ICCC — ISDN '87

EVOLVING TO ISDN IN NORTH AMERICA

INTEGRATED SERVICES DIGITAL NETWORK

Evolving to ISDN in North America

Proceedings of the
Regional Conference of the International Council for
Computer Communication, ICCC — ISDN '87
Dallas, Texas
September 15-17, 1987

Edited by

Irwin Dorros

Sponsored by
The International Council for Computer Communication (ICCC)

N·H
P C
1987

NORTH-HOLLAND - AMSTERDAM • NEW YORK • OXFORD • TOKYO

® 1987 by The International Council for Computer Communication

All rights reserved. No part of this publication may be reproduced, stored in a retrieval system, or transmitted, in any form or by any means, electronic, mechanical, photocopying, recording or otherwise, without the prior permission of the copyright owner.

ISBN: 0 444 70329 2

Publishers:
ELSEVIER SCIENCE PUBLISHERS B.V.
P.O. Box 1991
1000 BZ Amsterdam
The Netherlands

Distributors for the U.S.A. and Canada:
ELSEVIER SCIENCE PUBLISHING COMPANY, INC.
52 Vanderbilt Avenue
New York, N.Y. 10017
U.S.A.

The International Council for Computer Communication
P.O. Box 9745
Washington, D.C. 20016-9745

The views and the opinions presented in these proceedings are not necessarily endorsed by the International Council for Computer Communication.

Conference Commitee

Chairman
Dr. Philip H. Enslow, Jr.
School of Information and Computer Science
Georgia Institute of Technology

Program Chairman
Dr. Irwin Dorros
Executive Vice President
Technical Services
Bell Communications Research

Vice President
Dr. Peter E. Jackson
Vice President-Business Development
Northern Telecom

Publicity Chairman
Salvatore J. Barbera
Vice President
Strategic Planning and Development
Bell Atlantic Network Services Inc.

Local Arrangements Chairman
J. Larry Filmer
Director-Business Planning
Amdahl Communications

Program Committee

Bruce R. DeMaeyer
President
Ameritech Services

James P. Denvir
Partner
Akin, Gump, Strauss, Hauer & Feld

Robert H. Glaser
Vice President
Network Staff
Southwestern Bell

Paul E. Green
Research Staff
IBM

Paolo Guidi
President
Telenet Communications

Thomas J. Herr
Vice President
Network Systems
Universal Information Services & Strategic Market Planning
AT&T Network Systems

Leonard S. Hyman
Vice President
Merrill Lynch Capital Markets

Anne P. Jones
Attorney
Sutherland, Asbil & Brennan

Robert Kenedi
Vice President
Product Line Planning, INS
Northern Telecom Ltd.

James M. McCabe
Vice President
Donaldson, Lufkin & Jenrette

Robert Morris III
Securities Analyst
Prudential Bache

Kenneth Philips
Vice President
Telecommunications Policy
Citibank

Lawrence Sackett
Vice President
Gannett Company Inc.

John Thomas
Director of Development
Amdahl Communications

Williams F. Utiaut
Director
Institute of Telecommunication Services
Department of Commerce

T. Travers Waltrip
Vice President
Data Processing
Travelers Companies

Ward White
Vice President
Government Relations and Public Affairs
United States Telephone Association

Acknowledgments

Special thanks for making this conference and these proceedings a success are owed to the following: all of the ICCC-ISDN'87 speakers and session chairpersons, the ICCC-ISDN'87 Program, Publicity, and Local Arrangements Committees, Amdahl, Bell Atlantic, Bell Communications Research, Northern Telecom, and Southwestern Bell.

TABLE OF CONTENTS

Picturephone is a registered trademark of AT&T

KEYNOTE PANEL

Where Will ISDN Fit in the Telecommunications Strategy of Large Corporations?

T. Travers Waltrip

The Travelers Corporation

ISDN is emerging with tremendous fanfare. It is being touted as providing many things to many people. For large corporations, however, this might not be the case. The "S" in ISDN stands for services. Telecommunications managers must take a careful look at what these services are and how they can benefit from them.

The Travelers, like many large corporations, views DS1 and DS3 as commodity items. Cost-effective, high-capacity digital bandwidth is rapidly becoming ubiquitous. This capacity is enabling a myriad of new applications. Also, it provides connectivity through intelligent multiplexors to numerous corporate locations. This backbone network carrying integrated voice, data, and image, is fully customer-controlled and managed. It is clearly viewed as a corporate strategic weapon. Where does ISDN fit in this strategy?

The Travelers is basing its intra-building data communications strategy on local area networks and intelligent workstations. Included in this workstation is a voice terminal connected to a digital PABX. Peer-to-peer communications will logically tie the wide and local area network components together. To the worker, it will appear as though there is one network. Where does ISDN fit in this strategy for voice? For data?

Given this overall network strategy, ISDN will play its most important role for large users when they interface with the LEC to provide access for small or medium sized business or residential customers.

Challenges of ISDN: The Carrier's Perspective

Bruce DeMaeyer
Ameritech Services

ISDN has been the subject of countless publications, reports and studies since its first introduction at the 1972 Special Study Group session of the CCITT. Much of the focus, however, has been on the technical aspects of the architecture. The task facing today's local exchange carriers is to understand what the market for ISDN is, and what we must do to address that market. To accomplish that task will require adopting a thorough understanding of what customers expect from ISDN as the architecture evolves, and as their own communications needs evolve.

ISDN's evolution covers three phases. The industry is currently in Phase I, which is marked by "flavor of the week" ISDN trials and islands of ISDN availability. In Phase II, ISDN service will be possible across multiple exchanges. Phase III will be the period of full ISDN deployment and its interworking with other network architectures and services.

Each of these phases represents a marketing opportunity for local exchange carriers, and the market has not been silent. If ISDN prices are competitive, there is an opportunity to use ISDN to compete with LAN communications during Phase I. In Phase II, the market will demand that ISDN be widely deployed and interwork with packet and Public Switched Digital Service. In Phase III, ISDN must support the Intelligent Network and all the corresponding network services that it will entail.

To successfully market ISDN in each phase of its development will require local exchange carriers to do more than tout the technical wizardry of the Integrated Services Digital Network. The challenge for us is to build an ISDN deployment scheme that truly matches what customers want.

Challenges To Evolving ISDN In North America
The Equipment Manufacturer's Point of View

Tom Herr

AT&T Technologies

AT&T believes that challenges to evolving ISDN in North America may be also seen as opportunities. ISDN is an "equal opportunity" concept. Rather than one stakeholder making gains at the expense of another, everyone can come out a winner with ISDN.

ISDN is a logical evolutionary path in which communications technologies merge with computers to create an environment leading to what we call Universal Information Services (UIS). UIS will enable network providers everywhere to give any customer any voice, data or image service in any place, at any time, in any combination, with maximum convenience and economy. UIS supports open architecture which ultimately benefits all stakeholders.

We identify these stakeholders in network evolution to include:

- **End Users** are the ultimate beneficiaries of ISDN. Benefits will accrue to both business and residential users.
- **Network Providers**, both Local Exchange Carriers and Interexchange Carriers, will play a major role.
- **Enhanced Services Providers** (ESPs) stand to benefit greatly from intelligent networks. The ubiquity and power of intelligent networks deliver customers to the ESPs that would otherwise be inaccessible. Newly created or expanded industry associations are already directing their attention to formulating their telecommunication needs.
- **Standards Committee** standard interfaces and protocols within the public network and at network interface points are becoming ever more crucial.
- **Regulators.** The concepts of Open Architecture (ONA) supporting Comparably Efficient Interconnection (CEI) put forth by Computer Inquiry III (CI-III) reflect the most recent federal regulatory goals for evolution of the "national resource" of the communications carrier networks. The federal regulator goals to expand the U.S. role in the global information industry must be balanced with ongoing state regulator goals of universal service. Additionally, the role and opinions of the Department of Justice and Congress must be considered.
- **Equipment Manufacturers**, including both competing manufacturers of network equipment as well as customer premises equipment, wil reap the benefits of ISDN.

Less than a year ago, with our first office application on a production switch, ISDN became a reality. The barriers to ISDN are no longer scientific or technological. Rather, they are matters of public and industry policy. The real challenge, then, is to anticipate and support the standards upon which the full implementation of future networks will be based. As equipment manufacturers, we must approach the negotiating table with generously open minds. There is ample room within the open architecture for product differentiation based on proprietary applications and solutions.

We think the marketplace will grow as major end customers concentrate on being users of technology through the public networks rather than operators of their own communications systems. The openness of public networks creates a situation in which all parties can be winners.

Unquestionably, equipment providers are among the first to reap the sales benefits of ISDN. What we manufacture is needed to implement ISDN rapidly. We are commited to rapid deployment, because it will deliver the ultimate social and economical benefits of information productivity to the greatest number of users. We are looking forward to unparalleled opportunity and growth for the public network in the decade ahead. The key to seizing that opportunity lies in our ability to understand the issues, to engage in open dialogue, and to develop specific action plans addressing how we can work synergistically with each other.

Challenges to ISDN

Albert Halprin
Federal Communications Commission

From the government perspective there are two broad areas of concern connected with the introduction of ISDN. The first is cost—who will pay? Obviously, a tremendous investment will be necessary to achieve a meaningful level of ISDN penetration within a reasonable period of time. At the same time, the value of ISDN—even to the heavy users who are the most likely to benefit—depends upon achieving a high level of geographic penetration. In addition the breadth of consumer demand (across usage levels and types of users) is unknown and hard to predict. Will rate of return regulation—which removes both risks and rewards from stockholders for such a capital investment—politically or economically permit the rapid introduction of ISDN? If not, what is the alternative? What impact will introduction of ISDN have on rates (and costs) of POTS?

The second area of concern involves the legal and policy questions raised by Computer III and the MFJ. What impact will ISDN have on the line dividing regulated from competitive services established by the FCC? Similarly, what impact will it have on the U.S. District Courts' ability to separate permitted from prohibited activities under the MFJ? Conversely, what impact will these two orders have on the timing and manner of ISDN introduction? What role should the various government agencies have in making these determinations?

Currently, there is no proceeding at the FCC directly focusing on any of these matters. Consequently, questions exist about the manner in which these important questions will be addressed and, hopefully, answered. It is time to expeditiously address both the procedural and substantive questions.

ISDN—A Preliminary View from an Investor's Standpoint

Leonard S. Hyman

Merrill Lynch Capital Markets
New York, N.Y.

Investors want a return on investment. ISDN must involve products that can be profitably sold, which, in turn, allow communications carriers to furnish new services or cut costs, and regulators must let the carrier receive some of the benefits of ISDN, too, or investors will not get any return.

1. Introduction

Andrew Carnegie said "Pioneering does not pay." [1] The Integrated Services Digital Network (ISDN) is being brought to you by the same people who brought you Picturephone or at least attempted to. ISDN may be the recipe for full employment for telecommunications engineers and for those who wish to attend international conferences in such places as Brasilia or Dallas. Investors have watched high technology concerns go up and down like roller coasters: up during the euphoric days of development, when it looked as if the company had a great product and was ahead of its competitors, down when the product did not pan out or when the competitors caught up. There might be more profit in dishing out hamburgers than in developing new technologies.

2. What Investors Want

Investors want to receive a flow of funds that produces a return of the capital investment within a reasonable period of time plus a return on that investment (a profit). ISDN could satisfy those expectations for investors in manufacturers of equipment and in suppliers of software provided that new technologies or additional competitors do not supplant the original products and firms before the investors have reaped their rewards. Telephone companies can utilize ISDN to provide new services that should garner added revenues and cut costs, both of which could raise profits provided that others do not find a better way to do the job, and provided that regulators let the companies reap the rewards of innovation. Otherwise, the benefits of ISDN go directly to customers, and investors receive little or nothing unless it could be argued that ISDN is a defensive measure that helps the telephone company maintain its market.

The market for telecommunications hardware and software is a dog-eat-dog world. The initial surge of digitalization has subsided. Long distance carriers may strangle in an oversupply of optical fibers. PBX manufacturers batter each other and the resurrected Centrex threatens the PBX, too. Internationally, other countries have telecommunications plans which often seem aimed at protecting and promoting local monopolists and manufacturers as much as developing the telecommunications infrastructure. The United States does not have a plan. It has blind faith in free trade and free markets. Thus American investors in ISDN must not only worry about whether the technology is a good one, and whether the pioneers in it will make profits, but also whether American ISDN products can be sold abroad while foreign products can be sold here.

Perhaps the waves of innovation and destruction are coming too close together, so that producers of high technology might have trouble recovering costs before a new wave of product hits the market. Undoubtedly customers will benefit until we run out of those willing to take the gamble of investment. As one of the economics trade's great poets, Kenneth Boulding, wrote:

> We all agree that innovation
> Will benefit both world and nation.
> The question we must answer later
> Is, will it help the innovator? [2]

3. Prospects for ISDN

ISDN must have more than technological value to succeed. Will it lead to revenue producing services? New England Tel will launch an ISDN trial but only within the company, not with outside customers, because, "What we haven't been able to develop is a significant number of applications that we can go to our customers with yet." [3] A recent magazine story headlines British "Enthusiasm" but the text below the headline, about ISDN in the land of Prestel, doesn't seem to indicate that the British population is panting for ISDN as the technological replacement for high tea. [4] My skepticism about exotic services remains, but not my original skepticism about ISDN's ability to reduce costs and increase flexibility for both telephone companies and their customers. The evidence for that looks good. [5, 6, 7]

For the investor, the questions are whether the new service replaces old ones or generates additional revenues, and whether reduced costs of operation for the telephone company are passed on entirely to customers, or even whether reduced costs to customers simply means less revenue to the telephone company.

4. Conclusion

Kitahara speaks of the Japanese version of ISDN as involving:

> the development of a new comprehensive telecommunications network, made possible by the integration of computer and telecommunications technologies, which will achieve a higher efficiency than the present network. [8].

Efficiencies are commendable, but does the creator of the efficiency gain from it? Or must the hapless investor whose decision to finance ISDN made it possible settle for praise and no more?

The regulatory process will be the key to answering the questions. Incidentally, who is financing ISDN? The manufacturers are spending stockholders' money on R&D. That we know. Are customers of telephone companies paying for ISDN research through their rates? If so, investors in those companies might find themselves in the anomalous position that typifies innovative investment in a regulated business: if the innovation works the customer benefits, and if it doesn't, the investor foots the bill. Thus, if regulators use the standard procedures developed during the days of monopoly, then the telephone company will remain a conduit, earn a return on the investment in the conduit, and have little incentive to innovate unless forced to do so to protect its core monopoly. That outcome could slow the introduction of ISDN and its associated services and savings. America is threatened with a second rate communications system, if the present participants continue to operate in an atmosphere of Luddism, legalisms and pseudo-consumerism. Service economies, above all, need competitive communications systems. For that, ISDN is a necessity.

The old regulatory process, though, is breaking down, in my opinion. Regulators will free more of the system from rate of return regulation as long as residential customers are protected from price increases. That is good. In a recently published book, my co-authors and I wrote:

> ISDN ... will provide a general highway system to handle all forms of communication ... and thus accommodate all requirements for interactive communication. [9]

Investors view ISDN with puzzlement. What is it? Who makes it? Who uses it? Who gets the benefit? The concept is big. Many of the products to produce it seem to be add-ons to existing lines. If ISDN strengthens the network's position, then network system manufacturers could be winners and PBX manufacturers the losers. But who knows this early in the game? If ISDN comes along just as the regulators de-regulate, the telephone companies look like winners. Either way, the consumer wins, as long as the consumer has not already bet on the wrong equipment or systems.

My preliminary conclusion is that ISDN is the most important concept to develop in telecommunications since Theodore Vail invented universal service. In fact, it will bring Vail's concept to fruition. That has to be a recipe for success for investors.

References

[1] Morgan, T., *Introduction to Economics* (Prentice-Hall, Englewood Cliffs, 1956), p.21.

[2] Clark, J.M., *Competition as a Dynamic Process* (The Brookings Institution, Washington, 1961), p. 199.

[3] Watson, N., "New England Tel. Plans ISDN Trials", *Communications Week* (May 4, 1987), p. 12.

[4] Williamson, J., "ISDN in the U.K.: Enthusiasm Overshadows Skepticism as Private Applications Emerge", *Telephony* (April 27, 1987), pp. 56-57.

[5] Paul, F., ISDN Application Speeds AT&T 800 Operator Service", *Communications Week* (May 4, 1987).

[6]"AT&T Shows Off What ISDN Can Do; Bellcore Study Says ISDN May be Economical", *The Report on AT&T* (March 23, 1987), p. 1.

[7] *Bell Communications Research Annual Review 1986,* pp. 6-10.

[8] Kitahara, P. *Information Network System* (Heinemann, London, 1985), p. vii.

[9] Hyman, L., Toole, R., and Avellis, R., *The New Telecommunications Industry: Evolution and Organization* (Public Utilities Reports, Arlington, 1987), p. 58.

The information set forth herein was obtained from sources which we believe reliable, but we do not guarantee its accuracy.

TECHNICAL SESSIONS

ISDN - Will it Fit?

Roy Brunson

American Airlines

The Integrated Services Digital Network (ISDN) is expected to provide the network architecture for improved digital transmission. This paper examines how ISDN may be applied in the large voice and data networks which support American Airlines telecommunications requirements.

1. INTRODUCTION

American Airlines utilizes a variety of services in meeting its voice and data telecommunications requirements. In today's voice network AT&T 800-Service is used extensively for incoming reservation calls. However, voice private lines, Centrex, foreign exchange and public switched networks are also important elements in meeting the total corporate voice communication needs. On the data side, leased multidrop lines are the primary vehicles for connecting our many clients to the SABRE passenger automated reservation system. Public switched data network services are also used in limited applications. ISDN voice and data services will, undoubtedly, be added to the list. To what extent and how rapidly these new services will be integrated will depend on several factors with cost and availability being key.

The prospect of reduced costs and enhanced services is perhaps more significant to companies which rely extensively on long distance voice services and public data networks in meeting corporate information transfer needs. The implementation of the ISDN should provide additional flexibility over existing public switched services. Implementation will be straightforward. Existing business lines will be replaced with ISDN access lines.

For companies with large communication networks built to achieve maximum operational control and performance with the most cost effective service alternatives, the integration of ISDN is more complex. The use of ISDN based services may become more of an evolutionary process than one of wholesale conversion. Naturally, cost and capability of ISDN services as compared to the current base of services which form today's networks will be major factors in determining how rapidly such an evolution will take place.

I will examine some of American Airlines communication requirements and how ISDN may fit in satisfying these needs. However, first let's define ISDN. I perceive that ISDN may mean many different things to many people. For the purposes here I would like to define ISDN as a network that provides end-to-end digital connectivity.

Users access the network by two standard interfaces:

- The basic access is a 2B+D interface. The two B or bearer channels are 64 kilobit per second channels for carrying voice or data calls. The D or signaling channel is a 16 kilobit per second for handling call signaling status and possibly user data calls.

- The primary access interface is configured as 23B+D. Both the bearer and the signaling channels are 64 kilobit per second channels.

In addition to the two interfaces I would also like to include CCITT Common Channel Signaling #7 standard since it will make possible calling number identification which is one enhanced feature which has been much publicized. Other enhanced services are expected to evolve as a result of the ISDN implementation. Many different groups including local exchange carriers, interexchange carriers, equipment vendors and various user groups are involved in feature and service definition. We at American Airlines are participating in several ISDN definition efforts. Our objective is to influence the implementation of services which could be beneficial in meeting our information transfer needs. However, it is too early to speculate on what new services may evolve; therefore, I have not included them in this presentation.

2. VOICE REQUIREMENTS

Let's now examine some of American Airlines communication systems. The voice network must satisfy two diverse requirements. The reservation system must provide the connectivity between the traveling public and American Airlines reservation agents. The administrative voice network must provide the communication resources for employees to conduct the company business of operating the airline.

2.1. Reservations Voice

The key elements in the reservations network are the automatic call distributors located in Los Angeles, Fort Worth, Cincinnati, and Hartford. These systems handle more than 65 million calls per year. The reservations system relies extensively on AT&T 800 Service. In addition foreign exchange circuits are employed in many large metropolitan areas. The ACDs are interconnected with leased private lines to handle call overflow, as well as, call rerouting for several specialized reservation functions which are centralized.

How will ISDN fit in meeting reservations requirements? Already the AT&T 800 Service has been converted to AT&T Megacom service. Megacom has replaced the numerous individual analog 800 access lines with digital, D4 framed, 1.544 megabit per second lines which connect directly into the ACDs. In an ISDN environment the primary access, 23B+D would be a candidate to replace the twenty-four channel, D4 framed lines. Because Megacom is a digital service, no additional quality improvement can be expected in converting to the 23B+D format. Also because the ISDN signaling format is not compatible with the D4 format, either ACD hardware modifications or new signal conversion hardware would be required. Furthermore, in converting from D4 to 23B+D, one ACD port would be unavailable for use for each digroup changed from a twenty-four channel format to a twenty-three bearer channel format. This could represent up to four percent of the switch capacity. On the positive side, with a 23B+D access, the D channel would make possible common channel signaling for passing calling number identification from the network. Perhaps the most attractive use of this feature in a reservations system environment, is for accessing computer files to assist agents in handling calls. The calling number identification could be used to retrieve a file containing traveler related data. Such information as the address to mail the ticket and perhaps seating preferences could be made available. When the incoming reservation call is delivered to the agent, the supplementary information made possible with calling number identification is presented on the agent's terminal. The implementation of such a system would require an ACD, or combination ACD and interface device, capable of reading the calling number information, accessing a computer file and delivering the file to the agent assigned the call.

2.2. Administrative Voice

American Airlines' administrative voice requirements are more diverse. The voice network supports operations at over 115 airports, ticket offices and other locations. In addition, the four reservation centers, the Tulsa complex which is the center for computer and maintenance activities and the Fort Worth headquarters complex are major voice concentration points. The corporate voice network is primarily a leased, private line network. Users access the network via private branch exchanges or Centrex-type services. Most PBXs and Centrex locations are connected by tielines to one of five intertandem switches located at the four reservations centers and Tulsa. These five systems are interconnected with over 200 intermachine trunks. Each of the major switching systems and the PBXs are connected to the local central offices and the public switched network with direct inward dialing and direct outward dialing lines. The network is designed so that all on-net locations may be reached by dialing a seven digit number.

Many large companies including American Airlines continue to take advantage of new service offerings which further reduce the cost of private line networks while improving quality. The relatively new 1.544 meagbit per second, T1 offering is one such service. These digital systems are replacing analog circuits for both intercity requirements as well as for access to local exchanges.

How will ISDN fit in the administrative voice network? The most likely implementation of ISDN would be for digital access to the local exchanges. Many of the PBXs are connected to the local central office with individual analog inward and outward dialing lines and ordinary business trunks. These would be prime candidates for conversion to 2B+D access particularly if the pricing is competitive with individual lines. Analog tielines could be handled in a similar manner for access to the interexchange carrier facilities. At some low usage locations it is quite probable that leased private lines would be replaced by ISDN switched services. At locations already supported by D4 framed T1 service, the need to convert to an ISDN 23B+D format is less obvious. Compatibility with the next generation PBX, the nature of enhanced ISDN related services made available at central offices and, of course, cost will likely dictate the extent of conversion to an ISDN 23B+D format. Finally, the conversion of intermachine trunks from networks already employing digital services is even less likely. These links are part of a private digital transport network which derives at least 46 voice channels per T1 circuit using proprietary T1 formats incorporated within networking T1 multiplexers. While the transport network side will continue to operate with unique formats, it is realistic to expect that the interface between the T1 multiplexers and the voice switching systems will become a 23B+D format as ISDN switching systems are implemented. The packaging and unpackaging of signaling formats for transmission over the T1 network would be accomplished within the multiplexer. Many T1 multiplexers on the market today accomplish similar conversions in translating D4 format into the proprietary T1 formats which contain separate signaling channels as does the 23B+D ISDN format.

3. DATA REQUIREMENTS

The American Airlines' data network is a product distribution system which makes available the SABRE passenger reservation system to American Airlines agents, other airlines and travel agents worldwide. Over 14,000 discrete locations are served by approximately 1800 leased circuits. The majority of the circuits are multidrop lines which are operated at 2400 or 4800 bits per second. These lines connect over 90,000 terminal devices to the central computer system located in Tulsa. This system processes over 30,000,000 transactions per day. The system is query/response oriented with over 90 percent of the volume originating from the computer complex. Quick response is a critical factor and delays of over a few seconds are unacceptable. The current network of four primary polling concentration nodes is being replaced with fourteen packet switching nodes. These nodes will distribute airline reservation data from the computer to the remote terminal devices. These devices will be connected to the packet switching nodes by leased, multidrop lines as in the current network. The backbone transmission requirements to the central site will be supported by 56 kilobit per second digital trunks.

How will ISDN fit in meeting American Airlines data transfer requirements? The potential exists for 2B+D basic access lines to replace local access lines into the numerous terminal locations. The basic access could provide for higher data transfer speeds for most applications if a portion of the D channel could be utilized. I do not envision that total public switched network approach will be cost competitive with a hybrid network employing both switched and private services because of the high traffic volumes to be supported. However, the ISDN services could be ideal for local concentration and delivery to private long-haul switching and transport networks. Again, pricing will be key in determining the extent that ISDN services will be integrated into the data network.

4. CONCLUSION

American Airlines uses both leased private and public switched network services. The factors which influenced our network decisions include service availability, network control, quality, performance and cost. Given that no overriding reason to choose a particular service based on one of the first four factors, cost often becomes the key issue. In many of our analyses we find that the high traffic volumes typically support a private system approach instead of a switched network approach. In reviewing American Airlines communications systems and projecting how ISDN may be used to support our requirements, it is apparent that ISDN may fit very well in replacing switched service access. However, until pricing of new services is established, it is difficult to embrace ISDN as the ultimate solution to future communication needs.

ISDN - A CUSTOMER NEED OR VENDOR DESIRE

Sherman A. Murphy

Hartford Insurance Group
Hartford, CT

This presentation provides a customer view of ISDN. The position taken is that the customer recognizes (as have many others) that he has many of the planned features of ISDN already in place, or the potential to be made available, without ISDN. Further, it is felt that ISDN in the United States is being over-stated by suppliers without definition of customer business need or customer advantage.

I. Introduction

Today, major business firms in the United States have recognized Telecommunications as an important strategic asset in their business operations. The Hartford is no exception, and has deployed "state-of-the-art" data and voice networks and systems across North America. These are utilized to sell and service The Hartford's products, to record premiums, to report and pay claims, to supply intra-company business communications, such as messaging, and to support office automation.

Over the past several years, The Hartford has formed "strategic partnerships" with selected Telecommunications hardware and network providers. These arrangements have enabled The Hartford to avail itself of insight into vendor strategies; to position for product deployment; and to provide The Hartford with information concerning vendor product direction.

The Hartford views ISDN with a great deal of conservatism. It notes ISDN's currently announced capabilities as primarily those already available in various "pre-ISDN" offerings. Accordingly, it questions the efforts and costs being incurred by the vendor community without in advance cost benefit to the vendors and, more importantly, their customer. In effect, and as confirmed by consultants and customer associations, ISDN has come to be known as - "I still don't know".

This presentation will provide an insight into:

- Hartford Telecommunications services and deployment.
- How The Hartford is utilizing, or planning to utilize,"ISDN-like" features and functionality.
- The rationale for Hartford concern regarding vendor direction concerning ISDN.

II. Hartford Telecommunications Overview

The Hartford employs some 18,000 individuals located in approximately 250 locations in

North America. Its insurance and financial services products are sold by some 10,000 independent agents and brokers.

The Hartford has invested heavily in office automation and insurance and financial services proprietary applications software to support its business growth.

Two data centers support online and batch systems, with primarily IBM and Amdahl Processors utilized.

The data centers, and the networks that support them, are designed for disaster backup if a single data center should be lost. As well, The Hartford Online Telecommunications Network provides operational backup in the event of circuit loss to key locations.

The Hartford's principal networks (Wide Area, Metropolitan, and Local) provide data and voice support for a base user population of 11,000 data devices; 16,500 telephones; and 100+ facsimile devices. The data devices are a "mix" of 3270, PCs, and Work Group Processors.

The Hartford employs "state-of-the-art" transmission facilities and systems to support its Telecommunications requirements. Each of these is profiled below.

1. "HIGNET" Configuration

"HIGNET" is a multi-location Metropolitan area network which serves key Hartford Insurance locations in the Greater Hartford, Connecticut area. As noted in Figure 1, T1 Transmission Facilities are deployed in a "star" network configuration to provide data and voice services to some 7,000 employees. Data applications include Host-to-Host, Batch Transfer, Host-to-Print, User-to-Host, User-to-Work Group Processor, and Facsimile-to-Facsimile, using Group III devices.

Voice applications include an intra-company tandem network, shared Voice Mail, and "hubbed" mixed common carrier network supplier services to and from outside Hartford calling parties. The voice systems utilized are all digital and replaced former Centrex and Analog PBX products.

The T1 facilities (24 in total) are supported by "smart muxes" provided by two vendors which utilize CVSD and ADPCM for voice subchannels to and from the PBX systems.

Analog Dial Data is switched through the network at data rates to 4.8KBS, with dedicated data circuits in the network configured at 1.544MBS, 56KBS, 19.2KBS, 14.4 KBS, and 9.6KBS.

The Hartford handles its own design, provisioning, and operational control of this network, with vendor maintenance provided.

At present, The Hartford is working on plans to upgrade this network. Specifically, it recognizes "single point of failure" exposures in a star network; sees a requirement for data rates beyond T1; and is anxious to deploy additional fiber in its configuration. Accordingly, a "shared" city-wide network is being pursued with other major business users with DS3 in a "mesh" configuration likely to be deployed during 1988.

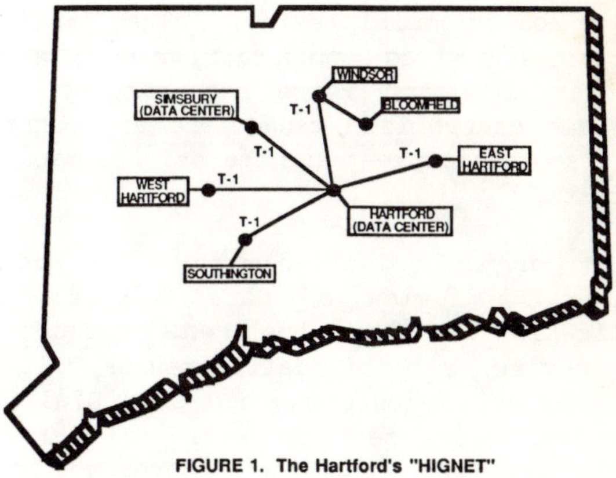

FIGURE 1. The Hartford's "HIGNET"

2. "Nodal" Network Configuration

The Hartford has recently concluded upgrades to its "backbone" transmission facilities which provide data support to its field locations and selected agent sites.

As noted in Figure 2, the backbone is now operating in a "ring" configuration served by T1 and 56KBS digital facilities.

At the communications network "nodes", the backbone facilities are sub-channeled to 32KB (via ADPCM for voice), and to various data tail circuit configurations operating at 19.2KBS, 14.4KBS, 9.6KBS, and 4.8KBS. The data tail circuits are point-to-point or multipoint to 242 company locations and 200 agent sites.

The design is supported by "smart muxes" which

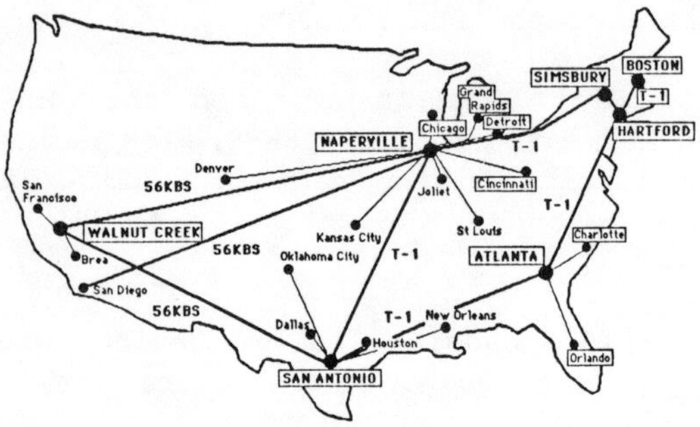

FIGURE 2. The Hartford's "Nodal" Network

provide automatic alternate routing. The 56KBS circuits are backed up via "switched 56". Automatic recovery time in the event of loss of a T1 is approximately six minutes via either route in the "ring". ATT's DACS/CCR is utilized in a disaster backup configuration in the event either of the data centers or network control centers should be lost. This capability is not used for operational recovery, nor are there plans to permit control via central office based multiplexing. The Hartford operates in a mixed common carrier environment for its data and voice requirements, and provides centralized management and control via its own equipment and operational management staff.

Due to dramatic increases in local access charges, a review of data tail circuit utilization has been completed. Based upon this review, selected tail circuits will be replaced by asynchronous and SDLC Dial VAN services. Such services will initially be public services, and then converted to a "hybrid" network configuration with X.25 PAD and switching equipment at selected locations.

3. LAN and Cablenet

The Hartford has utilized LAN Technology over the past several years, primarily Ethernet-based.

Such systems are in place in our Headquarters locations, with a few at our field sites. Approximately 2,000 devices are served including PCs and Work Group Systems. Planning and trial activity is in progress concerning Token Ring at this time.

An extensive installation is in progress at our Corporate Headquarters to replace coaxial cable with twisted pair utilizing baluns. This serves the horizontal office configurations of our office facilities. "Smart closets" containing patch panels are also utilized. The closets, riser feeds, and inter-building runs will utilize 62.5 micron fiber optic cable.

This concept is also being deployed on a selective basis at our major field locations and is tied to office relocations and office lease renewals and renovation. The objectives of the LAN and Cablenet activities are to reduce costs, position for LAN Technology deployment, enhance user office automation support, and to permit easy configuration change. The twisted pair balun configurations are supporting up to 4MBS data rates today and are expected, via planned vendor technology improvements, to support 10MBS within two years.

4. Telemarketing Centers

Multiple Telemarketing Centers have been established at various locations in the Far West, East, and Midwest regions of the United States. Such centers utilize ACD (Automatic Call Distribution) Systems, and "script sell" Customer Service Agents and customers direct.

On a given business day, some 50,000 "800" type calls are handled through these centers. Customer Management has access to various customer controlled features in the systems and in the network serving the locations to allow routing to various centers, "prompting" of calling parties, and reconfiguration of agent positions and trunks. Additional network features have been, and continue to be, reviewed for potential applicability.

III. ISDN in The Hartford's Environment

The Hartford has over the past several years followed ISDN developments and potential deployment closely.

At issue have been various overall questions and concerns:

1. Who will offer ISDN products and services?

2. What will the products and services be?

3. When will ISDN become available?

4. What will ISDN-based products and services cost?

5. What are the potential applications, and is there a positive cost/benefit relationship to be derived?

6. What will the impact be on existing products and services in place or planned over the next five years?

7. What conflicts exist in the vendor community, regulatory arena, and customer perception/market that will impact ISDN?

The Hartford's participation in major user groups, vendor advisory boards, and awareness of its requirements, has led to some preliminary conclusions and answers to some of these questions, with others still outstanding.

Prior to additional comment it is important to note that The Hartford is not "anti-ISDN". It believes there is merit to use some of the future planned service offerings, if the price is right, and our needs are served.

Given The Hartford profile of in place or planned Telecommunications network services and systems, the following comments and concerns are offered.

1. We view ISDN as a technology and standard to attain. All customers, The Hartford included, want solutions to business requirements, not technologies. Given what has been advertised to date as regards "pure ISDN", (e.g., conforming to international standards versus "ISDN-like" capability), we feel, so far, that no capability or functionality is being provided that can't be attained by other technologies. It is recognized that ISDN may accomplish required capabilities and functionality in a more efficient manner than is currently offered, but question is raised concerning the cost and timeframe to accomplish.

   For example, "basic" and "primary" ISDN (2B + D; 23B + D) interfaces are available in an equivalent now via our T1 and other configurations if we choose to utilize them.

2. To date, ISDN appears to be a vendor push, rather than a customer "pull", which is logically market driven.

   The history of ISDN is tied to its origin in Europe where the PT&Ts were and are anxious for standardization to assist in their monopoly intent - at tax payer expense.

   A different situation exists in the United States where there is increasing competition in both network and system offerings, customer private networks, and innovation in the use of such systems and services which is not seen in Europe. This has fostered "ISDN-like" services.

   In addition, bandwidth is being deployed in increasing capacities via fiber which of itself may lessen the requirement for ISDN.

   Customers in the United States have then a logical question - is there a "hidden agenda" in the pressure for ISDN activity? Specifically, are customers being asked to fund needed upgrades to vendor (not PT&T) facilities and systems; are existing installed customer products and services truly in need of change or upgrade; and is ISDN a vendor sales and marketing strategy to create new revenues at potential risk to customer management and control of their own business affairs?

3. ISDN trials are in progress at various locations. At question is the intent and purpose of such trials and the projected application of the results attained. Technical trials are an issue to prove capability. Application trials are another to prove customer acceptance and obtain solution to actual needs. At question may be how objective such trials are versus a comparison to non-ISDN solutions which could accomplish the same result. It appears that the key to all activity is vendors having decided that ISDN is a "given", and that applications must be found to drive an implementation.

4. ISDN appears to be a voice concept when customer requirements are driven by demand for increased data support and/or data/voice selective facility and system sharing. The classic issues of PBX versus LAN; Centrex versus PBX; Private versus Public Networks, etc., are all noted in this total situation. Most customers have come to grips with the fact that selected use of each system or network solution, based upon customer requirements and cost, is an ongoing task.

   The data user is of specific concern given the variables in data rates required, whether switched or dedicated circuit, and occasional versus interactive user profile. The Hartford does not use Centrex; does not utilize PBX facilties for interactive data users; and relies upon dedicated networks (LAN, Metropolitan, or Wide Area) for cost and performance reasons.

Accordingly, we do not see ISDN "competing" with LANs in high volume interactive customer requirements any more than PBX systems do today.

The business customer, in our view, will likely utilize the "primary" ISDN interface for Metropolitan and Wide Area connectivity, with the "basic" (2B + D) interface of little use except for possibly the home subscriber.

At this time, it is difficult to view how needed vendor investments to support ISDN (Telco, Common Carrier) can provide the system and software needed without a customer potential high cost when compared to what is in place or possible in today's technology. For example, T1, DDS, and LAN costs to the customer are coming down; data utilization is going up; and ISDN does not seem to fit given its switch-based architecture.

Obviously, there are "niche" requirements for small business customers, but the major business user will likely utilize dedicated services to support high volume locations.

Accordingly, ISDN may have missed the appropriate planning and deployment "window" with the major business customer given its limited capability at present.

A similar issue comes to mind in considering the vendor "conflict of interest" in determining who its customers are, and what its market direction should be. Certain vendors are in the PBX business, in the central office business, and in the network business, directly or indirectly. Accordingly, the business customer can potentially be marketed ISDN products and services from a Telco, switch supplier, network supplier, or all at once. "Putting the pieces together" will likely rest with the business customer, not the vendors. The so-called "value-added", "enhanced services" vendor offerings are still up for grabs and such issues as ONA (Open Network Architecture) will likely have an unknown impact on ISDN.

5. ISDN features and functionality are, in many cases, already in existance or available, without the need for specific ISDN investment by vendors or customers. A sample of some of those being touted as "applications", and suggested for cost/benefit, are as follows:

   A. Telecommunications access to information services - this capability has been available for some time. What's different is the "linkage" of network services with customer devices tied to "automated" accessed databases without manual intervention.

   Examples are "calling number services", "message switching services", and "data signaling".

   It is suggested that the technology which supports "911", DNIS, and various WATS offerings could provide equivalent capability to ISDN.

   A key concern of The Hartford is database integrity and security. The Hartford has strong security policies, procedures, and standards which require "live" individual password and ID access to any of our systems and databases. Any "automatic" system or network direct access interface to Hartford systems is going to be a significant issue and concern if/ when attempted.

   B. "Dynamic" change capability in Metropolitan and Wide Area Network Configurations - we have the capability to support this requirement today. This is obtained via "smart muxes" which allow reconfiguration of the T1 backbone to subchannel data and voice circuits as required; DACS/CCR which allows re-routing and "unbundling" of DS1 and DS0 subchannels; "smart WATS"; and planned offerings such as ATT's SINA, RINA, and DINA.

   Such capabilities do not require an ISDN "D" channel to activate and control which could lessen the overall current utilization of bandwidth to our requirements.

   C. Optimization of Local Telco or IEC Access Facilities - There is suggested advantage due to improvement in provisioning time, re-use of existing copper, "dynamic" allocation of circuits to varied Telecommunications requirements, etc.

When and where appropriate, we have considered access requirements via T1 level in our Metropolitan and Wide Area Configurations. Typically, this has been done to reduce local access expenses. We view the issue of Telco or IEC access and optimization of cable route as an advantage primarily to vendors to avoid new construction. At many locations, we have suggested fiber to position for the needs of both the vendor and The Hartford. As in Item (B), customer reconfiguration and allocation of access transmission media is available today, with additional enhancements coming shortly, without the need for ISDN.

D. Customer Premise Coax Elimination and Savings in Move and Change Activities - The Hartford has utilized PBX products for a number of years which permit customer activated software changes in station assignments. These are tied to "pre-wire" configurations installed at the time of the PBX installation. Our previously described ongoing cable network "project" utilizing baluns, twisted pair media, and smart closets is an additional example of this kind of approach. Inter-building and/or riser requirements between the closets will be supported through the use of 62.5 micron fiber. We feel comfortable in being positioned for today's requirements as well as those in the LAN arena, (e.g., 802.3 and 802.5). Again, ISDN is at question concerning logic for deployment.

IV. Conclusion

Previously, it was stated that The Hartford is not "anti-ISDN". Given some of our current uses of various pre-ISDN products and services, each of which was justified on its own merit, we feel "ISDN-like" features, functionality, and products can be deployed in the short-term if needed, (e.g., Group IV Facsimile). At present, we are not excited by what we see in "pure-ISDN" capabilities nor its potential cost for the products needed to support it. Accordingly, we do not see in the next five years a situation in which:

A. We will eliminate key private networks in favor of switched ISDN networks or their equivalent.

B. We will have need to upgrade or retrofit end user devices, systems, or cable plant for ISDN given its lack of proven need and business case.

If ISDN is to succeed in the major business customer environment, we see a strong need for the following:

A. Customer and vendor education.

B. Definition of logical problems, needs, and applications that can be addressed by ISDN.

C. Costs of all ISDN products and services must be specifically defined. Sales and marketing attempts to suggest "value-added" as a rationale for product acquisition is a customer determination, and will not be easy to attain.

D. Vendor support for the existing product set must be improved. Telcos and IECs are recognized as being in various states of difficulty in network problem diagnostics, quality and provisioning of services, and obsolescence in existing plant to support new services. Unless and until the current offerings work effectively under committed performance and availability standards (not FCC and Telco tariffs) customers will not "trust" their information transmission needs to "innovative" change. Given ISDN's order of magnitude, we feel this issue is quite mandatory. Accordingly, vendor investments in ISDN should be re-evaluated.

E. The customer perception of ISDN is in trouble and needs improvement. Various customer association annual sessions now evoke humor when ISDN is raised as a topic. The "image" of ISDN needs to be repaired, or an ISDN market demand will not take shape given that eventual requirements and needs are identified.

ISDN and The Conglomerate
An Operating Division Perspective

J. D. Wells

Gorham Division, Textron

This presentation offers a perspective on the potential uses of emerging technologies as they apply to an operating division of a major conglomerate. The relationship between vendors, the corporate parent, and the operating units are outlined, as well as the inter-relationships between operating divisions and the corporate office is presented.

1. Introduction

The continuing revolution in the use of communications services in business and industry has introduced new opportunities and challenges. Within the service segment of the economy and in centralized and homogeneous organizations the utilization of integrated communications services have been more evolutionary. Spurred by the increased use of data processing to manage on-going operations organizations offering products and services which require a continuous flow of accurate, up-to-date information capitalized on developments in computing hardware and took a leadership role in the demand for increased communications capabilities. The response to these demands have been an almost constant release of new products and services in which real-time, on-line data communications have become an integral part of these businesses.
For the decentralized organization communications have followed a somewhat different path. Within the classical conglomerate, the diversity in the use of data communications, and in fact data processing has been dictated as much by the constant change in the mix of businesses as by the diverse nature of these operating companies.

This paper will describe the two primary aspects of advanced communications confronting the conglomerate, specifically:

- An overview of Textron as a business entity and the relationship enjoyed among Textron business units and vendors and between Textron and its operating divisions.

- An examination of the role of data processing in a modest operating division and how communications, specifically ISDN might possibly be used.

Finally a few comments will address current positioning of Textron for potential use of ISDN .

2. Textron, A Corporate Profile

Textron, as a major diversified corporation employs over 72,000 employees worldwide. With headquarters in Providence Rhode Island, Textron controls over thirty five operating divisions throughout the United States.

As a multi industry conglomerate Textron can be viewed as three distinct segments to the same organization.

- The traditional Textron operating companies.

- The newer Avco business units which joined Textron as a result of a merger in 1985.

- The Ex-Cell-O companies acquired in 1986.

These segments are engaged in Aerospace Technology, Commercial Products and Financial and other services. Recent acquisitions have resulted in an on-going restructuring of Textrons primary businesses and sales increasing to an excess of seven billion dollars in 1986. Continued growth of the corporation and its business units is seen to be in the following areas:

-Helicopters; Bell...

-Propulsion; Lycoming...

-Systems; Avco Systems...

-Commercial Products; Homelite...

-Finance; Textron Financial...

-Insurance; Paul Revere...

-Management Services...

2.1 Communications Profile

Corporate

While Textron's businesses are dispersed over much of the United States, from California to Georgia, voice communications continue to be through a private leased line network which link the major business units and corporate headquarters in Providence through a digital voice network. At the current time this network does not carry data nor are their any plans for corporate wide data communications in the future.
The central node of the Texcom network is a System 85 located in Providence Rhode Island. A secondary system for Voice Mail is connected to the Texcom Network or may be accessed through pulse tone phones. The Voice Mail node is located in Stratford Connecticut.

All other communications between divisions and or the corporate offices are via Telex, or Facsimile. Data communications between different operating units are practically non existent and between the operating units and headquarters are limited.

2.2 Communications Profile

Operating Divisions

The major business units, as mentioned previously, are directly connected to the Texcom communications network. However, the smaller business units are not for the most part. Each operating division must justify the type of communications equipment needed and its use. Consequently, there is a great deal of diversity among the various business units in the type, capabilities, and configuration of telephone switching equipment and data processing communications control units. Again, as previousley mentioned, communications among the various divisions tends to be more informal with little central control or direction over equipment, software in the case of data communications, or network services in the case of voice or image processing.

3. Gorham A Business Profile

Gorham, as a division of Textron employs 800 people located in three manufacturing plants and a warehouse and distribution center. Sales average 100 million dollars per year through retail outlets located in the United States. As an operating division we are accountable to Textron for:

- Return on Investment

- Cash Flow

Gorham is engaged in three primary businesses offering multiple product lines to our distribution network. These product groups are:

- tabletop; Sterling, China, Crystal

- Gift;

- Bronze; Memorials, Statuary, Architectural

Conducting business in a mature market offers a somewhat different set of challanges than are found in an expanding business environment. As a multi-site manufacturing and distribution operation the challange we have been confronted with is to position our division to be capable of taking advantage of technological advances as they become fully available, at a cost consistant with our marketplace and our accountability to Textron. To Accomplish this Gorham has over the past two years embarked on a major upgrade to all of our facilities and the services they perform or provide. Currently Gorham is in a new silver manufacturing plant located in Smithfield Rhode Island, a new Warehouse and Distribution facility located in Williamsport Maryland, a new Bronze manufacturing plant located in Aiken, South Carolina with another plant located in City of Industry, California. We also have direct retail outlets in Pidgeon Forge, Tenn and North Conway, New Hampshire. Plans have been drafted to open five additional outlets in the next two years. In addition, we have showrooms strategically located in major cities in the Mid-West, Southeast, and on the West Coast.

With the major upgrade to facilities Gorham has also undertaken a significant improvement to all data processing and communications capabilities required to support our business. Again, each of these changes had to be made with a firm understanding of the cost benefit relationship and an eye toward where our business is going and how technology can best be used to service that business.

3.1 Communications Profile

The changes made to our data processing and communications environment over the past two years have all been made to provide on-going service to the division now and yet be in a position to evolve into a more heterogeneous climate as our business changes or as technological advances become more readily available.

Currently we have a telecommunications network linking all major facilities with our general and administrative offices located in Johnston, Rhode Island. Each facility is connected via a 9600 baud dedicated lease line with multi-drop capabilities. Johnston is connected to Smithfield, which in turn is connected to Williamsport, Maryland. Williamsport is in turn connected to Aiken, South Carolina which drop to the showrooms located in the Southeast and Southwest. Californian is connected to Johnston, Rhode Island with drops to showrooms located in the mid-west and the west coast. While this is not a complex network it does serve our purposes and does not require significant expenditures for communications personnel or software. Redundancy has been built in through the use of in bound WATTS lines. Concurrent with the establishment of the data network we also moved to install Facsimile equipment in all primary locations as a means of improving our ability to communicate documents. While short of true image processing, it is again a practical solution to a common problem. Voice communications presented yet another problem. With a long term contract on our Dimension telephone switch we were confronted with either having to add expensive direct in dialing or find another solution. The Apex Call director satisfied this requirement. Each in-bound call is answered by an operator, one of which may be the "phantom".

While all of these changes have made significant improvements in our daily operations non would have been practical without significant additions to our hardware, software and applications environment.

3.2 Data Processing Profile

Over the past two years Gorham has also gone through a major upgrade to its computing machinery, systems software and applications portfolio. Evolving from an IBM 4341-11 computer with native DOS/VSE for an operating system, BTAM supporting remote communications, VTAM for local communications, 3375 disk drives, two 3420-6 tape drives and two 650 LPM printers and a 3705-M81 communications controller we now have installed:

-4381-P2 computer
-3380 AD & BD disk drives
-3705-M83 communication controller
-3420-8 tapes
-4245-12 printers
-VM/VSE with two guest machines
-SNA/SDLC
-DL/1, CICS
-3174 Cluster Controllers
-Series One; Shop Floor Data Collection
-CMI bar code terminals
-System 36 in Aiken, Williamsport

A similar change has taken place in our applications portfolio. Starting with a tewenty five year old payroll, a seven year old Order Entry, and an Accounts Receivable,and Accounts Payable system of indeterminate age and finally a General Ledger we have made significant progress in the development of new systems to support our business. First was a major upgrade to all systems software to bring it up to currnet release level. Next was the initiation of concurrent projects to design, develop, and implement using database technology an integrated series of systems. Among these are:

- Payroll/Personnel

- Warehouse/Distribution System

- MRPII

- Procurement

- Shop Floor Data Collection

- Order processing/Customer Service Bronze

- Shop Floor Controll Bronze

Each of these application have been developed and are in the final implementation phases i.e loading Bills of Material, Routings, etc.
Of greater importance each are being used in the daily conduct of our business. The integration between the various sites has been dependant on the use of communications to tie all facilities togather and database to provide commonality.

4. ISDN in a Conglomerate

Despite the fact that we as a division of a major corporation enjoy major account status with practically every vendor with the attentive discount structures we are often unable to take advantage of the product or services being offered. This is due to two factors. One, the autonomous and decentralized nature of the corporation obviates the need for many of these products and services. Two, the lack of centralization of purchasing and direction results in multiple level pricing structures. This results in buy/lease decisions and vendors being determined by the operating divisions. Essentially they determine what to get, who to get it from anf when to get it. Corporate reserves the right to review the financial justification, legal wording of all contracts and the need assessment for the equipment or service being requested. The lack of centralized reporting again suggests that ISDN, in a conglomerate is a solution looking for a problem.

To the operating division there is a lack of need, to say nothing of financial justification or cash. For the corporation, with the potential for restructuring through the purchase/sale of operating business the need to be able to assimilate or disconnect with minimal effort dictates that while a concept such as ISDN is attractive it is not a necessity now.

5. Conclusions

While Integrated Services such as ISDN will undoubtedly play an increaingly important role in the future for the conglomerate they are not essential now.

A State Public Service Commission Staff View Of ISDN

Robin Pischel Ancona
Robert Bryant
Ronald Choura

Michigan Public Service Commission

Lansing, Michigan

1. INTRODUCTION

Government regulation has been a way of life for many years although many industry participants are proposing to reduce government regulation. The Federal Communications Commission (FCC) claims they are reducing regulatory burdens, but in fact have implemented more regulations [1] affecting more local exchange carriers than ever in the history of telecommunications. We did not study Roman civilizations while in school because it was Rome. We study past history to shed light on the present. The impact of regulation is pervasive, and all of us must learn to deal more effectively with government as a major participant in the telecommunications area. To quote Vern J. McGinnis, former president of the American Marketing Association, as someone from outside a regulated telephone company, ". . . to an increasing degree, decisions formerly under the control of marketers are limited by or dictated by powers outside of their own organizations."

In fact, the current federal administration and the FCC believe that government has no positive role to play in domestic and international communications. The past Chairman of the FCC, Mark Fowler, has launched a crusade to deregulate all local exchange service rates (that which is in the states' regulatory authority and not the FCC) leaving captive consumers to the mercies of their monopoly local exchange carriers. This "back to the future" program sounds adventuresome but ignores the historical lessons that monopolists must be regulated in the public interest in order to provide essential services at reasonable rates. [2]

This policy also eliminates the level of oversight that exists today to criticize and limit the one-sided views of the FCC and big business in setting future policies for all of us to follow.

The Communications Act of 1934 was passed when only 30 percent of all households had telephone service. As a result of the Communications Act of 1934, which put into place government controls on cost-sharing, monopoly markets and rate averaging, we now have 93 percent of the nation's households equipped with telephone service. We have virtually no integrated information switched services in today's households. Similar congressional action may be needed to achieve Integrated Services Digital Network (ISDN) in every household so that consumers can be a part of the information network of tomorrow. However, there may be room today for multiple providers, competition and ubiquity in the industry. State regulators still need to protect ratepayers/consumers from unnecessary and unfair rate increases through the use of innovative rate structures which will assure the affordable basic services that exist today. [3]

We are rapidly entering a new market and business environment where information becomes a commodity, and the means to move information becomes a vital business activity. The telecommunications industry must realize the need for end users to command many types of information quickly and efficiently while minimizing costs. Thus, the information age has yielded many methods and schemes to move information. Each method of communication has its own form of transmission and each is autonomous. Finally, each individual way of communication has a unique set of expenses.

The computer, telephone and video services of today are valuable services. If industry can bring these services together, clearly a far more valuable and needed service will result. [4] This will clearly be a case where the sum of the parts is not greater than the integrated whole.

Even with advances in telecommunications, today there is not yet on a ubiquitous, nationwide basis any feasible way to incorporate all of the communications needs of one user without great expense. What is needed today is a system that could carry signals of voice, data, two-way video and a multitude of other services for people with increasing communication needs. [5] Leaders in the telecommunications industry are observing that current technology could bind all needed services together through one transmission medium. There was no one that had done it economically, and there were no ideas on how to let such a system use today's technology and yet have the capacity to embrace new services in the future. However, the industry is developing interim steps and technologies to begin meeting these increasing communications needs. Some of the steps are Switched 56, Switched Digital Networks (SDN), Packet Switching, Signaling System 7 and Integrated Services Digital Network (ISDN).

The idea of ISDN is still developmental. Marketing and advertising offices splash glossy ads of ISDN switches and software applications in trade magazines. Except for a few trials, there is no ubiquitous ISDN system in place. A working universal ISDN system is still at least a decade away.

2. ISSUES OF ISDN

Regulation and policy decisions are very reactionary. They come only after a problem in technology has passed a point of grave concern. New technology presents problems with regulatory policy in terms of foreseeing possible negative effects. Many times regulators are not in the position to allow new products because of the simple lack of information and increased costs. It is for this reason that the topic of ISDN should be addressed now, so that current and plausible effects are anticipated.

ISDN may or may not be practical or cost-effective. Its implementation calls for a "revolution" in the public exchange network. It requires every user and carrier to submit to its standards of connection in the public exchange in order to be truly effective. ISDN services will have a great effect on business and society once it is in place, and those overseeing its implementation should not make light of its effects.

The decision to go along with this revolution is a key issue in and of itself. There are those who are wary of hedging on the public exchanges going to ISDN; others see ISDN as the zenith in communications capability and are willing to forge ahead with this new idea. What must be considered along with both views is that, eventually, advances in telecommunication will make the current system obsolete. Therefore, choices have to be made now as to whether or not to go along with this "all-encompassing" technology or to wait for a new form of integrated network that is better than ISDN. This is a real dilemma. But what many fail to recognize is that Integrated Digital Networks (IDN) are a prerequisite to ISDN, and so are Signaling System 7 and tariffs for regular service. Even then, all we may see is narrowband ISDN. These prerequisites are not without their problems, and the industry alone is not solving them amicably.

3. STANDARDS OF ISDN

The concept of ISDN is the ultimate goal for many exchange carriers to survive competitively in their attempt to wire the world with communication capabilities. However, this concept is still a fleeting one. Some of the critical standards have yet to be agreed upon. While many carriers are unwilling to commit fully to ISDN until universal standards are accepted, others are gambling on certain standards and moving ahead.

The final authority rests with the International Telecommunications Union's Consultative Committee on International Telegraphy and Telephone (CCITT). Their job since 1972 is to formulate ISDN technical and policy standards for use internationally. They plan to give the final decisions on ISDN at the CCITT's Ninth Plenary Assembly in October of 1988. [6] It is expected that the decisive final definition cast in 1988 will set ISDN apart from IDN. This final decision will be greatly shaped by the government and industry vying to set the standards of ISDN in the most favorable light for themselves. For example, the FCC, in its pro-competitive deregulatory stand, stated that it:

> ". . . tentatively concluded that it is important that the overall ISDN concepts which evolve appropriately accommodate the Computer II policies; that a multiple-vendor environment in the United States not be foreclosed through technical recommendations or concepts directed toward a single-vendor environment; and more generally, that sufficient flexibility be built into the broad range of different market structures which exists in various nations." [7]

While the FCC recognizes that nationalized exchanges like the Post, Telegraph & Telephone authorities (PTTs) will have monopolized control, they still want the CCITT to incorporate a pro-competitive stand in the ISDN formula that specifically benefits the FCC. The FCC, to promote this stand, recently said that when the North American industry develops a standard for the ISDN numbering scheme, the Commission will use the industry recommendation as the basis of a rulemaking proceeding. [8] It is for this reason that telecommunications companies and state regulators should exercise caution before committing themselves to ISDN in both manufacturing and policy. The plans for implementing ISDN calls for the system to advance to the primary rate of interface once the basic rate is in full use. [9] However, at this time many questions remain regarding this standard. The key questions are: When will the switch to broadband be made? How much additional costs will there be to upgrade the system? And what are the network termination, terminal equipment, interface, compatibility and interconnection standards? [10]

These remaining technical standards for ISDN have an impact in three areas: strategic planning in business, regulatory issues in government; and, finally, new technology in manufacturing digital equipment. Those who are most concerned about these remaining technical standards are those companies who have gone ahead with ISDN compatibility in their products. They realize that their investment might be wiped out if standards vary from original projections. Thus, the influence they try to exert on government and standard-setting is very great.

4. SERVICES OF ISDN

Through standards of ISDN, the network will be able to offer services that have never before coincided with one another. This prospect has led to intense excitement by carriers who are eager to offer the ability of integrated audio, video and data services. Carriers also realize that the capital outlay of offering these services will only happen once; then each user will be able to order their own unique selection of services. The network will also open up new service opportunities including human-to-human communication, encryption, real-time configuration of leased lines and calling party identification. [11]

Carriers are going to have to review how they offer these services in relation to billing practices now used by common carriers. New methods of tariffing will have to be developed that set a fair price for a true degree of service while enabling the carrier to survive profitably.

5. GOVERNMENT AND INTERNATIONAL GROUPS

In order for the ISDN technology to be an evolutionary process rather than revolutionary chaos, extreme participation by various governmental groups and international organizations must be maintained. One thing that we all must keep in mind, however, is that customer needs create markets; scientists and engineers invent new technology; and regulators and other governmental organizations formulate rules. All participants must be compatible, supportive and work together to achieve the increased opportunities of an integrated information age. [12]

The first and foremost group is the CCITT. Composed of members from many countries, they set the standards that are followed by all United Nations countries. The second group that plays a role with ISDN is the T1 Committee's T1-D1, ISDN Subcommittee. They develop standards for U.S. telecommunications and have contributions forwarded to the CCITT via the State Department and U.S. CCITT study groups. The T1-D1 is composed of working groups made up of members from carriers, resellers, manufacturers and government agencies. The third major group is composed of the Federal Communication Commission and the National Telecommunications and Information Administration (NTIA). This group regulates, forms policy and greatly affects the CCITT.

The fourth group has yet to affect ISDN, but they will. The public utility commissions (PUCs) in the United States will observe the evolution of ISDN in the public exchanges of their states. In turn, the PUCs will have to deal with how implementation affects the citizenry and how new tariffs for all of the services will be handled.

The role of each group is important to the success of ISDN. Each group affects the outcome of the others, and all of these groups will leave their mark on the network. Their effects on the final network structure is of paramount interest.

5.1 The Role of the CCITT

The world's advanced nations felt that international communication was an important issue for international commerce, law and peace. The International Telecommunications Union (ITU), an agency of the United Nations, was established to provide worldwide standards for communications. The Union's Consultative Committee on International Telegraphy and Telephone is the medium for recommendations for international communications systems. [13] The recommendations set by the CCITT serve to let all U.N. member countries achieve communications capability with telecommunications systems, thus making international communications possible.

All recommendations for standards are formed at a study group level. Two important study groups are:

- Operation and Quality, and
- General Tariff Principles.

Although standard-making groups, such as the CCITT, set technical standards for the best integrity of the ISDN, they have been accused of being biased. However, most of the problems are not technical in nature, but rather focus on when those technologies have a bearing on regulation and policy. [14]

Policymakers in the U.S. should also be aware that the CCITT is responding more to European PTTs due to the fact that ISDNs in countries such as Germany and France are at a very advanced state. Therefore, standards at this point may reflect an effort to allow the European ISDN to advance.

There are still uncertainties as to how the United States will cope with these decisions.

5.2 The T1 Committee

The T1 Committee is an independent group which develops interconnection standards for telecommunications services in the United States. It was founded as a post-divestiture standard-setting organization. The Exchange Carrier Standards Association (ECSA) is the T1 Committee's secretariat and administrative body. [15] The T1 is accredited by the American National Standards Institute (ANSI), who, since 1979, was chosen by Congress to represent the United States in dealing with non-governmental standard-setting organizations abroad. [16] Membership in the T1 is open to exchange carriers, interexchange carriers and resellers, manufacturers and vendors and those with general interest in telecommunications standards, such as government agencies, consultants and user groups. [17]

The T1 represents a great force on the U.S. CCITT with recommendations that are then passed on to the CCITT in Geneva, Switzerland. However, like the CCITT, the T1 may be somewhat "captured" by big telecommunications companies. Others express concern that the T1 is ignoring the effects and standards on terminal equipment, notably private branch exchanges (PBXs). [18] But the fact remains that the T1 is only one cog in this international struggle for the ideal set of ISDN standards. The T1's effect will be measured by how many of their recommendations actually hit the table in Geneva.

5.3 The FCC

The Federal Communications Commission will have a great bearing on the success and environment in the U.S. in which ISDN will evolve. Moreover, the FCC is taking steps to ensure its position is accounted for internationally. Concern about this was apparent when, on April 2, 1984, the Commission adopted a First Report in Docket 83-841. For the first time, the Commission took action on a topic that was yet undetermined by the CCITT. This unprecedented move started a new era of regulation; an era many feel will lead to further federal deregulation in the telecommunications industry. [19]

The reason the Notice of Inquiry was filed is twofold. First, the FCC prepared a positional paper in 1982 that was submitted to the CCITT expressing concern about whether ISDN standards will permit differing national systems to offer ISDN and if consumers of the service can choose the routing of messages. The second and prevailing reason was that the FCC perceived a shifting role of the CCITT from purely international technical standards to that of standardizing techology that would avoid the duplication in systems of the PTTs. [20]

Another reason for concern by the FCC was that ISDN was being lead by the Europeans, notably the Deutsche Bundespost of West Germany. The pre-divestiture posture of AT&T was not in a position to recognize problems relating to competition, and therefore, could not resolve them. Thus, the FCC felt that they should step in and at least seek comment on the FCC's role in ISDN and to stimulate public interest. [21]

The Notice of Inquiry was open for comment and thirty-four parties filed comments. Some had never had experience with ISDN before. The first report that resulted included the following points:

> "It is not the function of the Commission to plan or design carriers' or others' ISDNs . . .
>
> "The ISDN numbering plan must have sufficient flexibility to allow customers to specify carrier choice in those countries that choose to implement such an option . . .
>
> "ISDN recommendations should not limit the number of satellite hops allowed on an international connection, but should provide for customer choice as a natural option . . .
>
> "Customer provision of the NT1 interface device must be a national option since the United States views this as competitively supplied customer premises equipment . . .
>
> "CCITT recommendations must be sufficiently flexible to accommodate the U.S. basic-enhanced dichtomy." [22]

The FCC views its role as more of a guide than taking a major part. Indeed, the FCC has decreased its role in radio and the provision of customer premises equipment (CPE) to encourage competition in the industry. This trend may or may not continue with ISDN.

No matter what trend the FCC will follow regarding ISDN, one major obstacle was the Computer Inquiry II (CI-II) which did prohibit local exchange carriers from offering "enhanced" services or providing CPE except through a separate subsidiary. [23]

In June 1983, the Commission clarified that Network Channel Terminating Equipment (NCTE) is customer premises equipment within the meaning and policies of the CI-II and that it therefore is to be provided on a competitive unregulated basis no differently than other CPE. [24]

Pacific Bell has put their ISDN test project "Victoria" on hold while asking the FCC to reconsider a Common Carrier Bureau (CCB) ruling that Victoria's multiplexer is NCTE and thus CPE.

However, the FCC eliminated its CI-II restraints, and in July 1985 initiated the Third Computer Inquiry (CI-III) and issued its decision in June of 1986. [25] This decision does not address the concerns of project "Victoria".

FCC Commissioner Patricia Diaz Dennis, in a speech to the International Communications Association in New Orleans on May 21, 1987 said the FCC has a role to play in advancing national competitiveness through facilitating the provision of advanced telecommunications and information technologies to all users.

Commissioner Dennis said that "regulatory policies that tightly curb or compartmentalize telephone company growth and development may result in an out-of-date, high cost, low performance communications system." She questioned policies that "may prevent a significant number of customers from benefiting from possible Bell Operating Companies' (BOCs) efficiencies and that "may have the unintended and negative effect of depriving small users access to the kind of rich network features, functions and services currently available to large users . . . through . . . PBXs and private networks."

She said that the well-being of large corporations "is linked to that of small users, because our economy depends, in large part, on the entrepreneurial spirit of small firms that generate the largest number of new jobs as they both supply and buy from large corporations . . . . If these small users cannot benefit from advanced telecommunications and information technologies, we all will be worse off."

Commissioner Dennis questioned whether FCC rules "designed to separate basic and enhanced services or competitive from non-competitive equipment and services fully serve the public interest. . . . The distinction between basic and enhanced services is near meaningless," she said. Because protocol conversion is integral to regulated packet switching services, Commissioner Dennis disagreed with the Commission's recent decision in Computer III, Phase II, to define protocol conversion as enhanced, rather than "as a neutral function or option which could be provided on a tariffed basis." She said that "fundamentally, I come out on the side of allowing technology to be used."

The FCC CI-III decision is properly moving to provide a national framework which encourages the upgrading of local public networks. The thrust of CI-III is to correct some poor regulatory policy and allow the public to benefit from new technology. Although standing in the way of this progress are the restraints of the Modified Final Judgment (MFJ) on the BOCs. [26]

Out of basic issues that involve CI-III, one involves ISDN specifically and two deal with ISDN indirectly. The first of these issues is the plausibility of locating other carriers' equipment in the central office of a dominant carrier. This rather unusual suggestion may solve some of the interconnection and location problems of the ISDN while giving the opportunity for dominant carriers the chance to earn additional income from renting out space. Regulators may be needed in this area, however, to insure fairness in the provision of these types of services. CI-III also proposes two new regulatory structures, Comparably Efficient Interconnection (CEI) and Open Network Architecture (ONA). These structures require exchange and interexchange carriers to provide their own enhanced services operations and those of customers and competitors with equal access to the carriers' underlying transmission facilities.

A final question left to be answered is what stand the FCC will take in the regulation of ISDN. In the end, proponents of ISDN hope that it will be an end-to-end worldwide network. It is felt that the concepts of local access transport areas (LATAs) and inter/intrastate calling will be moot points in tariffs and regulation of the network. However, that time has yet to arrive, and until it does, the FCC may have to let the states take a greater role in the initial stages of ISDN development--the topic of our next section.

5.4 The PSCs

The role of the public service commissions (PSCs) is to control the published rates, regulations and descriptions governing the provisions of ISDN services. [27] The PSCs are expected to protect the consumer, yet create an environment that fosters the development of new technology and consumer options. This is no easy task. Many are critical of the PSCs due to the wide range of unique regulations that each state has. In addition, the dividing line between federal and state regulators is often hazy. The NTIA expressed concern that:

> "The line between state and federal jurisdiction is not always clear because the small facilities may often be used for both types of service. The transition to an ISDN may be delayed in the United States unless these jurisdictional disputes are resolved. New firms may be reluctant to enter the integrated services market unless they know if, by whom, and how they will be regulated." [28]

PSCs develop and coordinate interconnection of competitors and consumers at the grass roots level. PSCs resolve disputes with technical support staff and legal arbitrators who are knowledgeable in the industry (a feature that courts don't offer).

In mid-January 1986, a survey was taken of the nation's PSCs by the Michigan Public Service Commission staff (MPSC) to find if there was any policy being drawn up on the topic of ISDN. The survey was in the form of a questionnaire requesting views on many different issues.

States replying to the survey wonder if the ISDN regulation or tariffs they suggest will be preempted by federal control. But more often, state PSCs express that it is too early for any consideration of ISDN regulation to take place.

6. IMPLEMENTING ISDN

The task of implementing ISDN will be one that will involve significant cost and time. Most agree, however, that it will not be put into place all at once. Some see it as being implemented in stages which will spread the cost and labor over time. How carriers will approach this implementation schedule will depend on the demand for the services that ISDN offers, coupled with the retirement of outdated switching equipment in a geographic area. However, it is viewed that this switchover to full ISDN will occur in three large phases--the change to ISDN switching at the central office level; the implementation of narrowband ISDN; and finally, the change to broadband ISDN. [29]

Because ISDN will only occur at significant cost, some carriers are first demonstrating what the network might do in terms of service and options, then signing customers up for the service. The carriers are opting for this strategy rather than building the network and finding it won't sell. Some accuse the network providers and vendors of overselling the possibilities of the network, thus leading to inflated hopes and customers unsure of whom to trust. [30] On the bottom line, the real success of ISDN lies with those who will want, use and ultimately pay for the services.

7. COSTS OF ISDN

In the MPSC survey of the PSCs, the most expressed concern was over "premature retirement of the existing plant". This dilemma is evident for the approximate 75 percent of the switches in the U.S. which are analog in nature. [31] However, if ISDN is to evolve at an orderly, cost-efficient rate, it is hypothesized that the replacement will occur only in those areas with a base of willing customers that can bear the initial cost of the service. To update the typical switch may cost up to $6 million. [32] This cost, hopefully, can be spread out over the users of the service and not the users of the basic local service. While regulators have begun to help modernize the industry by moving toward a more cost-based structure, [33] basic safeguards need to be established to ensure that ISDN new ventures are not unfairly subsidized by basic services. [34] Paul Popenoe noted in a NARUC Subcommittee on Communications meeting in July of 1985 that:

> "Any replacement program requires extremely close scrutiny to assure that it is truly cost effective and directed to those areas where greatest benefits can be derived from the digital technology." [35]

It is expected that the regulation issues of replacement will be addressed on a case-by-case basis until a clear direction on the issue is reached. State regulators should keep communication between them open on the issue in order to establish a uniform policy. [36] Doing this will mean less work for all PSCs and a unified base of policy in the United States. Also, this will help assure that cost allocations are proper, otherwise ratepayers may subsidize the competitive ventures of the industry. [37]

8. TARIFFING ISDN

Tariffs presented to the PSCs on ISDN may be based on totally different principles than with the current local exchange carriers. Currently, principles of rate-of-return pricing and rate averaging are widely used throughout the country. [38] With ISDN, a more objective stand may be taken with tariffs.

The traditional methodology of cost and rate of return must try to match the objective side which is value of service and public interest. The public interest is the added factor in trying to determine the value of service. [39] Thus, value of service may have many individual components. However, fairness is a necessary requirement to overall policies in assigning costs. [40]

Study Group III of the CCITT has outlined questions on the tariff issue of ISDN. They are broadly based, but are key in determining the objective side to our equation.

- How can tariffs for voice and non-voice services be reconciled?

- How should the tariffs on similar-to-traditional analog services be established?

- What is the efficient means of accommodating the fact that costs are the same for every user but benefits differ among user classes?

- Can the effect of the Local Area Network (LAN) be accounted for in tariffs?

- Can the services provided by satellites be subjected to standard tariffs?

- What should determine optional access fees for interconnection of ISDN and non-ISDN networks?

- Should the fact that, according to national policy, ISDN service is available only in particular regions affect the tariff setting policy?

- Can the fluctuation of costs, dependent on the degree of expansion of service areas, be represented in the tariff structure? [41]

State regulators can use some of these same questions to help determine these objective "value of service" questions.

On the other hand, tariff setting for the value of service may be very subjective in terms of handling. Billing by the amount of traffic handled is one possibility. Doing this effectively cancels out any "value of service" consideration. Arguments against doing this with communication are these: "An ISDN is essentially a network of networks and will involve a variety of conditions which will present problems starting with the lowest common denominator. That denominator is that all services are digital and that they pass through the same interface such as an NT1 or NT2 terminal. [42] Thus, digital services can be lumped together in the "billing-by-the-bit" method even if their service is of the highest importance. Thus, voice would cost more than facsimile, ignoring that the current value of facsimile service is put on a higher level." [43]

Value of service involves six component features:

- Distance
- Time of Day
- Type of Service
- Amount of service handling required
- Bit-load of the service
- User (residential or business)

Each of these categories has a bearing on the value. In voice communication, the information is unchanged and the most basic; thus having a low value, while services such as data base access involve great handling and are placed at a higher level of service.

9. REGULATION AND ISDN

The structure of this new network may create the need for new thoughts and ideas in common carrier regulation. There will be a need for new policy in three areas:

International
- transborder data flow
- intelligence ownership
- international provider, manufacturer and government cooperation

National
- competition with the network
- interstate traffic
- FCC and PSC tariff, cost and policy coordination

State
- interlata/interstate traffic, cost and policy coordination
- tariff standards
- consumer relations - arbitrator in complaints on problems of implementation

These areas affect the roles of the others in that international regulation/control will affect national regulation/control and so on. However, in each of these areas a new policy must be developed that will be adaptable to the new technology, structure and services of ISDN. As stated by James E. Olson, Chairman of AT&T, regulatory oversight is needed--oversight which is more focused and goal-oriented, more flexible and less cumbersome.

For most state regulators, international problems should never be a problem. However, ISDN knows no geographic boundaries with respect to the structure of the network. Therefore, recognition of some key issues is necessary to understand the possible regulation that could transpire out of international problems.

Many other nations in the world do not share our policy for the marketplace approach to regulation. The most obvious reason, among many, is that many countries do not have a free and open economy. The economic climate in Third-World countries is continually dominated by only a handful of industries. Many countries view information and data flow as an "instrument of state policy" and require greater management of the medium. [44]

The FCC will have a great impact with regard to competition in the ISDN system at the national level. Currently, the FCC is at odds with the international standard-setting committee of the CCITT in the areas of network termination and technical standards of the network that have resulted in non-technical policy considerations.

Domestic policy with regard to competition will continue to hold the ISDN implementation scheme to levels of field testing and business applications unless these policies are altered. Dominant carriers, such as AT&T and the BOCs, are generally prohibited from offering "enhanced" handling of communications without a separate subsidiary or waiver from either the FCC or federal district court. This condition will means that the dominant carriers are unable to use the technology and resources that are needed to implement the network. Since these dominant carriers serve 90 percent of the market, which includes many rural areas, the technology may not reach many customers for a long time unless the policy is changed. In an article written in March of 1981 by Manley R. Irwin and John D. Ela in <u>Telecommunications Policy</u>, they stated that current policy:

> ". . . identified the forces acting on industry or assessed the implications of those forces on the alleged concept of national monopoly? Has policy been focused on the private rather than the public section in search for future prescriptions? Policy makers are currently attempting to wed the past, the present and the future to marry monopoly to competition to deal with technological change and uncertainty." [45]

The FCC and other involved policymakers are trying to ascertain how to regulate in a competitive environment. Solving this problem is paramount if ISDN is to be successful. Dominant carriers must be free of the constraints that hold advances back. With the vast amount of revenue that could be earned in the service sector of our economy, it could prove to be economically disastrous to continue to hold down technological implementation with the false argument of competition. In the simplest terms, we may find ourselves behind other countries in telecommunications technology. With the structure of ISDN, there is simply no need for structural regulation. Instead, the FCC needs to develop a policy to allow integration of services without unfair cross-subsidization.

The FCC and other national policymakers need to work very closely with both international and state policymakers as proposed by The National Association of Regulatory Utility Commissioners (NARUC) in St. Louis, Missouri in 1983. Each of these regulators needs to assess their goals and objectives so that all levels of policy setting are working in harmony. The FCC needs to establish broad principles for cost allocation, tariff structure and policy direction, but only after coordinated input and consideration of state concerns. Then the FCC must relinquish day-to-day oversight to the states and industry so that only one rate structure for joint state, national and international services can be developed. This would eliminate many of today's multi-regulatory steps which are not only time-consuming but costly and would, at the same time, retain control and needed policy direction that both consumers and industry want.

For the state PSCs, the issue of what positions to take in regulation is still not in the clearest of terms. However, two practices can make regulation at the state level easier--cooperation and interexchange of thoughts and ideas between the FCC and the PSCs, and state communication amongst themselves. ISDN may be heavily regulated by the FCC because of the end-to-end worldwide architecture that it is supposed to achieve. On the other hand, regulation may be left to the states, because outside of interstate transmission, all processing is done within the geographic boundaries of the state. [46]

State PSCs have the front-line responsibility for creating mechanisms to bring the benefits of new technology to the public via upgraded local networks. [47] This is particularly true in light of the recent Supreme Court decision rejecting the FCC usurpation of state jurisdiction over intrastate communications. [48]

The Communications Act of 1934, 47 U.S.C., et seq., provides for a dual federal-state regulatory scheme and reserves jurisdiction over intrastate communications to the states. Specifically, it provides:

> "Nothing in this chapter shall be construed to . . . give the Commission jurisdiction with respect to 1) charges, classifications, practices, services, facilities, or regulations for or in connection with intrastate communication service by wire . . . of any carrier . . . ." (47 U.S.C. 152 (b))

A separate issue will arise regarding any attempt by the FCC to bar state regulators from developing their own cost allocation rules for purposes of intrastate regulation. Federal and state regulators acting independently, may arrive at different conclusions regarding the appropriate procedure for allocating costs associated with a specific dually regulated activity. The FCC, for example, may use all minutes-of-use for allocating all access services as was proposed for equal access costs associated with providing interexchange toll facilities for equal interexchange carrier access service. On the other hand, a state commission might conclude that only the minutes-of-use associated with services providing interexchange access is a more appropriate surrogate. A similar result might occur with respect to composite allocators. The FCC may establish a composite allocator of state interlata and interstate/interlata minutes-of-use for allocating the cost of interexchange access service, while a state commission may select a different composite allocator.

It is beyond dispute that development and implementation of a single set of cost allocation rules will be an enormously complex task, imposing significant demands upon the resources of regulators and the industry. The complexity and demands will only increase if two separate regimes of state and federal cost allocation rules are developed. Thus, the development of consistent state and federal cost allocation rules must be encouraged.

The FCC, therefore, lacks the authority to command state regulators to adhere to its cost allocation rules. Under existing law, state regulators may develop their own cost allocation rules for intrastate services regardless of the FCC's action in this proceeding for interstate services.

In <u>Louisiana Public Service Commission v. FCC</u>, 54 U.S.L.W. at 4511 [slip op. at 20-21], the Supreme Court recognized that state and federal regulators could require communications companies to maintain separate books of account. While the implementation of separate state and federal accounting requirements would surely be burdensome, the Court's opinion suggests that such a system would pass muster under the Constitution and the Communications Act of 1934. Thus, FCC preemption of state authority is not available to prevent development of separate regulatory requirements.

It is believed that the best method for developing cost allocation rules for dually regulated services lies in the use of a cooperative federal-state working group akin to the Section 410 (c) Joint Boards.

Such a procedure would have many advantages. It would bring to bear the combined resources of state and federal regulators to address the complex and difficult issues raised by cost allocation for dually regulated services. The FCC would also obtain the expertise of state regulators familiar with services deregulated at the federal level. Moreover, development of these cost allocation rules in a cooperative procedure will promote agreement on consistent federal and state rules. Finally, a cooperative procedure would eliminate some of the costs associated with development of separate federal and state regulatory schemes. This procedure is not without flaws and will not work unless both state and federal regulators work together.

Accordingly, the FCC is urged to use this or some other cooperative procedure, not a legally untenable attempt to exercise preemptive authority, to develop a body of consistent federal and state cost allocation rules. The MPSC staff in FCC Docket 86-111, the Joint Case proceeding, has offered its recommendations for the accounting procedures to follow in allocating common costs. The MPSC staff recommendations state that all costs should be directly assigned whenever possible, and then common costs should be allocated using a "common denominator" approach as a general rule. There is no possible way for the FCC to develop proper allocations of common costs between regulated and non-regulated activities or between dually regulated activities without knowing all of the components involved. There will be jurisdictional revenue shifts involved if the cost allocations are made either before or after the separations process. The FCC needs to set up test models and studies to see what these effects are before they attempt to develop cost allocation methodologies. This process must include state regulators' involvement in order to prevent future jurisdictional conflicts.

The FCC should not mandate the alternative measures for allocating common costs. On a case-by-case basis, the companies should petition both the state and federal regulators with their cost allocation plans for review of the accounting and allocation mechanisms prior to setting up an ISDN system. The FCC could set up a 30-day filing period and a 30-day review period for both state and federal approval. This should be done on a test basis for a few years so the companies could develop the alternatives themselves. After a few years of testing, a uniform methodology could possibly be developed. If there are differences between what the state regulators, federal regulators and the companies feel is the proper method of allocation, then a joint committee process similar to the Federal-State Joint Board process could be set up to resolve the differences. The main emphasis here is that of mandatory state participation in this whole process. Anything that is to be required to be filed with the FCC must also be required to be filed with the appropriate state commission, and any reporting requirements must meet both state and federal concerns.

States should continue to have the authority to regulate all intrastate telecommunications operations, since the state regulatory authorities are optimally positioned to determine and evaluate local conditions while protecting the needs and interests of consumers. Historically, the states have been the only regulatory bodies to extensively audit and investigate the activities of the nation's telephone utilities to uncover and correct any areas where the cross-subsidization of regulated and unregulated activities has occurred. The states are better equipped to handle this process as they are involved in the local ratesetting process where the disallowances of these unwarranted costs occur. The states should be allowed to continue to monitor this process through the use of audits, investigations and any other means necessary to prevent cross-subsidization. State and federal governments should work to develop national guidelines, but states should be allowed to administer the procedures. The actual procedures to prevent cross-subsidization at a national level need to be flexible and dynamic for various geographic differences.

Federal procedures should preserve existing state authority to determine the records and accounts of carrier operations subject to state rate jurisdiction. Although the states generally voluntarily adopt FCC-prescribed records and accounts, state-regulated communications services frequently involve operations significantly smaller than those regulated by the FCC. That fact, coupled with regulatory concerns unique to each state, suggests that FCC accounting rules may not always be the most appropriate. Accordingly, we respectfully urge that the states be encouraged, but not required, to adopt any FCC accounting rules, and that an administratively final Federal-State Joint Board be set up to develop these rules. This board should also establish the allocation procedures of these accounts between the state and federal jurisdictions in order to establish the respective jurisdictional revenue requirement.

10. CONCLUSION

Changes in technology are one of the hardest areas to conceive in today's world. However, the United States is blessed with a system of government that can incorporate these advances as they come. Society, too, has become well adapted to change and advancement. The public often seizes such changes and molds them quickly into domestic use. ISDN represents an opportunity like this; however, unlike advances in products, ISDN is a new concept in our communications system. It will bear a multitude of new products and services in our economy. It will change work habits and the amount of time people devote to certain tasks. In short, ISDN will change any activity associated with the telecommunications industry.

ISDN will only achieve a degree of success if it is allowed to be implemented in a long-term time-frame, at a reasonable cost and in a conducive regulatory environment. The regulatory environment is most critical. If the United States is to catch up with the current rate of advancement now experienced in Japan and Europe, then dominant carriers must be allowed to operate without discriminatory regulation. If the government truly wants to view communications such as telephone as competitive, then it should stop regulating it like a utility. The services of ISDN will not be used by everyone, especially the average rate-payer, but should be available to everyone. This does not mean that universal service is threatened. The market for this new network will be limited to those customers with computing and data handling needs, and only when economies of scale are realized will the market broaden. Policy should reflect an effort to achieve economies of scale.

Regulators must become better educated on the topic if suitable policy is to be set. Regulators must not deregulate and run away from new technology such as ISDN. The FCC and the PSCs must discuss the dividing lines between them so that less confusion will result. States may want to think about similar objectives as those the FCC and CCITT are working on. ISDN may be the start of a "global village" as touted by Marshall MacCullen some twenty years ago. To have dissimilar policies would threaten this concept.

We urge those who strive to formulate policy to learn more by studying past mistakes and positive directions. We cannot stress enough the effects this new network will have on the average consumer and on commerce once it is in place twenty years from now. Setting policies for this industry and for ISDN is too vital to be left only to parties trying to achieve narrow ideological agendas. Policymakers need to set objectives: 1) develop universal switched information services to meet consumer demands; 2) ensure fairness in the distribution of costs to services; and 3) promote equitable competitive standards for international communications. Only through dialogue, hard work, cooperation, flexibility, imagination and fortitude will we have a chance to achieve these objectives.

REFERENCES

[1] W. Page Montgomery: "Empiricism in Telecommunications Regulatory Policy", IEE Communications Magazine, Vol. 25, No. 1, p. 5, (January 1987).

[2] John D. Dingell: "Communications Policy for the Future - And for the Present", IEE Communications Magazine, Vol. 25, No. 1, p. 11, Committee on Energy and Commerce, U.S. House of Representatives, (January 1987).

[3] Paul F. Levy: "The Swing of the Pendulum", IEEE Communications Magazine, Vol. 25, No. 1, (January 1987).

[4] James E. Olsen: "New Policies for a New Age", IEEE Communications Magazine, Vol. 25, No. 1, p. 32 (January 1987).

[5] Mimi Weyforth Dawson: "The Reverticalization of the Telecommunications Market", IEEE Communications Magazine, Vol. 25, No. 1 (January 1987).

[6] Masanao Tanase: "Behind the Telephone Debates - 5, Integrated Services Digital Networks (ISDN): Concepts and Issues in the United States and Japan", p. 11, Program on Information Resources Policy, (November 1985).

[7] FCC Docket No. 83-841, p. 3, (April 2, 1984).

[8] David Meyer: "Companies Search for ISDN Telephone Number Compromise", Communications Week, p. 1 (May 11, 1987).

[9] Judy Getts: "ISDN, Network of the Future", PC World, p. 225, (January 1986).

[10] Karen E. Fitzgerald: "Be Prepared for ISDN", Telecommunications Products & Technology, p. 14, (August 1985).

[11] Charles Baker and Robert M. Wienski: "Getting Ready for ISDN", Business Communications Review, p. 3, (November-December 1985).

[12] Irwin Dorros: "The New Future-Back to Technology", IEEE Communications Magazine, Vol. 25, No. 1, p. 57, (January 1987).

[13] John and Mary M. Blyth: Telecommunications, (Indianapolis: Bobbs-Merrill Education Publishing), p. 52, (1985).

[14] "International Telecommunications Standards; Issues and Implications for the 80's: A Summary Record of a July 1982 Workshop", (NTIA Sp-83-15), pp. 113-116, Quoted in Tanase, p. 78.

[15] Megumi Komiya: "Integrated Services Digital Networks in the United States and Japan: A Comparative Analysis of National Telecommunications Policies", p. 15, (mimeographed) East Lansing, Michigan, (1985).

[16] Ibid.

[17] Interview with Donald Hoes, Northern Telecom Inc., Troy, Michigan, (February 18, 1986).

[18] Interview with James Splear, Electronic Data Systems; Southfield, Michigan, (February 17, 1986).

[19] Michael J. Marcus: "The Regulatory Point of View: ISDN in the United States", Telephony, p. 32, (March 11, 1985).

[20] Ibid.

[21] Ibid.

[22] Marcus, p. 34.

[23] Gary W. McBee: "Regulation Restricts ISDN Implementation", Telephone Engineer & Management, p. 98, (June 1, 1985).

[24] First Report in FCC Docket 83-841, p. 22, (April 2, 1984).

[25] "FCC Initiates Third Computer Inquiry", FCC News, Report No. DC-166, July 25, 1985, p. 1, Amendment of Section 64.702 of the Commission Rules CC Docket No. 85-229, FCC 86-252, (Released June 16, 1986).

[26] Dawson, p. 16.; Alfred C. Sikes: "Tomorrow's Communications Policies", IEE Communications Magazine, Vol. 25, No. 1, p. 27, (January 1987) and Richard E. Wiley and R. Michael Senkowski: "Computer III: Breakout of the BOCs", IEEE Communications Magazine, Vol. 25, No. 1, p. 49, (January 1987).

[27] Blyth, p. 53.

[28] U.S. Dept. of Commerce, NTIA/Staff: "A Primer on Integrated Services Digital Network (ISDN): Implications for Future Global Communications", Washington, D.C., (NTIA Report 83-138), p. 57, Quoted from Tanase, p.11, (1983).

[29] Getts, p. 218.

[30] Editorial: "Bandwagons Roll", Telephony, p. 27, (August 5, 1985).

[31] Getts, p. 222.

[32] Ibid.

[33] Olsen, p. 30.

[34] Brian R. Moir: "Post Divestiture Telecommunications Policy Making", IEEE Communications Magazine, Vol. 25, No. 1, p. 47, (January 1987).

[35] Paul Popenoe, Jr.: "POTS or ISDN? Whither the Local Telephone Exchange", 1985 Report of the Committee on Communications, 97th NARUC Annual Convention, p. 129.

[36] Levy, p. 27.

[37] Dawson, p. 17.

[38] Tanase, p. 95.

[39] Levy, p. 25.

[40] Sikes, p.27

[41] K. Narano: "Ryokin N. Taisuru Kangae-Kata", (CCITT N. Okeru ISDN Ryokin No Keukyo), "How to Set Tariffs: Study on ISDN Tariffs at the CCITT", ISDN Seminar, Tokyo, Japan, Quoted by Tanase, p. 97, (November 14, 1983).

[42] Philip Black: "ISDN: A New Challenge for Test Instrumentation", Telecommunications, p. 49, (July 1984).

[43] Komiya, p. 43.

[44] Corydon B. Dunham: "A Workshop on New World Information Flow and Other Issues", Vol. 2, p. 25, Washington, D.C., (1982).

[45] John D. Ela and Manley R. Irwin: "U.S. Telecommunications Regulation, Will Technology Decide?", Telecommunications Policy, p. 24, (March 1981).

[46] Leland Schmidt: "Future in Telecommunications", 1985 Report of the Committee on Communications, 97th NARUC Annual Convention, p. 16.

[47] Dingell, p. 12.

[48] Louisiana Public Service Commission v. FCC, 54 U.S.L.W. at 4511 (slip op. at 20-21).

Regulatory Involvment in ISDN

V. Louise McCarren

Vermont Public
Service Commission

The degree and type of state regulatory involvement in the future of ISDN will depend on several factors, including: what the generic term ISDN is, the regulator's fundamental view of the role of economic regulation in the marketplace, and the ability of the mixed (regulated/ unregulated) communications industry to deliver the promise of the "communications age."

Whether ISDN is, or is perceived to be, a specific set of new services or an architecture to which the network evolves will be important. If ISDN is a specific set of services, competitive cost allocation and pricing questions are more important. If ISDN is a standard to which the architecture of the network will evolve, questions of integration, investment and availability will be more important.

Faced with a changed and rapidly changing communications market, state regulators will respond based on their fundamental view of the role of economic regulation. The rationale of economic regulation - to provide a replacement for the market where a monopoly supplies an essential service - may no longer exist. Regulator's goals, however, are the same: to ensure reasonable price, consistent quality, and availability of new services. The minimalist will respond differently from the market manager in an effort to achieve these goals.

The minimalist will remove _de jure_ and _de facto_ barriers to entry, remove return/price constraints on competitive services, and take other actions to replace traditional economic regulation with the tensions of the marketplace. New regulatory structures will be developed.

The market manager will continue to exercise traditional economic regulation, and take on antitrust concerns. Control of entry by use of the franchise/exclusive right to serve will be continued. Market managers will retain price setting control over all, or most, services and rely on detailed cost allocation studies. Concern with the relative market share of all entrants and antitrust issues will be great, and some may use regulation to redistribute income.

Ultimately, the ability of the communications industry to deliver collectively the promise of the communications age will profoundly affect the degree of government involvement at all levels. Government has a strong and legitimate interest in the essential communications infrastructure, for economic, social, and defense reasons. If new services of high quality, integrated into the network, are not delivered, government will become more involved, particularly if there is a perception that the United States is lagging behind other countries.

The Promise and Problems of ISDN
A State Regulator's Perspective

Terry L. Murray

San Francisco, CA
California Public Utilities Commission

The paper presents the benefits and difficulties associated with ISDN from a state regulator's viewpoint. The benefits of ISDN derive from the service offerings it makes possible (e.g., audiotex, remote meter reading and access to on-line databases) and more cost-effective provisioning of existing services (e.g., multiple lines to a customer's premises). In the short run, large businesses seem the primary beneficiaries of ISDN.

The transition to ISDN poses problems of: (1) timing of investments and (2) mismatching of costs and benefits. The latter may occur if current ratepayers pay more than their fair share of the costs of ISDN, or if one class of ratepayers, e.g., residential ratepayers, pays a share of costs greater than its share of the benefits of ISDN. These and other problems can be solved by cooperative efforts between regulators and telecommunications utilities. Ultimately, ISDN should bring Information Age benefits to all telecommunications users.

State regulators have had a lot of negative press lately. Although a few state commissions are singled out for praise, the outside world's view of most commissions seems highly critical. One of the most frequently voiced complaints is the allegation that state regulators impede technological progress through arcane, backward-looking regulatory processes. In this view, state regulation of telecommunications is particularly onerous, because the telecommunications industry is experiencing more rapid technological change than any other regulated industry.

Integrated Services Digital Network (ISDN) represents the hope on which the telecommunications industry has pinned its future. Without the cooperation of state regulators, ISDN deployment could be delayed, diminishing the benefits that are expected from this new technology. Thus, it is important to ask how state regulators view the promise and problems of ISDN.

This paper represents the perspective of only one state regulatory staff analyst; however, I suspect that many state regulators share many of the enthusiasms and concerns expressed regarding ISDN. I see ISDN as being merely a subset of regulatory policy issues raised by local exchange company modernization.

The benefits of such modernization are well known to most industry observers; therefore, I describe these briefly. The brevity of the discussion should not, however, be interpreted as a measure of the importance placed upon what are very real and significant benefits for all ratepayers. The hazards of the transition from today's network, which is still oriented primarily to POTS (Plain Old Telephone Service), to an ISDN environment receive more attention. Again, this is not intended to indicate an assessment that the risks outweigh the benefits that deployment of ISDN technology will bring.

The remainder of this paper is organized as follows. First, I discuss the promise of ISDN for local exchange ratepayers. Second, I consider the problems that state regulators will encounter in managing the transition to an ISDN environment. The two major problems addressed are: (1) timing of investment in ISDN technology; and (2) matching of costs and benefits of ISDN over time and between classes of ratepayers. I also speculate briefly on problems that may arise in meshing ISDN development with network architecture changes stemming from the Federal Communications Commission's (FCC's) Computer III order, which calls for AT&T and the Bell Operating Companies (BOCs) to implement Open Network Architecture (ONA) plans.

1.0 The Promise of ISDN

Pundits hail the upcoming Information Age as a revolutionary lifestyle change. ISDN is supposed to be the vehicle through which the Information Age delivers many of its services to end users. Thus, the benefits of ISDN are inextricably linked to the benefits of the Information Age services it makes available.

Imagining what those services will be and how they will benefit consumers is not always easy. We have limited experience at this time with information services delivered over the telephone network. "Dial-It" services using the 976 prefix are just a sampling of what may be offered once transmission costs drop and greater bandwidth becomes available via ISDN. If today's 976 offerings and France's experience with videotex are any guide, the most widely sought information services are "Dial-a-Porn" services. Presumably, the information services made possible by ISDN will be more varied and more valued by consumers. The usual list includes access to on-line data bases, remote meter reading for utility services, home energy management, banking and shopping by phone, and many others.

Until those applications are developed, which will likely lag behind deployment of ISDN technology, the greatest promise of ISDN will be the cost savings it make possible for existing network services. The obvious area of application is business data transmission services. Cost savings may also be realized in the provision of multiple lines to residential and business customers, using the greater bandwidth of ISDN to eliminate the need for additional copper pairs.

2.0 The Problems of ISDN

The potential benefits of ISDN are tremendous, but most will not be realized until many years into the future. Yet to obtain those benefits, telephone companies must undertake substantial upfront investment in new network technology. State regulators are understandably concerned about the impact of such investment expenditures on basic rates during the transition to ISDN. Two key problems emerge. First, what is the correct timing of investment in and deployment of ISDN technology? Second, how should the costs and benefits of this technology be matched in ratemaking?

2.1 Timing of ISDN Expenditures

In California, network modernization expenditures have been hotly contested, especially in the current Pacific Bell general rate case. The debates centers on whether POTS customers truly benefit from retiring equipment that is still physically useful, such as some electromechanical switches, in order to deploy new digital technology.

The general framework within which modernization expenditures should be analyzed is not in dispute. Both company experts and the Commission staff and consultants agree that some form of cost-benefit analysis should be used. The categories of costs and benefits are also largely undisputed. Beyond these generalizations, however, neither the specific methodology nor the quantification of costs and benefits seems susceptible to agreement between the parties.

The stakes are high. New investment for Pacific runs around $1 billion annually. Some of this investment is needed to serve growth in demand; much, however, goes toward network modernization. To some degree, maintenance cost savings may justify modernization. A significant aspect of the justification of most modernization expenditures, though, is their revenue-generation potential.

Forecasting the revenues to be gained through new service offerings is typically the most controversial aspect of any analysis of modernization expenditures. The more the service departs from existing offerings and the further into the future one projects demand, the more uncertain revenue forecasts become.

The temptation for regulators to put the brakes on network modernization expenditures can be great. The outlays are immediate and readily measurable; the short-term rate impact of these expenditures is clear. By contrast, the benefits of the new service offerings and even the potential cost savings are distant and ephemeral; today's regulators may not even be around to accept the kudos if modernization programs succeed.

Yet most state regulators recognize that too little and too slow modernization is also risky. If large business customers leave the network because POTS simply does not meet their needs, the remaining ratepayers must pay higher rates to cover the fixed costs of the network. State regulators are also sensitive to the fact that small business and residential ratepayers may be left out of the Information Age unless these services can be offered through the local network. Unlike large business customers, these groups cannot afford to build their own private telecommunications systems. Finally, regulators realize that a POTS-only network could put their state at a disadvantage in attracting and retaining major businesses that rely heavily on the high-tech services a modern telecommunications network can provide.

State regulators must strike a careful balance between the extremes of too rapid and too slow network modernization. Similar issues of investment timing have already been addressed with considerable success in electricity resource planning. Unfortunately, planning in telecommunications is far more complex, and the state of costing and demand forecasting is far less advanced than in electricity. Astute telecommunications managers share regulators' concerns about the optimal timing of investment in new technologies such as ISDN.

2.2 Matching Costs and Benefits of ISDN Investment

Even if the problems of investment timing are solved, state regulators must still develop ratemaking treatments that appropriately match the costs and benefits of network modernization expenditures. Both economic efficiency and equity dictate that the costs of new investment should be borne by those who benefit. Translating this truism into ratemaking is a major challenge.

2.2.1 Intertemporal Equity

Most of the discussion above suggests that the costs of ISDN will be incurred long before most of the benefits of that investment are obtained. This is a typical experience in a capital-intensive industry. Unfortunately, conventional cost-of-service ratemaking "front-loads" costs into rates, so that ratepayers in the early years bear the brunt of investment-related costs. This ratemaking treatment presents little problem when there is a relatively constant level of investment expense over time. Deployment of a new and long-lived network technology such as ISDN, however, is likely to cause an initial surge in investment expenditures, with a significant dropoff in later years.

It would be unfair to make ratepayers today and tomorrow pay higher rates to provide newer, cheaper services for future ratepayers. The issue of fair matching of costs and benefits between ratepayers in different time periods is sometimes referred to as intertemporal equity. Intertemporal equity would best be served by a ratemaking approach that placed more of the burden of ISDN-related investments on future ratepayers, who will benefit most from the services that ISDN will make possible.

One such ratemaking treatment is **trended rate base.** This approach has been widely discussed with respect to electric utilities, in part as a solution to the rate shock associated with bringing expensive new generation facilities on line. In California, the Commission's staff has also put forth a proposal for trended rate basing for Pacific Bell's new investments. Under this proposal, a portion of capital recovery is deferred, resulting in an upward trend in revenue requirements over time for a given investment. Trended rate basing would <u>not</u> prevent companies from recovering the full cost of their investment over the economic life of the plant in service. It simply restructures the pattern of revenue requirements over time to better match the stream of benefits accruing from the investment.

2.1.2 Interclass Equity

The question of who pays for ISDN is not just a matter of current versus future ratepayers. Regulators are also concerned with fairly allocating the expenses between customer classes, particularly between residential and business customers. This issue is in part related to the intertemporal problem because large business customers are the likely initial beneficiaries of ISDN.

The matching of costs and benefits for ISDN between customer classes is a variant of the thorny cost allocation problems that have confronted telecommunications regulators for decades. Nearly everyone now agrees that there is no one "right" way to allocate joint and common costs within a telecommunications network. Nevertheless, some approaches seem "righter" than others to regulators who seek a fair apportionment of costs between customer classes.

For example, consider a new switch installation that is justified on the basis of expected revenues to be obtained from customized business applications. Suppose further that the anticipated levels of revenues from these new services are not achieved, and that overall utilization of the switch is lower than originally projected. How should the cost of the switch be allocated between customer classes?

Cost allocation rules that focus on the <u>actual</u> relative use of the switch may assign a higher fraction of the cost of the switch to residential customers than would rules that rely on the <u>projected</u> relative use of the switch built into the cost justification for the investment. An actual use rule places customers other than the intended beneficiaries of the investment at risk if demand differs from that anticipated. Given an unbiased demand forecast, the opportunity for gains and losses by these other ratepayers should be equal. State regulators are probably not risk-neutral on behalf of residential ratepayers, however, so this outcome may not seem sufficient to justify the risk of the new investment.

The FCC's proposed Part X accounting rules appear to handle this problem fairly well. Nevertheless, ratemaking reality may compel regulators to place the burden of failed investment opportunities on residential and small commercial customers. These customers have few alternatives to service from local exchange companies; the inelasticity of their demand for basic telecommunications services makes them a tempting target for making up any revenue shortfalls.

The difficulties inherent in cost allocation between classes have led regulators to search for alternative regulatory approaches that would eliminate the need for such allocations. Some states have already adopted variants of a "social contract" method. The hope is that local exchange companies and their regulators can agree on formulas for setting affordable basic rates without developing detailed service- and class-specific cost estimates. The California Public Utilities Commission will soon be evaluating the desirability of various alternative regulatory frameworks.

Moving away from cost-of-service ratemaking may in the long run reduce regulators' concerns over interclass equity. But in the near term, the question of who pays for modernization is central to the debate over ISDN. Even if every state were to adopt a social contract approach immediately, there would still be the problem of setting initial rate levels for each customer class and of determining how those rates should be adjusted over time. Expectations of levels of modernization expenditures and levels of benefits associated with those expenditures are sure to influence the terms of any social contract.

2.2 The Impact of ONA

As if there were not already enough uncertainty over ISDN, the FCC's Computer III order has added a layer of complexity. That order calls for implementation of ONA by the BOCs. Enhanced service providers have already indicated their belief that access to the ISDN "D" channel is an essential element of any ONA plan.

However, the international standard-setting process for ISDN is well underway, and it appears that little emphasis has been given to developing standards that would facilitate an open architecture in an ISDN world. Nor is it clear that the other countries that are involved in ISDN standard-setting will view ONA as a priority item since many of these countries still have nationalized telecommunications networks.

Why does this matter to state regulators? After all, state commissions have virtually no involvement in these standard-setting bodies. The potential problem is that any difficulties in meshing ISDN and ONA may result in increased costs for implementing both programs. And increased costs have an unfortunate habit of landing in the intrastate jurisdiction, where basic ratepayers often end up footing the bill. So, the relationship between ISDN and ONA merits monitoring by state regulators in their role of protecting basic ratepayers.

3.0 Conclusion

Despite the concerns raised above, I share the confidence of industry observers who see ISDN as the natural foundation of a thriving Information Age economy 20 or 30 years from now. Just as clearly, I see today's network as still being predominantly oriented toward the POTS user. Getting from here to there without a lot of growing pains in between is a challenge that must be borne by regulators and operating companies alike. That transition can be most effectively managed if ISDN's proponents are quick to acknowledge the legitimate concerns of state regulators and to work with the state commissions to find workable solutions that secure the benefits of new technology for all consumers.

ISDN: Beyond Jokes to Payment

Stanford L. Levin

Southern Illinois University at Edwardsville
Edwardsville, Illinois

ISDN looks like it will finally be a reality. It also appears, however, to be driven by technical, supply considerations, rather than by market demand. The former may have been appropriate in the old regulated regime, but the latter is necessary in the competitive markets in which ISDN will be offered. ISDN providers will find that they can no longer price ISDN services based on the type of use customers are making of the network. Pricing will have to reflect market realities. In addition, only those customers who actually use ISDN services will pay for them. Regulators are unlikely to allow ISDN costs to be recovered from ratepayers in general. Indeed, they may attempt to charge other costs to ISDN services. To be successful, ISDN providers will have to come to terms with both the regulators and the market.

1. INTRODUCTION

The only thing about ISDN more tedious than the continued promises and the wait for it actually to arrive have been the constant ISDN jokes. The jokes have sometimes been amusing, but they have, for the most part, been anti-ISDN. They have also attempted to dismiss ISDN by suggesting that no one really wants it rather than confronting the very real regulatory problems which it raises. While the industry may have oversold ISDN, and certainly it has engaged in premature promises, the current trials and customer interest suggest that this time ISDN really will arrive soon, although probably not with the pervasiveness its supporters once claimed.

This paper discusses two key state regulatory issues raised by ISDN. While on the surface these may appear to be soley regulatory issues, they are also, in fact, issues that must be confronted by telecommunications providers as well. First, the way in which tariffs are designed will have to change as a result of ISDN. Second, who is going to pay for ISDN should be resolved before ISDN becomes widespread. Failure to address these issues promptly at the state level raises the possibility of significantly increased risk for regulated telecommunications companies and may serve as a barrier to the spread of ISDN itself.

2. TARIFF DESIGN

Without offering criticism or a defense, one can note that state regulators, with the acquiescence if not the open support of local telecommunications companies, have implemented a wide range of tariff designs and pricing policies which have served not to relate price to cost or to market conditions but rather to achieve other social goals. For example, businesses usually pay more for the same service than do residential customers. Access is often overpriced and usage underpriced or free. Various types of voice transmission are usually priced differently under different tariffs, and voice and data services may be priced differently, even though both use the same facilities to the same extent.

ISDN will change this system radically. With current technology, telecommunications companies know, for the most part, who customers are and what they are doing with their telecommunications facilities. With ISDN, telecommunications companies will just be selling generic transport. They may or may not know who the actual customers are; they most certainly will not know what their customers are doing with the telecommunications facilities. This will completely change the ground rules for pricing ISDN services from those currently in effect for similar telecommunications services.

Customers will simply refuse to identify or

pay a higher price for a particular service if there is another service available at a lower price. For instance, if data transmission is tariffed at a higher rate than voice, customers will simply claim that all of their transmission is voice. The telecommunications provider will have no means to verify whether or not this is true. The problem is analagous to the intrastate/interstate traffic division reporting requirements for long distance telephone companies who are not offered equal access.

A simple example using current technology may help to make this clearer. Southern Illinois University at Edwardsville is in the process of installing a new telecommunications system. Each office wall outlet will have a "voice" jack and a "data" jack. The plan is to charge more for the data outlet than for the voice outlet. Of course, everyone needs a voice outlet, but few users will pay the additional charge for the data outlet. It is much simpler to buy a device for a couple dollars and plug a computer modem into the "voice" outlet. ISDN availability will make even this example seem innocent.

Pricing will be both simpler and more complex. It will be simpler because there will be many fewer tariffs. Each service can be priced individually, if desired, but these prices cannot vary by the type of use the customer makes of the network. It will be more complex because in some cases there may be significant changes in prices for services after ISDN is available than before. This will probably result in price reductions and revenue decreases from ISDN customers, more difficulty in forecasting revenues, customer adjustment problems, and upward pressure on prices for services that have received the benefit of subsidies in the past.

3. PAYING FOR ISDN

The major problem threatening the success of ISDN is likely to be regulatory and not technical. A couple observations are enlightening.

First, ISDN should be considered in a competitive context, as the best that ISDN providers can hope for is that regulatory realities will cause it to succeed or fail on a competitive basis. In a competitive environment, customers demand products or services, and suppliers, seeing an opportunity to earn a profit, attempt to satisfy these demands. The market is demand driven, and new products appear only if they can meet some real or perceived demand of customers.

Regulated markets, on the other hand, have been driven by supply considerations. Engineers have developed new technologies and service improvements, and these have been put in place. Regulators were eager for the best possible level of service, as this is one important measure of judging regulatory success, and consequently they were willing to pass on the cost of these improvements to all customers. Indeed, if there was a criticism of companies by regulators, it was often that the improvements were put into place too slowly. New services or improvements did not pass any market test; they were not driven by demand considerations.

At this point ISDN looks like the latter model of supply driven enhancements. It is technically possible, but little analysis has been done, apparently, to see if anyone wants it enough to pay for it. While this procedure has worked in a regulated environment, it may not be successful in the competitive market environment which will be the appropriate framework for evaluating ISDN.

This leads to the second point. While regulators have supported ISDN experiments, as they have supported experiments for other developments in the past, companies must recognize this for what it is. Regulatory support means that regulators will allow the company to put the cost of the experiment into their general expenses, recovered from all ratepayers. This does not mean that regulators will permit companies to engage in widespread ISDN implementation and recover the cost from the general ratepayers. It will be a serious and costly mistake to generalize improperly from the regulatory treatment of ISDN experiments.

So who is going to pay for ISDN? Regulators have been taking a lot of heat recently for increases in local exchange prices and for the subscriber line charge. It is certainly unlikely that they will allow the costs of ISDN to be passed on to general telephone customers. At best, all ISDN costs will be recovered from those using ISDN. This requires a marketing approach, rather than a regulatory or engineering approach, to evaluating the financial viability of ISDN.

It is probably in the providers' best interests to have ISDN treated in as competitive a fashion as possible. Regulation will not offer any protection from market forces because regulators will not allow ISDN costs to be recovered from ratepayers in general. Even if they wanted to, in any case, competitive pressures in other telecommunications markets will probably not allow them to do so. Indeed, the opposite is more likely. If regulators keep ISDN under regulation, they will be tempted to use fully-distributed-cost allocations or other equally arbitrary and artificial schemes to justify higher prices for ISDN. Since there is a real risk that the prices which result will be above those sustainable in the market, the ISDN provider could be significantly at risk without the flexibility to succeed in a competitive market.

While this strategy may seem unfair, one has only to look at the history of regulation to see that it is nonetheless likely. Thus there is a real incentive for the company to work with regulators to have ISDN services treated with the lightest regulation possible. Because there are substitutes for ISDN services, it is competitive whether or not everyone wants to admit it. ISDN prices cannot be above market levels, as regulation might be tempted to attempt. The best that ISDN providers can expect, therefore, is for ISDN to be provided with no regulatory oversight, subject to the vagaries of the market.

The telecommunications provider who is bearing all of the risk has a great incentive to achieve a successful regulatory framework. Regulatory and pricing strategies need to be separated from the technical, engineering aspects of ISDN feasibility. ISDN must also be a success in the market. It is unlikely that it will be allowed to be a "success" in the traditional regulatory sense of providing it and charging its costs to all customers.

4. CONCLUSIONS

ISDN has the possibility of greatly improving telecommunications services for many customers. At the same time, telecommunications providers must understand clearly the regulatory context in which they will be providing ISDN in order to minimize their risk and to insure that ISDN can be offered successfully in the market.

ISDN regulatory issues will probably be resolved primarily at the state level. Many of the services will be local exchange services under state regulatory jurisdiction, and recent pre-emption decisions have limited the authority of the Federal Communications Commission in this area.

While the issues raised here appear to be regulatory problems, in reality, however, they actually are not. They are problems confronting telecommunications providers who must offer ISDN under regulation or who must succeed in getting regulators to allow them to offer ISDN in a deregulated manner.

First, the companies themselves must understand that ISDN will be offered in a de facto market environment, regardless of the regulatory treatment. At best, the costs of ISDN will be recovered from ISDN users and not from customers in general. At worst, significant overhead costs may be assigned to ISDN. This may pose certain pricing problems for providers, but they are difficulties which will have to be overcome if ISDN is to be offered.

Second, the telecommunications providers are going to have to work with the regulators to find an acceptable method of handling ISDN. This will vary from state to state, from complete deregulation to full regulatory treatment. Companies should start educating and working with regulators now. Special programs for regulators may be appropriate, and companies should invite regulators to their ISDN programs and presentations. In this way, everyone will understand as well as possible all of the ramifications of the introduction in ISDN.

It is important for companies to begin solving the regulatory issues surrounding ISDN quickly, before ISDN is widespread. Failure to do so will simply increase the risk to providers and the costs to users. In addition, as has been seen over and over again in telecommunications for the last few years, it is almost always easier to solve these regulatory problems earlier rather than later when there is more investment, more cost, more dislocations, more vested interests, and a greater probability of price increases for some customers.

ISDN may be the area of telecommunications where currently regulated companies are confronted fully with the implications of markets. Regulators will be unlikely (and unable) to provide protection, and customers will have alternatives for satisfying their demands. ISDN will succeed or fail depending on how well companies can deal with regulation and how well they understand the market for ISDN. Not only will companies have to limit regulatory burdens which can not be supported in a competitive market, but ISDN itself will have to pass a market test. There are many pitfalls along the way, but it is extremely important that ISDN providers find a successful path.

ACKNOWLEDGEMENT

The author is indebted to Bruce L. Egan of Bell Communications Research in Livingston, New Jersey, for some of the ideas expressed in section two of this paper.

ISDN - GTE's View

Kent B. Foster

GTE Service Corporation
Stamford, Connecticut

This paper presents GTE's ISDN perspective and provides an overview of ISDN activities currently underway and planned within GTE.

INTRODUCTION

GTE is a diversified organization, a major business focus of which is telecommunications. Its operations in this area range across a broad spectrum. They include local exchange, interexchange, cellular mobile, satellite, and custom network applications, as well as data processing and manufacturing based services and products.

As a large, dispersed provider of local exchange service throughout the country, its domestic telephone companies serve 11.1 million access lines in portions of 31 states. GTE Telephone Operations (Telops) generated $11.3 billion in revenues during 1986. The GTE Telephone Operating Companies (GTOCS) allocated $2.9 billion for network expansion and modernization during 1986, and have budgeted similar amounts for these programs in 1987. GTE has maintained an aggressive network modernization program over the years and as a result, approximately 45% of our central office access lines are now digital. This aggressive modernization program is being driven by customer demand coupled with sound business practices.

GTE's View of ISDN

GTE is committed to supporting ISDN. This commitment encompasses GTE's telephone operations as well as the company's central office switching system manufacturing, CPE manufacturing, and data processing subsidiaries.

CCITT defines ISDN as, "...a network, in general evolving from the telephone network, that provides end-to-end digital connectivity to support a wide range of services, including voice and non-voice services, to which users have access by a limited set of standard multi-purpose user-network interfaces." Therefore, ISDN provides a framework in which the telecommunications network can serve total customer communications needs for voice, data, and image services. With ISDN, customer access to multiple communications services is integrated at a single point of interconnection with the transport network. ISDN has five major attributes which are attractive to GTE:

- End-to-end digital connectivity
- Integrated access to all services
- Standardized customer-to-network interfaces
- Out-of-band signaling
- Customer control

These five attributes allow the network to achieve higher performance levels, better service quality, improved customer and carrier economies, and enhanced functionality.

GTE views ISDN as a network architecture with the explicit purpose of satisfying our customers' needs and demands. The increased functionality of ISDN makes it the most appropriate near term approach to fulfilling customer needs for simplified integrated network access, dynamically allocatable bandwidth, lower costs per service and enhanced service content.

WHY ISDN?

The telecommunications industry has experienced tremendous upheaval in recent years - brought about by deregulation, divestiture, increased competition, and accelerated technological developments. Along with this upheaval we have seen a rapid proliferation of data processing capability in conjunction with the ongoing evolution of the information age. Our customers are displaying an expanded knowledge and understanding of available telecommunications options. As a result, competition has increased dramatically in recent years. Users of telecommunications services are faced with a wide range of alternatives to satisfy their telecommunication requirements. These alternatives include local area networks, digital PABX's, satellite communications, digital termination systems, private microwave radio, private fiber optic facilities, and, of course, local exchange and interexchange carriers.

As a result of this increased awareness of options, most business customers are planning to meet their communications needs with the latest available "state-of-the-art" technology. These customers perceive a real value in end-to-end digital communications and recognize opportunities in the midst of industry upheaval. They are seeking alternatives that will reduce their communications cost, while enhancing their companies' productivity. They are searching for ways to manage only one network - capable of providing a full set of services instead of multiple, specialized networks and discrete, independent service solutions.

GTE views ISDN as a means to satisfy our customers' needs. Through the application of ISDN network architecture, the local exchange carrier (LEC) is capable of providing these customers with a single digital interface that will come closer to ultimately satisfying their communications needs than any system that has been developed to date. ISDN provides the customer a more comprehensive and flexible communication capability than previous network architectures. We recognize that many of our customers will continue to require highly customized and unique solutions. However, the control structures of ISDN and other emerging standards will permit significant customization that we believe will often satisfy unique customer needs via the public switched network.

ISDN is an important stage in the evolution of the telecommunications network to meet the evolving needs of the world-wide information society. ISDN provides the bridge linking present network capabilities to a future flexible network infrastructure for low cost basic and enhanced services. Our desire to evolve GTE's segments of the public switched telecommunications network to ISDN is motivated by market, economic, and technical considerations.

MARKET DEMAND

The market demand for ISDN-supported services beyond basic voice communications will be driven by the impact that computers and microprocessors are having on the work and home environments of customers, and by the increasing use of digital technology in telecommunications networks. Increased capabilities will become available to telecommunications users over the next decade as networks evolve to ISDN. To assess the need for ISDN, it is necessary to understand the changes that will take place in telecommunications usage over this time frame.

In recent years, non-voice communications has grown about 20% annually, and this trend is expected to continue for the next ten years. While non-voice communications is growing very rapidly, voice communications has been experiencing average annual growth rates in the range of only three to five percent. Despite this disparity, voice communications is expected to remain the dominant service offering into the foreseeable future.

Basic needs of customers such as reliability, responsiveness, price/performance, flexibility and control will always have to be satisfied by the LECs. Voice communications will remain a primary service and will be the major source of revenues for the local exchange carrier over the planning horizon. Therefore, to achieve widespread acceptance, ISDN must provide services that can satisfy customer needs in both the voice and the non-voice market.

The American economy is experiencing a growing emphasis on information management that is largely attributable to the expansion of microelectronic technology. Computer storage capacity and processing power will continue to increase, while costs are expected to decrease geometrically. The computer terminal market is growing rapidly with a trend toward more sophistication (e.g., faster terminals, new protocols and more intelligence). These changes reflect increased customer interest in a network that can accommodate voice and non-voice services in an integrated and convenient manner.

Virtually all PBX's sold are now digital and a majority of PBX lines will be digital in 1990 (compared to only 18 percent in 1980). Significant penetration of T-1 service is emerging within the PBX market. With the introduction of ISDN, a substantial amount of this traffic may be extended to the network in an integrated digital format as afforded by the ISDN primary rate access.

In the residential market, homes are now targeted by a variety of service providers for advanced technologies such as security systems and home computers. Many customers now want second access lines for business and personal use. An increased emphasis on education and work in the home is also anticipated. We expect these applications to increase traffic demands on the network and thereby increase the demand for ISDN access.

ECONOMIC IMPLICATIONS

The economics of evolving the present local exchange network to ISDN are very complex because of the dependency on current network configurations, the perceived markets for new services, and the unknown costs of ISDN hardware and software. However, two very distinct economic advantages of ISDN will result in lower cost for customers and increased margins for participants in the information "value chain" -- equipment manufacturers and suppliers, and service providers (local exchange, interexchange, and auxiliary services). These benefits are:

- Capital and expense savings due to integrated network access

- Feature/cost benefits due to ISDN access capabilities and economies of scale.

Providing a variety of services to the customer today requires distinct network channel terminating equipment at the customer's premises, separate facilities to the central office, distinct terminating equipment in the central office (including maintenance spares in both locations), and separate network operation, administration and maintenance (OA&M) functions. Simplifications and their resultant capital and expense savings are afforded by integrated access in which a single piece of NCTE, one access link, and a single piece of CO terminating equipment enable a customer to gain access to a range of network services. Thus, the economics of providing ISDN access should be significantly more cost effective than the provision of discrete access to an array of separate services.

The second economic advantage arises from the improved feature/cost ratio when compared to access capabilities provided by existing services. The fundamental capabilities of switched digital capability and packet switched data network access are equaled by ISDN transport capability. However, ISDN offers significant additional features above and beyond these basic transport capabilities (including customer interface to network signaling and call control, internationally standard interfaces, computer readable call progress messages and fast call set-up). These additional features include calling number forwarding, closed user groups, and customer control features. The availability of both basic and advanced features allows ISDN's greater functionality and new services at much lower incremental cost.

TECHNICAL CONSIDERATIONS

In addition to market economic drivers, a number of technical considerations are driving the telecommunications network towards ISDN. Evolution of the U.S. telecommunications network from a primarily analog voice network to end-to-end digital connectivity began with the introduction of T-carrier digital interoffice transmission in the 1960s and stored-program-control central office switches in the 1970s. Since then, many more digital or digital-capable systems have been developed and deployed in response to technical and economic advantages associated with the transmission and processing of telecommunications signals in digital form.

The increased deployment of digital facilities has led to increased interconnection with other digital facilities. This has created a totally digital transmission environment, with resultant elimination of costly terminal equipment, improved transmission quality, simplified maintenance, and greater justification for additional digital facilities. In the mid-1970s, the point at which signals are digitized was moved closer to the customer premises through the development of digital loop carrier systems. In addition, direct interface of DS-1 (1.544Mb/s) rate facilities, tariffed as "T-1 service," has extended the digital interconnection point to the customer premises itself.

The latest steps toward end-to-end switched digital connectivity are being provided via the deployment of packet switching networks and switched digital capability access. Therefore, the local exchange networks are evolving toward end-to-end digital capability. This evolution is a major force behind the development of the ISDN concept. ISDN provides for the standardization of interfaces, signaling, and other technical aspects to achieve intercompatibility. ISDN also provides for a cohesive, integrated network evolution to digital capability, as opposed to independent digital evolution of discrete network components.

ISDN TEST BED ACTIVITY

The technical aspects of ISDN are quickly becoming understood, but the quantified market demand for the services associated with ISDN has yet to be fully identified. At this point in the development of ISDN technology, there are many unknowns regarding customer acceptance and demand for services that could be provided through ISDN functionality. As result, GTE chose to deploy a number of ISDN test beds designed to help determine what is the appropriate mix of services to meet the various customer needs. Many ISDN based services are being implemented for the test bed activities; additional services have been simulated or simply described to obtain customer feedback. To gain further knowledge regarding the relationship between technology/functionality and services/customer application, the initial ISDN sites have been structured as test beds rather than as field trials or evaluations. This approach will allow GTE to gain customer input prior to final product and service development and deployment, and will help determine which ISDN services will best serve the customer needs.

As a point of clarification, GTE defines the term "test bed" as an experimental technology in which a prototype of an innovative technology, architecture or service is tested. During the test bed activities, alternatives are evaluated and refined in the context of real users fulfilling business functions. In the case of the ISDN test bed activities, a combination of simulated capability, prototype, and commercially available hardware and software have been employed.

The GTE Telops ISDN test beds have been carefully planned and coordinated with the explicit purpose of fulfilling a well defined set of objectives. Generally, the test bed objectives fall into four categories. The first - and perhaps the prime - set of objectives deals with the market aspects of the ISDN capability and related services. The test bed efforts are designed to more clearly define ISDN services and customer applications in terms of customer benefits and savings. In addition, the test beds are providing GTE with the research data to better assess the related market opportunity. In the category of OA&M, the objectives are intended to identify the operations, administration and maintenance issues associated with the deployment of ISDN. The third category of tests is technical in nature and is intended to determine the functional characteristics of the various switching units, including intercompatibility and verification of the established ISDN standards. Finally, the fourth set of test objectives deals with the identification of costs associated with the deployment of ISDN - including investment requirements, associated expenses, and potential savings for both the LEC and customers.

GTE's ISDN test bed activities are producing the desired results. Though the projects are not yet completed and the data gathered has not yet been completely analyzed, it already appears that the ISDN test bed program will completely satisfy the desired objectives.

GTE has included several central office switch and CPE vendors in our ISDN test bed program in order to have access to a variety of markets and customer applications. Test bed activities are currently under way at four sites - each managed by one of our telephone operating companies - and utilizing three central office switch vendors. To supplement the applications provided by the switch vendors, GTE Data Services (our information management subsidiary), has developed several unique prototype applications for these test beds. The test beds are described below.

General Telephone Company of California

The General Telephone Company of California (GTC) ISDN test bed utilizes GTE's GTD-5 EAX central office switching system. The GTC test bed is scheduled to run for approximately one year and is being conducted in two phases. The initial phase began in June of this year and presently links the Company's headquarters office with three regional sales offices. The second phase began in August and is scheduled to be conducted over a nine month period. During the second phase, ISDN based services will be demonstrated to approximately 1,000 GTC customers.

General Telephone Company of the South

The General Telephone Company of the South (GTS) ISDN test bed will employ a Northern Telecomm DMS 100 central office equipped with 50 Basic Access (2B+D) ISDN lines. The test will be conducted in the Research Triangle Park (RTP), a major research and manufacturing industrial park located in North Carolina. The test will initially utilize 20 lines internal to GTS. The lines will be reassigned later in the test to individual customers for evaluation. Northern Telecom itself will be the major customer during the latter phase of the test bed with a total of 30 lines assigned to multiple locations within the RTP area. A demonstration center will be equipped to display ISDN applications and capabilities to major GTS customers. The GTS ISDN test bed will begin October 30, 1987, and is scheduled to conclude in July 1988.

General Telephone Company of Florida

The General Telephone Company of Florida (GTFL) ISDN test bed was initiated, late this past summer and will continue through year's end. The GTFL test bed is structured around an AT&T

5ESS host switch with two optical remote modules (ORM) equipped with sixty Basic Rate lines. The ORM's will be linked to the host via fiber optics. One ORM is located in downtown Tampa - while the other is located 15 miles north at Telecom Park. ISDN Basic Rate services will be provided internally to GTFL and to several GTE Data Services locations for evaluation. In addition, a product demonstration room has been equipped to provide GTFL customers an opportunity to experience ISDN capabilities first hand.

General Telephone of the Southwest

The General Telephone Company of the Southwest (GTSW) test bed is being initiated late this year with Texas A&M University at College Station as the customer. The ISDN capabilities are being provided in conjunction with digital CO-centrex service. Texas A&M and GTSW will be testing integrated voice and data applications, utilizing basic rate ISDN access. The services are being provided via the GTE GTD-5 EAX located adjacent to the University campus. The tests will include standard ISDN capabilities, ISDN applications for university administrative processes and engineering functions, such as Computer-Aided Design/Computer-Aided Manufacturing (CAD/CAM).

DEPLOYMENT PLANS

As indicated earlier in this paper, GTE has viewed an aggressive modernization of its network as absolutely essential to satisfying the demands of our customer and ensuring the long term viability of our telecommunications business. In that light, ISDN may be seen as an extension of GTE's ongoing network modernization program. However, ISDN will provide a direct interface with the customer and will impact the customer in a far greater manner than any recent network modernization effort.

ISDN provides the customer greatly increased network and service flexibility and control. It equips the customer to better control communications costs. It provides the framework for new customer applications and rapid roll out of new services. It allows the customer a single integrated network access as opposed to multiple overlay networks for special applications. ISDN gives the LEC the ability to meet the demands of the telecommunications customer in a timely, integrated, and cost effective manner.

It is the intent of GTE to pursue the deployment of ISDN capability in response to customer demand as rapidly as possible. The velocity of the ISDN conversion will be tied directly to forecasted customer demand. The initial demand for ISDN based services is expected to be greatest from the business community, with an eventual migration to the residential community. As a result, the initial deployment of ISDN capability will take place in the areas of highest business concentration as the demand develops.

Therefore, expect our key market areas to experience the earliest conversions because of their high business customer concentration. Eventually, ISDN capability will be made available on all digital central offices to ensure a uniform availability of services and functionality from all GTE serving locations. Again, market demand will determine the velocity of GTE's ISDN deployment.

CONCLUSION

ISDN will play a critical role in the future of the telecommunications industry at all levels. No segment of the industry will be isolated from the impact of this new technology. Customers will demand ISDN-related services, and the LECs will have to provide the capability or face customer dissatisfaction and perhaps a migration of customers to an alternate carrier. Similarly, the IXCs will have to establish a compatible interface with the customer and LEC or face competitive erosion. As a result, we believe ISDN will ultimately become the non-optional network modernization program of the telecommunications industry. GTE is committed to a thoughtful, scientifically supported approach to this important evolution.

Southern Bell's ISDN Rollout: Narrowband and Broadband

Richard K. Snelling, Kathleen Wilson Kaplan

Southern Bell

1. INTRODUCTION

Southern Bell is committed to the deployment of ISDN, and firmly believes, as did the IEEE ISDN-85 conference, that it is THE network architecture of the future. That commitment has lead us to begin our ISDN deployment with a first application or initial service, not a trial. This will begin our rollout of ISDN to all our four states with viable commercial service offerings. Southern Bell's first application is heavily supported by BellSouth Services.

2. KEY FEATURES

The major components of our ISDN first application in Atlanta, Georgia, testing in 1987 and service in early 1988 are as follows:

1. Multiple Offices
2. Multiple Vendors
3. Signaling System 7 (SS7)
4. Basic, primary and Low Bit Rate Voice (LBRV) option rates
5. Business and residence service

2.1. Multiple Offices

Atlanta installed an ISDN test machine - a 5ESS, 5E4.1 - in Toco Hills in May, 1987. This machine is being used for internal testing of operations, CPE and interworking.

The first ISDN service office will be Dunwoody which will service in March, 1988 (5ESS, 5E4.2). Dunwoody is a rapidly growing area located on the northeast portion of the "beltway" highway encircling Atlanta. The Dunwoody area could possibly have as much office space as downtown Atlanta by the 1990's, and has many nice residential areas as well. This area has potential business and residence ISDN customers. Southern Bell's Georgia operations are headquartered in the Dunwoody wire center.

The next offices in Atlanta to be equipped with ISDN capabilities will be Courtland Street in mid 1988 (DMS-100, BCS 24) and Peachtree Place in 1989. Courtland Street, in the heart of downtown Atlanta, also has many businesses including the state government and Georgia State University. A portion of the Southern Bell Headquarters building is served from this wire center also. Peachtree Place is in the north downtown area of Atlanta, sometimes referred to as Midtown. Peachtree Place has many businesses located within its wire center, not the least of which is the Southern Bell Headquarters building. The new BellSouth Headquarters building opened in this area in mid-1987. A major university, the Georgia Institute of Technology, is also in the Peachtree Place area.

These and other offices in our four state area which will see early ISDN have high growth rates, customers with a likely willingness-to-pay who are interested in "advanced" telephone services, packet switching and fiber facilities, for example. All will have digital switches and Signaling System 7. SS7 deployment begins in the third quarter of 1987. The Signal Transfer Points (STPs) were installed in the third quarter of 1986.

2.2 Customers

On May 19, 1986, Southern Bell announced its first two ISDN customers, Prime Computer and Trust Company Bank. On June 18, 1987, two more customers - Hayes Microcomputer Products and AT&T - were also announced. All are located in the Dunwoody area, and are interested in higher efficiency, lower cost voice and data services.

Prime Computer, Inc. designs, manufactures, markets and services general purpose computer systems for a wide variety of applications in principal world markets.

Trust Company of Georgia is one of the state's largest banks. It is part of SunTrust Banks, Inc., a major regional banking group with assets of $20 billion.

Hayes Microcomputer Products is a leader in data communications products, and recently demonstrated an ISDN PC card and a terminal adapter.

AT&T is a supplier and customer, and has a long history of involvement in the development of ISDN. AT&T provides equipment from switches to terminals.

2.3. Multiple Vendors

The real world network, at least Southern Bell's digital switch network, uses more than one vendor's digital switch. Our network currently has two vendors' digital switches deployed, a third to be added this year, and the probability of adding at least one more in the near future. Thus, a multi-vendor environment is a reality which is here to stay, and with which all will eventually need to deal. So, rather than putting off the inevitable, we decided to tackle the real world of interworking now.

The Toco Hills and Dunwoody offices will use AT&T 5ESS's; Courtland Street and Peachtree Place will have NTI DMS-100's. Additionally, there is a third vendor involved in our ISDN efforts in Florida - a Siemens switch will provide Florida's first ISDN presence. Challenge is apparently one of our favorite words.

How do we proceed to make our multi-vendor application work? We began by making history in 1985. On July 23, 1985, all three vendors - AT&T, NTI and ITT (who was at that time providing the STP's. That, of course, is no longer the case.) - met with Southern Bell and BellSouth Services for the first time to begin working together toward making this first multi-vendor ISDN service a success. A common underlying motivation for us all is that a successful venture will definitely benefit everyone involved. Each company agrees with that goal.

August 1985 saw two more multi-vendor meetings; issues and questions were boarded and discussed. The initial formal responses from the two switch vendors (AT&T and NTI) came in on October 1, 1985. Since then, further questions and answers have been exchanged. Additionally, many individual vendor meetings between Southern Bell, BellSouth Services and a vendor have occurred and will continue to do so.

On December 16, 1985, we established an Officer Steering Committee for ISDN which involves all companies. Officers from all three vendors (the third is now Siemens), Southern Bell and BellSouth Services met with the working committee, the aforementioned group who had been meeting during previous months since July, and Bell Communications Research. This meeting's purpose was to ensure commitments from each company involved, to ensure consistent communications among all of our positions and the major unresolved issues at that time, to formulate action items toward issue resolution, and to establish an escalation avenue for matters which may need officer level attention. A second of these officer meetings was held in August 1986 separately with each vendor; they have been and will continue to be held quarterly.

Standards issue. The major problem facing this multi-vendor forum and effort is the lack of set standards in all the needed interface areas. For example, standards are yet to be fully defined for the supplementary services and FCs (Functional Components) to name a few. If we waited until all the standards are in place to begin ISDN, we would indeed be beyond the initial market window, and considerably beyond 1987-88.

Therefore, we have to go forward with the standards that are defined, participate heavily in the standards process and expedite it as much as possible, and be prepared to begin with a MINIMUM amount of non-standard equipment. Any non-standard equipment will be replaced by standard equipment as soon as that version becomes available.

We are also leading and encouraging agreement among the vendors on some of the yet unresolved standards issues in an attempt to expedite the standards formation process. Meetings to that extent and to work on other issues resolution will proceed throughout the implementation years.

Signaling System 7

Common Channel Signaling System 7 (SS7) is an inherent part of ISDN's foundation; true ISDN does not exist without SS7. As the January 6, 1986 issue of MIS Week (p. 26) stated, "ISDN is absolutely worthless if isolated pockets of customers cannot be connected. The bridge is SS7." Since Southern Bell completely agrees with that statement and concept, SS7 is a major portion of our ISDN rollout and ISDN services offerings. Signaling information transport is one of the primary purposes of the D-channel.

We began our CCS System 7 deployment in July of 1986 with the installation of a pair of STPs in the Atlanta area. STPs are very powerful packet switches which act as concentrators for switches and data bases. They are the CCS network nodes which provide efficient message transport among all signaling nodes in a CCS network. After SS7 trunk setup is operational in our network, SS7 deployment begins in the third and fourth quarters of 1987 in Atlanta, Orlando (Touchstar) and a few other cities. Some of our highest priority, initial SS7 services are City Wide Centrex and Touchstar, our Service Marked name for CLASS (Custom Local Area Subscriber Services), and User-to-User Signaling which requires ISDN and SS7.

2.5. Services

Some of the services we want to provide will include:

Touchstar. Touchstar (CLASS) is a collection of many advanced custom calling type services. Southern Bell has been trialing a number of these services since July 1984 in Orlando, Florida. These services are activated with the * button on a Touch-Tone telephone and two digits.

Automatic Callback (Call Completion to Busy Subscriber) allows the subscriber to activate a continuous check on a busy number for up to 30 minutes. When both lines are idle, the subscriber receives an alerting signal that the call is being placed. (This includes Repeat Dialing - redialing the last number dialed.)

Call Return allows the subscriber to return a call to the last number that called him or her without knowing that number.

Call Block (Selective Call Rejection, or more affectionately known as Reject-A-Jerk) allows the subscriber to cause a call to be routed to an announcement and not alert the subscriber.

Call Tracing allows the subscriber to automatically trace an incoming call. The trace may be originated during or after a call.

Distinctive Ringing allows the subscriber to assign a different ringing pattern to a limited number of incoming call numbers.

Calling Number Display (Call Monitor/ Suppression allows a subscriber with a display device to view the incoming call number. The calling party can prevent (suppress) the display of their number at the called party's premises.

Preferred Call Waiting allows the subscriber to receive call waiting indicators only if the incoming call number has been placed on a list generated by the subscriber.

Preferred Call Forwarding allows the subscriber to have forwarded only those calls whose numbers are on a separate subscriber generated list.

Call Tracking permits all incoming numbers to be provided to the subscriber along with some call detail information such as time of day.

City Wide Centrex (CWC). CWC extends centrex usage across multiple central offices within a LATA. (We will also provide a gateway to extending CWC on an interLATA basis). With CWC a multiple location business can have access to the centrex line features, attendant features, and dialing capabilities available today only within a single central office location. For example, four-digit dialing is a desirable service to offer a CWC customer on an interoffice as well as an intraoffice (enjoyed today) basis.

User-to-User Signaling allows short messages to be sent from one ISDN terminal to another ISDN terminal at the beginning of a call or during a call. The network will transmit the message without storing the information. A subscriber will only be permitted to send user-to-user messages if CCS connectivity exists between origination and destination and if both parties are on the ISDN network. These messages are transmitted on the D-channel.

Other desired services to offer with ISDN include Message Desk and existing services such as the traditional Custom Calling Services Call Waiting, Call Forwarding, 3-Way Calling, and Speed Calling; Conference Calling, Call Transfer, Call Pickup and Call Hold. We are working industriously with the vendors involved in our ISDN rollout to have the capability to offer as many of these services as early as possible in our rollout. We will grow our interoffice ISDN services capability as our vendors' switches ability and compatibility increases. That is a vehement and common goal but not an easy one in the present world of late or missing standards!

2.6 Basic and Primary Rates

Further "services" in our ISDN rollout will take the form of multiple access rates. We will offer both the basic access rate of 2B+D (144 kilobits per second) and the primary access rate of 23B+D (1.544 megabits per second).

Additionally, we would like to offer our customers another ISDN access rate option which is based upon Low Bit Rate Voice (LBRV) using 32 kb/s rather than 56 kb/s for voice transmission. Our 32+(16+3.2 or other data rates) kb/s proposal would give the customer voice transmission, the 16 kb/s D-channel for signaling/data transmission, and possibly 3.2 or X kb/s left for some low rate data or maintenance bits -- all within a single 64 kb/s bit stream. We are proposing this 32+(16+X) rate simply as an additional option for accessing our ISDN network. (Southern Bell has been trialing 32+19.2 kb/s simultaneous voice and data transmission in our loop plant in an Atlanta suburb).

2.7. Business and Residence Service

Another distinctive aspect of our ISDN application is that we plan to serve both business and residence customers. Southern Bell is as yet in a minority of those who believe that there is indeed a market for ISDN in residences prior to the end of the decade. The increase of those in the work force who do and will work at home is one of the reasons for our efforts to serve residences. Some studies have shown that perhaps a third of our customers will work at home at least part time by 1990. The Information Age users demand for more information at their fingertips, and added conveniences will extend to residences as well as businesses.

3. ISDN TERMINAL

On July 1, 1985, BellSouth Services issued a Request For Information (RFI) on ISDN terminals. This RFI's purpose was to survey the terminal manufacturing industry and identify interested terminal vendors who could manufacture ISDN terminals which would be compatible with our offering. The RFI communicated to the terminal vendor world our ISDN plans and outlined the parameters required to meet our ISDN needs from a terminal perspective. A total of 110 vendors received our RFI; approximately a fourth of those in receipt responded. From that group of respondents, five at this time are seen as strong candidates being looked at more closely. We hope to find several terminals which we can recommend to our ISDN customers for their use with our services. We have formed a CPE subcommittee which concentrates more thoroughly on terminal portability, and looks more in depth at pricing, quantity and timing availability.

3.1. Broadband

Full ISDN includes both narrowband and broadband capabilities. We are starting now with the narrowband capabilities which are available and are moving toward broadband applications as quickly as market demand dictates and vendor capability can meet that demand. All fiber network plans are underway - Southern Bell recently announced the nation's first all fiber network residential development now operational in Florida - as are broadband customers discussions. Fourth generation switches will start appearing in the 1989-1990 time frame. We are strongly pushing for broadband standards to be agreed upon in 1987, so that broadband's standards and services are more timely with the market demand.

4. CONCLUSION

Southern Bell believes there is a market for ISDN for all categories of customers from residences to small, medium and large businesses. Many customers have contacted us; we have four announced and several additional potential ISDN first application customers. We will deploy ISDN throughout Southern Bell on a market willingness-to-pay basis, which means that deployment will probably not be done in mass ubiquity or homogeneity. It will happen in different places at different times where there is market demand. ISDN's fruition will come when simultaneous voice, data, and video transmission is a reality, fulfilling ISDN's promise of providing the entire spectrum of exciting services to customers' homes and offices.

ISDN - A Carrier's View

Daniel E. Crawford

MCI Telecommunications Corporation

Washington, D. C.

As high-performance private and public switched data communications networks progress beyond ISDN standards, it is probably a mistake to assume that the operators of those networks will be willing to lower performance demands to suit ISDN standards that attempt to accommodate the low-end voice market. Data communications users can be expected to spark demand for ISDN services if carriers debundle existing ISDN technologies and offer high-performance, cost-effective integrated services in advance of a full ISDN roll out. Such customer demand is essential in the U.S., where a technology-driven implementation of ISDN is difficult due to the competitive nature of the telecommunications industry.

There's an old story about three blind men describing an elephant in three totally different ways -- according to the body part they happen to be closest enough to feel -- that seems custom made for this discussion of ISDN. All of us, I think, are still trying to get our arms around this beast, still trying to figure out why there are so many descriptions of what is supposed to be the same animal. And we continue to avoid touching one of those, well, those sensitive spots that might cause both us and the ISDN elephant a lot of unnecessary embarrassment.

So at the risk of being accused of molesting our subject here today, I'd like to talk about the evolution of ISDN from several perspectives. My basic perspective, obviously, is that of a competitive U.S. inter-exchange carrier. MCI was the first company authorized by the FCC to compete against AT&T in the long distance market, and operates the second largest long distance network in the world. In recent years that network has evolved to the point where we can concentrate on offering high-quality, high-performance services to large, sophisticated users. MCI now has a large international presence as well, with voice service to more than 50 countries. Our telex, electronic mail and data services have an even greater reach overseas.

MCI's efforts to penetrate the large user and international markets has given us another perspective -- that of our customers. So our view of ISDN is greatly influenced by what our most sophisticated customers tell us. After all, we're in business to serve their needs. And we look on our network not as an engineering artifact , but as a problem solving tool.

This perspective has led us to view ISDN with a peculiar combination of excitement and concern. We're excited about the potential for ISDN to deliver all the promises of the Information Age. There can be very little doubt at this point that ISDN will be most significant force in telecommunications in the next decade.

We're concerned, however, because at least in this country, there doesn't seem to be the kind of push for ISDN on the part of customers that would lead to an early roll-out. We're also concerned because the limitations of the ISDN standards to address the disparate requirements of voice and data networks. In addition, I think customers understand that there is no clear consensus yet on how this industry will overcome the significant political obstacles that block the road to ISDN.

Up until this point, the users' interests have not been given proper attention. Business organizations don't want "pie in the sky" promises about what ISDN will do. They want to know what ISDN will cost, and what it will do to justify any added expense. They want to know if there is any way to get the same kind of bang for a lot less buck

One thing you have to remember is that ISDN was hatched by the Bell System 25 years ago as a way of extracting greater revenue from the public switched network. So ISDN was originally a monopolist idea, and still carries with it a good deal of monopolistic baggage -- for example, the control and administration of the ISDN numbering plan.

I think it is also significant for those of us in this country to realize that the vision of ISDN as a single, standardized network connecting all the voice and data terminals on earth fits the monopoly model a lot better than it does the multiple, competitive network model the U.S. has adopted in recent years. Private networks, or even public switched networks, which often operate at higher levels of performance than ISDN standards, seem even more problematic for the one-world view.

We need to recognize that there are rather fundamental differences between the ISDN envisioned by the key players in the CCITT, which happens to be dominated by the state-owned monopolies, and the ISDN our customers dream about. They see ISDN as a transparent digital stream -- a way to get universal access to an infinite variety of information through a single point.

My understanding of the monopolist dream of ISDN is radically different. Where we dream of a digital stream, oblivious to borders, through which information is allowed to flow unhindered, the monopolist dreams of ISDN as a huge, self-contained ocean -- an ocean with a lot of small self-service islands, and possibly, one huge, new continent called data processing.

In some countries, ISDN is viewed not only as a way for the state monopolies to enter data processing, but also as a way to beat back calls for American-style competition. The Europeans, for example, are co-operating within their Conference of Post and Telecommunication Administrations to advance their own vision of ISDN. If the CCITT is unable to establish a consensus on international standards within a reasonable period of time, the Europeans will probably implement their own standards, which could then be offered to the rest of the world and become a defacto standard.

As you may be aware, the West German company Siemans, backed by the Bundespost, has moved ahead in this area. ISDN devices, including multi-purpose terminals, plugs and jacks -- like nothing ever seen on these shores -- are already on the market, and ISDN is commercially available in two German cities.

This is an example of how a monopolized industry can institute ISDN from the top down. Here in the U.S. that is simply not possible. There are just too many players with opposing views and differing interests in the North American market. Does this mean the U.S. market, which is not monopolized, is doomed to a second-class ISDN future?

As long as ISDN is technology-driven, as opposed to market driven, our reliance on market forces does seem to put us at a disadvantage. But the market has some hidden strengths. In fact, I think we are already seeing a market that is moving quickly toward ISDN capabilities. So perhaps we ought to stop looking for a top-down evolution to ISDN, and concentrate on developing our own ISDN market from the bottom up.

Let me give you an example from MCI's own experience. As you know, we recently introduced our own 800 service, offering the first competitive alternative to AT&T in that market. The database technology necessary to offer that service is not ISDN, but is analogous to ISDN in ways I think will become apparent. The competitive carriers were told by the local exchange industry that in order to offer 800 service, we would have to wait until the necessary databases were installed, and that we should expect to wait several years.

MCI's reaction to this situation was to start looking for ways to offer 800 service as quickly as possible. And we were able to find those ways. Now, years in advance of the dates given to us by the local exchange carriers, our 800 service is up and running.

I think this example proves that you don't need to have all the standardized configurations in place to deliver or receive the benefits of advanced services associated with ISDN. Forward-thinking organizations today realize that ISDN is already here, both as a way to solve problems today, and also as a way to prepare for the future.

MCI and other carriers already offer services that allow customers to integrate their voice and data functions, which is of course the definition of ISDN. This is typically accomplished through an evolution that starts out as a digital private line network upgraded with customer-premise equipment capable of supporting ISDN. The next step in this evolutionary process will be virtual digital networks offerings by carriers such as MCI. Our network already contains clear channel digital cross connects and common-channel-signalling-system-seven-based network control systems. This allows us to offer our virtual private network service, Vnet. So all the essential elements are in place for MCI to provide ISDN. As soon as the local exchange carriers can provide the last remaining element -- the digital local loops -- ISDN will be in place for our customers.

That same network control system, now being implemented through common channel signalling, is being configured as the focal point for integrated control. We believe that in a world moving toward integrated networks and services, we should also move toward integrated management of those services. It makes no sense to us to integrate the functions, yet leave the management of those functions to separate systems.

Consequently, much of the thrust of our network engineering program at MCI is now focused on what we think are some innovative steps to help us integrate control of the services we offer. In fact, we're now spending a significant portion of our capital investment dollars on network management and control. And we've formed a new group within MCI dedicated solely to the integration of our network management tools into a single, unified interface.

MCI's digital reconfiguration service, or DRS, which you may have read about recently, is being implemented with clear channel, 64 kilobit per second ISDN capabilities. These and many other network managment features are going to grow out of the centralized, software-driven network control technology that MCI is developing in anticipation of ISDN.

Recently announced products such as DRS and MCI's Netview PC interface show clearly how we intend to help customers not only integrate the various services carried on their networks, but also integrate network management functions. But what, you may ask, do all of these advanced ISDN features have to do with plain old telephone service? To tell you the truth, not a hell of a lot. And that's the point I'd like to leave you with here today.

Our experience in serving the real needs of the real markets that are out there tell us that there are two very different and distinct telecommunications markets -- the voice market, and the high-performance, interactive data market. In simple terms, both of these markets can be accommodated on the same transmission net-work. But the outlook, at least as far as the high-performance market goes, is anything but simple.

When you look at the high-performance data market, compared to voice, and you look where it is going, you begin to realize that there is a tremendous disparity in the performance need of these two markets. They have different service characteristics, different needs in terms of set-up times, holding times, error rates and so forth.

As your high-performance data communications market continues to race forward to ever-higher levels of performance, you can't help but wonder if it's a mistake to assume that the Plain Old Telephone Service network needs to be brought up to the levels of performance associated with large-scale data applications. Certainly, no one is really prepared to incur the kind of expenses that would bring the entire national voice network -- from end to end -- up to these kind of data standards.

By the same token, it is probably a mistake to assume that the high end data market will be willing to lower its performance demands to suit ISDN standards that seek a middle ground -- at least until there is some compelling strategic or economic reason to do so.

So ISDN, as it is currently conceived, may not be able to bridge the gap between these two worlds. It would seem to me, then, that the best way to help move ISDN forward would be to look at decoupling the various technologies involved: The standardization of interfaces and protocols...The standardization of digital local loops...The cross connect and switching functions...The intercarrier signalling.

If we could fully take advantage of the subsets of the ISDN technology that is now or soon will be in place, we not only better meet the needs of data users today, we could help spark the kind of market demand for these services that we need to move ISDN forward in this country.

For example, we could use the digital local loop technology -- right now, independent of ISDN -- to replace much of the analog equipment needed today for digital services. This would greatly enhance the performance and cost-effectiveness of most data networks.

MCI, as an example of the interexchange carriers, has put the essential elements of ISDN in place. Now we're ready to work with you and your local exchange carrier to deliver the benefits of ISDN to the computer communications com-munity. Eventually, all of telephony will realize the benefits of the ISDN concept. But you -- the digital communications customer -- will be the first to enter this exciting new era.

Let's work together to make it happen as soon as possible.

Thank you.

Federal Regulatory Challenges: A DOJ Perspective

B. B. Grossman

DOJ

Paper will be provided at the conference.

Federal Regulatory Policy and Information Services: An Economist's Perspective*

Thomas C. Spavins

Office of Plans and Policy
Federal Communications Commission

No aspect of current regulatory policy has caused such an intense mixture of hope and frustration as the treatment of information services provided by dominant telecommunications carriers. Hope stems from a vision of new services that can be uniquely provided by those firms that will enhance customer choice, and the vitality of the telecommunications networks generally. Frustration stems from a seemingly endless set of rules, procedures, approvals and conditions that need to be satisfied before even limited progress can be made in actually offering these services in the marketplace.

This paper explores the sources of both of these features of the current discussions of the provision of information services. A key source of the hope and the frustration in the current debate is the inherent limitations on the ability of the regulatory process to provide an empirical assessment of the economics of dominant carrier participation in the information service market, a difficulty that will not be resolved anytime soon.

Another source of strain stems from the need to develop an effective framework to assess, and where necessary, control the monopoly power of dominant carriers, a framework that needs to recognize the great difficulty any regulatory system has in making difficult empirical estimates of both the prospective and retrospective success of any firm in these markets. In this latter area, while much work needs to be done, recent economic research seems to provide an institutional design that will do the job.

The Sources of Hope

Let me first explore the sources of optimism about the outcome of dominant carrier participation in the information services market. The first source of hope stems from some general observations about technical change. In general, the achievements in physics, electronics, mathematics and computer science that are included in the "computer revolution" have been a major source of productivity growth. These technologies and their application are generally viewed as a strategic asset for all business firms. If a merger of telecommunications and computing has occurred, there is also a merging of computing and transportation, computing and banking,

*The views expressed in this paper are not necessarily those of the Office of Plans and Policy, the Federal Communications Commission, or any other member of the Commission's staff.

and computing and a host of other industries. Therefore, arbitrary restrictions on the ability of communications carriers to provide services that incorporate these features and sell them to the public seem out of step with everyday experience.

More importantly, the telecommunications industry has historically been a leader in the application of technologies stemming from research in applied physics. Over the long term these advances have enabled the telephone industry to systematically do better than the economy as a whole. It has been significant that such a large and mature industry has been systematically able to achieve a long-term decline in its prices, relative to all other prices in the economy, of about two percent per year. While the number is not large at first glance, it is on reflection impressive.

It must be remembered that the telecommunications industry, like other industries, includes a conglomeration of many tasks. Some, like long distance transmission, billing services, and switching, are subject to quick applications of new techniques. Other activities, like construction of the loop plant, may not be so easy to improve. In order to achieve an average improvement of two percent greater than the entire economy, telecommunication firms have to achieve a very rapid pace of improvement indeed, within their targets of technological opportunity.

Recent developments with the telephone industry appear to indicate that these long-term trends continue. The recent repricing of telephone services has enhanced network efficiency. Corporate reports stress continued productivity gains. Trade literature on technology indicates that deployment of computer technologies through the network is increasing.

Thus, everyday experience, backed up by the consideration of long-term trends, causes observers of the telephone industry to focus on application and deployment of information services techniques.

The Sources of Frustration

The frustration concerning the deployment of new information techniques has several sources. One is a continuing regulatory concern about the abuse and extension of monopoly power by dominant carriers. These concerns encompass both fears that information services markets might be subject to monopolization, and separate concerns that participation in these adjacent markets will cause prices to rise in monopoly markets.

Compounding this long-standing problem is the very divided nature of the regulatory regime for the telephone industry. One-stop-shopping is not the rule of the day in telephone regulation. Information Services provision is potentially subject to regulation by: (1) the Federal Communications Commission; (2) the applicable state regulatory agency; (3) the federal-state joint board process (where separations are involved); and (4) the U.S. District Court (for the BOCs and GTE) as the result of two antitrust consent decrees.

Each of these regulators in turn is "regulated" by varying degrees of legislative and judicial oversight. This means that every decision by each regulator must not only dispose of information services issues on the merits, but must be accompanied by a careful analysis of the relationship among the four controlling jurisdictions. Each decision must also be crafted as part of a judicial strategy to receive appellate approval and a legislative strategy to be sure that actions do not cause legislative reversal. Thus, the process of reaching a decision in the information services is extremely complex, but from the perspective of any one participant unavoidably so.

Root Causes of Our Dilemma

The roots of both the hope and frustration in the information services include extreme empirical weakness on both the benefits from more extensive dominant carrier participation in regulated markets, and the dangers which accrue therefrom.

Let me consider first the factual foundation or the dangers of dominant carrier participation in related markets. The historical record in this area is really nothing more or less than the seemingly endless disputes that characterized the great unpleasantness that ultimately led to the AT&T divestiture. It is clear that, at best, history seems to speak with a forked tongue, in terms of the record of either cross-subsidy or predatory pricing. The service with strongest monopoly power - local exchange telephony - was the recipient of a net contribution from the long distance market, an area often thought of as being subject to more competition. The repricing of local telephone service to correct this has been a major struggle. On the other hand, despite deregulation, the Bell System's embedded CPE business has not been able, when faced with a market test, to earn anything like an economic return.

There exists a similar dichotomy concerning anticompetitive access denial problems. It is probably true that relatively simple interconnection rules destroyed the ability of the traditional telephone companies to control the CPE business. On the other hand, even with extensive and complicated equal access costs, the non-AT&T carriers have had their hands full trying to establish themselves. While the issues can be developed with never-ending complexity, the policy debate in this area is marked by the gravest weakness in our ability to make precise estimates concerning prospective harms from dominant carrier participation in related markets.

The same factual gap covers any assessment of the benefits of dominant carrier participation in related markets. Let me take the most extreme case. Immediately after divestiture, many observers, including this author, were very optimistic concerning the ability of the new AT&T, freed from much regulation and the 1956 Consent Decree, to contribute to related markets. The unchallenged role of Bell Laboratories and the high reputation of Western Electric for manufacturing excellence, together with the established base of CPE and the long distance network, suggested that if any communications firm could succeed in related markets, the new AT&T could. While the verdict is not in yet (and perhaps the inability of the federal regulatory process to fulfill the deregulation side of the divestiture/deregulation combination has not helped), the new AT&T has found it very difficult to do extremely well in these markets.

Thus, public policy for information services markets rests on a very shaky foundation. While it is certainly possible to provide more bulk to the discussion, the weakness of the factual debate is endemic.

<u>Is There Light at the End of the Tunnel?</u>

I have had little to offer so far by way of a happy ending to my survey of the state of public policy toward information services. Yet, despite the very complex and frustrating problems we face, there are two separate strands of hope.

1. The Third Computer Inquiry.

First, as the result of the Third Computer Inquiry, the telephone industry is developing a set of standards for open network architecture and cost accounting that should dispose of the major concerns regarding local exchange carrier entry into these related markets. Solid cost accounting standards should push back our concerns about cross-subsidy. ONA should provide sufficient unbundling such that access denial will not be a cause for monopolization.

Two big problems remain with this approach. First, neither is in place yet. Cost accounting manuals are just beginning the approval process. The resulting audit process is not yet up and running. Until this is done, doubts about the exercise will not entirely go away. We have the same problem with ONA.

Second, more fundamentally, the great uncertainty concerning the impact of regulation, and the need for each step of the change to be part of a broader strategy, makes for regulation that tends to be conservative, that errs on the side of those whom we seek to protect, at the expense of the carriers we seem to unleash. It is very easy for this process of strategic caution to result in a set of safeguards that are no better or are even worse than the restrictions they displace. While everyone at FCC is optimistic, accuracy requires that this concern be noted.

2. A New Regulatory Framework.

For the longer term, recent economic research provides a road map that should create a stable regulatory regime to permit full participation by the dominant carriers in the information services market.[1] This longer-term, new view of telephone regulation is not motivated solely by the need to solve the substantial regulatory problems involved in dominant carrier participation in information services markets. Indeed, this new regulatory framework could be justified independently of the problems, if only because the Third Computer Inquiry and its offspring may have provided a workable approach for the information services problems.

There are two key elements to the new regulatory framework. First, regulation must be modified to break the direct link between the rates a dominant firm can charge and its own costs. It is this link that causes much of our concern about cross-

---

[1]See Willig, Robert D., "Designing Regulation for AT&T and the BOC's in Information Services Markets" (Division of Research, Harvard Business School, 1987); Spavins, Thomas, "Federal-State Regulation of Local Exchange Telephony" chapter 15 in Proceedings of the Telecommunications Deregulation Forum (College of Business Administration, University of Arizona, 1986); Haring, J., and Kwerel, E., "Competition Policy in the Post Equal Access Market" 2 FCC Record 1488 (1987); and Levitz, K., "Loosening the Ties That Bind: Regulating the Interstate Telecommunications Market for the 1990's" 2 FCC Record 1495 (1987).

subsidy and waste from dominant carrier participation in the information services markets. When rates are determined independently of the firm's own costs, powerful economic incentives work to curtail our concerns about dominant carrier participation in related markets. If such participation is unsuccessful, as it very well may be, the firm's stockholders, not ratepayers, will bear the burden. (Although the converse must also be allowed to happen, if the new ventures are successful the stockholders need to be able to keep the associated profits.) Predatory pricing behavior becomes much more expensive when the dominant firm cannot rely on ratepayer financing of the behavior.

It is possible to break, or at least substantially weaken, the link between a firm's own costs and its rates. The basic theme of this approach is to start with an initial set of satisfactory rates, then allow adjustments based solely upon general trends independent of the firm, such as inflation and industry, not the firm's productivity.

Several federal and state regulatory proceedings are underway at this time, seeking to implement variations of this new approach.

The second aspect of the new regulatory framework is an effort to reclassify the service offerings of dominant carriers. This apporach suggests that if we can isolate a set of key or core communications services, and precisely control the prices of these core services, then carriers should have great pricing freedom concerning other new communications services. This approach should greatly simplify the regulation of new hybrid information communications services. These services are almost always not likely to be core services at the time of introduction. While this two-step classification of services by no means eliminates the need for unbundling and equal availability of the essential transport services, it indicates that there is no need at all to fragment each service such that it can be pieced out at the discretion of the dominant carriers' competitors.

In Conclusion

The problems faced by dominant carrier participation in information service markets stem from the most difficult questions of regulatory policy and technology. The problems are not the result of simple failures of ability or motivation by any set of individuals or institutions. The Third Computer Inquiry process appears to have set in motion a set of interim solutions to our dilemma. The new regulatory framework offers a real long-term and stable solution to our concerns.

Remarks on ISDN

M. MacCarthy

U.S. House of Representatives
Committee on Energy and Commerce

The evolution of ISDN in the United States poses technical and regulatory issues for telecommunications policymakers at both the state and federal levels. In this paper, however, I want to address some underlying political realities that shape the more technical debate.

Policymakers are under enormous pressure to bring the benefits of the new computer and communications technology to as many Americans as possible. Members of Congress face angry constituents who are outraged by the quality and price of telephone service in a post-divestiture, deregulatory environment. These complaints have been largely substantiated by a recent report from the U.S. Labor Department which noted that local rates have increased by 40 percent since the AT&T divestiture.

Long distance rate reductions counterbalance these increases for some customers. But for most customers there has been no decline in overall price -- and no improvement in overall service.

The new technology makes possible the provision of a vast array of information age services -- bank at home, shop at home, electronic yellow pages, video retrieval, pay-per-view, digital music, and so on. And yet the number of Americans who receive these new information services today is miniscule.

What are we doing wrong? Part of the problem is an overemphasis on antitrust, at the expense of a telecommunications policy committed to bringing service to the public.

The effort to create a competitive industry structure for the new computer and communications technology is in accord with the historic U.S. commitment to pluralism and economic decentralization. The antitrust tradition treats all concentrations of economic power with dark suspicion, and seeks through divestitures and restraints on lines of business activity to create competitive markets. The heart of the belief of this tradition is that a workably competitive market is the best check on abuse of economic power.

This antitrust approach in the telecomnunications industry reached its zenith in the AT&T consent decree and the Computer II inquiry.

U.S. traditions have provided policymakers with an additional way of dealing with economic power. Instead of destroying monopolies or oligopolies through antitrust action, this approach accepts the need for large-scale economic units, but insists that they be regulated in the public interest. This is the approach embodied in the Communications Act of 1934. It emphasizes the need to bring service to the public and to set behavioral regulations to protect the public.

Americans would like to have both the new services be made widely available at affordable prices and a pluralistic industry structure with a diversity of suppliers to assure a diffusion of economic power.

It is becoming increasingly clear that we cannot have it all.

For instance, in order to prevent all risk of anticompetitive activity, the Bell Operating Companies have been barred from providing information services that are well within their technical capacity to provide. The public is denied the benefits of new technology.

The consequences are clear: in the United States we have a limited videotex market; in France, the market is thriving.

For instance, under the Computer II regime the FCC prevented the Bell Companies from providing Custom Calling II, on the theory that competitive suppliers would fill the void. In its Computer III notice the Commission noted that this had not developed. Again the public loses.

Lifting the restrictions in the Consent Degree that inhibit Bell Company provision of new services and abandoning the antitrust separate subsidiary requirements of Computer II are proper reactions to the development of new technology and push forward the program of bringing new services to the American people.

As the nation moves away from prohibitions and separations and toward an ISDN environment, the distinctions of the past -- between basic and enhanced service, regulated and unregulated, monopoly and competitive -- will be increasingly difficult to make. Any regulatory policy that puts excessive weight on such distinctions will be doomed to failure.

In moving away from the past overemphasis on antitrust, two extremes must be avoided.

One is laissez-faire. This would diminish antitrust safeguards (that is, prohibitions and separate subsidiaries) and replace them with no governmental mechanisms at all. Regulators would do nothing to oversee the service offerings, facility construction, and pricing practices of telecommunications firms. The public interest would be entrusted to the unchecked conduct of these firms. Unfortunately, this approach may be taken by the current FCC, and some state regulators.

A second approach is simply to return to old-fashioned economic regulation. Regulators would enforce a limited rate of return, require detailed cost justification for tariffs, and in general struggle to keep up with an industry spinning out of control under the impetus of dramatic technological change.

Both extremes are in error. Laissez-faire is hopelessly naive -- or hopelessly cynical -- about the power of large companies to use their economic power to the public's detriment. Old-style economic regulation is hopelessly optimistic about the effectiveness of government oversight of the conduct of large companies.

In this context, a search for new ways to regulate makes good sense. Policymakers must devise new ways to identify the public interest and to provide incentives for large companies to attain it.

The FCC is embarked upon this right now and has proposed a "price cap" mechanism to replace the current regulatory regime. Without additional safeguards, however, this proposal would merely allow companies to reap unregulated windfall profits.

In one other area, the Commission could move forward on the program of bringing services to the American people. In its Computer III decision, the FCC treated protocol conversion as an enhanced service, and established additional antitrust protections for providers of enhanced service, including a requirement of providing customers with multiplex access to their facilities. In so doing, it rejected the alternative of treating protocol conversion as basic when attached to a basic service such as packet switching.

This policy makes it more cumbersome for the Bell Companies to provide this service, and sends an important message: integration of facilities to provide an ISDN will be hemmed in and limited in a fashion that will make it difficult to provide this service at all. Large users will not suffer from this because they have the resources to obtain those services themselves through private networks or through customer premises equipment. But residential and small business customers will not be brought into the information age.

There is a danger that in pursuit of an efficient ISDN system our pluralistic goals will be compromised. But the greater danger is the two-tier society. Universal service no longer means plain old telephone service. It means the kind of information services needed to be a functioning citizen in the twenty-first century.

ISDN is not just a technical issue. Behind the technical disputes lurk conflicting visions of what a democratic society needs to flourish in the next century. Policy makers must strive to reconcile these conflicting visions.

They must balance the need to bring new services to all Americans at affordable prices with the need to maintain an effective check on potential abuses of economic power by large corporations.

ISDN - A Central Office Supplier's Optimistic View

R. L. Weindruch
Siemens

ISDN - A REASON FOR OPTIMISM

I am very optimistic about the telephone industry. The trends towards deregulation combined with the powers of new technology and aggressive marketing capabilities of the telcos make the outlook bright. ISDN is an important part of this optimism. ISDN is a technology that will streamline the existing telecommunication network and take advantage of the embedded twisted pair outside plant.

For more than a hundred years, advancements in telecommunications technology were inspired primarily by the requirement to reduce the costs of providing basic telephone voice services. The need to develop revenue producing services took a back seat. Separate networks for voice and data evolved independently, because of the technology available and the regulatory conditions that existed. Each was suited to particular purposes and neither readily integrated with the other. The result today is the quilt work of incompatible connections, software and customer premises equipment. ISDN has the promise of reducing these many interfaces to only a few. With ISDN, separate networks can be integrated and new services can be conceived. These services can be offered at improved value and more convenience to the users with additional revenue possibilities and less administrative costs to the telephone operating companies.

What is required to make these good things happen? First, true standardized interfaces (not defacto standards) are critical. Second, development of an operational support system strategy for ISDN is a must to keep costs controlled. Third, more innovation in the manner in which both suppliers and network providers market network services is required. Suppliers and telcos must do more in working together (and with third parties) to determine what rings the network cash register. The demand for new information services is on the rise in our businesses and higher income urban residential communities. The demand for those services that prove to be most useful will spread rapidly in the early 90's to rural and less affluent areas. The image of public network services is changing from one of a follower of PBX'S to one of a leader in providing information services. Centrex is being positioned as a vehicle for delivery of value added services. The ISDN capable public switched network will have significant power to handle the increased traffic and intelligence that these services will require. End users will buy ISDN services because they will get more useful information services per dollar spent on communications.

Three types of equipment are critical for the introduction of ISDN. First and foremost is the central office with SS#7 capabilities. Second, terminals and other forms of ISDN compatible CPE are required. Lastly, the introduction of Broadband will require the extensive use of fiber in the loop.

Most industry observers agree that Narrowband ISDN Islands will gracefully evolve into Narrowband networks and further evolve into Broadband ISDN networks. This assumes that ISDN is a modular addition to Digital POTS capability, and further, that Broadband ISDN capability will be a modular addition to Narrowband. This view of ISDN deployment conflicts with the inability of most existing switches to evolve gracefully from POTS to ISDN.

This paper will explore the driving forces behind the requirements for ISDN introduction, the reasons why end users will buy ISDN services, the equipment required for ISDN and a likely scenario of ISDN implementation.

REQUIREMENTS FOR ISDN INTRODUCTION

The foundation for the ISDN capable network is a set of uniform internationally accepted standards. Good progress is being made on standards that will allow the network to make available to subscribers all forms of information including voice, data and ultimately full motion video with the ease and flexibility of today's telephone call. Both Narrowband and Broadband ISDN standards are integral parts of this evolving network. It must be a smooth process for the operating telephone company to implement and maintain ISDN lines. The switch must be capable of quickly provisioning ISDN. Wherever possible the existing OSS facilities must be enhanced to allow ISDN lines to be tested and administered. Physical testing of the facility to the subscriber network terminator should be no more complex or difficult than what is used for analog lines. The

capability of the D channel enhances the maintenance opportunity by being able to report bit errors even when a subscriber is not actively transmitting. The challenge before us is to integrate the capabilities of the existing OSS infrastructure without creating complicated and expensive parallel systems.

The third and perhaps the currently most important requirement for successful ISDN introduction is Marketing. We, as an industry, now need to put even more emphasis on the process of defining more precisely what is useful to our customers. We should encourage as many companies and individual end users as possible to conceive, define and trial communication services. We need to find out what end users want and what they are willing to pay for, how they are willing to pay for it and the impact of these new services on network capacity. The bottom line is that the technical capabilities of ISDN technology and the public network do not ensure ISDN success. The keys to a successful ISDN introduction are held by the marketeers. We must all encourage them to get on with the important job of determining how to ring the network cash register. Any supplier or telco who claims that they can define ISDN services and how they should be priced does not understand the complexity of the marketing task. The risk is that we will allow ISDN to become an improved transport medium stripped of its value added potential.

WHY END USERS WILL BUY ISDN

Users will be able to receive electronic mail messages on their personal computers, talk with the individual at the other end and receive hard copies all at the same time using only one line. They will be able to access multiple voice and data networks and services. Users will be able to work, shop and bank from anywhere they have a terminal, home or office. Office features will be transportable to the home. MIS managers will become increasingly familiar with telcom managers. Political walls between these two groups will be demolished. Instead of separate networks with separate management, an integrated network will emerge with integrated management. Telcos will become much more acquainted with the MIS Department to enable them to sell ISDN/Centrex as the premier vehicle to deliver value added services.

This starts to sound like an ISDN revolution is near but, in my opinion, nothing could be farther from the truth. ISDN will evolve in the late '80s with the introduction of a few common sense services. Special services will go through switches and be administered with network software, not screwdrivers. Additions of data circuits will not require a backhoe or a carpenter. ISDN residential service will displace the rudest feature of the 1980's (call waiting). Lastly, the introduction of ISDN will be managed in a way that the POTS subscriber will not pay more so that his neighbor can reap the benefits of an ISDN line.

The key advantage to ISDN is that users will be able to dedicate their resources more to solving their own communication problems rather than designing multiple ways of accessing information. The users will be able to select equipment from among many competitive products and design applications suited to their communications requirements. ISDN will sell because it offers more cost effective choices to end users with new and different products and services.

ISDN combined with SS#7 will enhance PBX'S by providing both new capacity and feature options. All of the noise about battles between Centrex and PBX will be replaced by a new spirit of interworking. ISDN Centrex will allow end users access to a more advanced set of network services without the requirement of managing their own networks, but with the opportunity for the end user to choose how much network control he requires. Integrated access over existing telephone plant will eliminate the requirement for different kinds of wires, cable and connectors.

Higher transmission speeds, especially with Broadband capabilities using fiber optics will allow for high volume applications such as file transfers, database updates, graphics design, and even full motion video. In the business areas, the demand focuses on high-speed data communications. The increasing processing power of computers and the growing use of distributed databases will utilize significant bandwidth. The increasing bandwidth requirements of local area networks and the trend of greater interconnection among them will account for the demand of Broadband data communications. After Broadband is deployed, it could also be used to offer economic video communication services, such as video conferencing. In residential areas, the initial demand will center on the home entertainment services, such as film retrieval. This is indicated by the success of video stores and movie rentals. The use of Broadband networks for movie distribution can provide much better video quality than that currently provided by consumer VCR's and video tapes. In addition, when Broadband communications become popularly priced, full interactive video communications will eventually become commonplace in residential areas. Once again, though, the key will be in marketing the potential services. The doubters should consider that most of us pay more each month for cable TV than we pay for local telephone service.

We must dedicate ourselves to discovering what services the end users want and what is their willingness to pay under each circumstance. The success we achieve in building the ISDN infrastructure will position telcos to become more market oriented in providing value added services. Telcos that choose to wait for regulatory approval prior to planning the introduction of ISDN services face a high risk of merely being low cost providers of improved transport.

EQUIPMENT REQUIRED FOR ISDN INTRODUCTION

The ISDN capable central office switch with SS#7 capability will become the heart of tomorrow's communications networks. New terminals and terminal adaptors to existing PC's will be available to facilitate the introduction of new ISDN services. Lastly, fiber in the loop will allow for wider bandwidth services.

A few years ago, the PBX was portrayed as the office controller. Now the C.O. has the potential to be a network controller of 1990's. The central office will provide access and enhancements to PBX's, LAN's, packet networks and residential subscribers. Access to these services through the ISDN capable office switch means a whole new approach to meeting the needs of the users. The ISDN capable central office coupled with the power of the embedded network give the telephone operating company a unique opportunity to reposition Centrex from a PBX substitute to an information pipe.

The key to integrating the various networks and thus successfully implementing ISDN is the central office switch. The "ideal switch" has the following attributes. The switch architecture must be capable of providing today's analog features and interfacing with existing OSS's. The switch must also provide a smooth transition to ISDN, both Narrowband and Broadband, in a cost effective manner. There should be little or no cost impact on existing analog POTS subscribers who do not want ISDN. The existing subscribers on IA's should not have to relearn how to use the telephone. SS#7 will be mandatory because ISDN cannot be implemented in isolation. Interaction among various switches in the network is necessary and there should be feature transparency from supplier to supplier. Subscribers should not have to be retrained just because their central office was "improved". There should be minimal impact on the real time of the switch and the response time perceived by end users when it is equipped with ISDN and Signalling System #7.

Many telephone features do not need to be provided by the switch itself. They can be provided initially by independent nodes. These nodes are add-on subsystems that are connected via a defined interface to one or more ISDN switches and offer a particular service or feature. They may be owned and operated either by the telco or entrepreneurs. The switch architecture must be flexible enough such that services can be integrated back into the switch or more service nodes can be added. This begins to sound like the intelligent network.

ISDN terminals are the second type of equipment required for ISDN introduction. Various suppliers are taking different approaches towards the development of terminal equipment. Both telcos and suppliers must encourage terminal suppliers to undertake immediate steps to ensure that cost effective ISDN compatible devices will be available. It is clear that the development of terminal adapters to interface with a huge existing embedded base of CPE apparatus is essential. The terminal adapter will be required for interface capability to asynchronous terminals, personal computers, synchronous data networks, fax machines, telex devices and many others.

There will also be ISDN telephones. The capabilities of the digital telephone will include an LCD Display, programmable buttons, and hands free operation. This device will connect directly to the S bus without the need for any terminal adapters. Intelligent chip cards could be used to export services or simplify accounting.

It is my belief that although new ISDN terminals will be available, standardized terminal adaptor products will dominate the market. These products will interface into existing PC's, which I predict will be the future standard CPE. These PC's will offer a variety of input/output options. This protects the customers investment in his currently installed equipment and minimizes ISDN getting started costs. New software based services available in or through the network will let users take advantage of network functionality without expensive application design and development or changeouts of terminal equipment.

Fiber is the third type of equipment required for ISDN Introduction. It is not true, as some believe that ISDN stops at 23B+D, Primary Access. ISDN is a bandwidth independent concept. We should all be planning for the introduction of fiber into the ISDN Loop network. While twisted pair and pair gain systems provide access for integrated voice and data services, the introduction of fiber will extend the Narrowband capabilities to include image and video services. Connectivity to LAN's via ISDN at these data rates makes the Broadband ISDN a natural backbone for the Metropolitan Area Network (MAN).

ROAD MAP FOR ISDN INTRODUCTION

The introduction of ISDN will likely be in three phases. The first will be providing Narrowband ISDN network transport solutions to special customers in ISDN islands. These initial ISDN islands may be extended with remotes to build overlay networks. The initial ISDN island implementations may consolidate subscriber transmission requirements, integrate voice and data networks, provide additional ease for inter-department data communication capability or provide additional end user control. The second phase will involve providing Narrowband capabilities throughout the network to anyone who wants it. In this phase, ISDNs will be both overlayed on existing networks and used to replace older technology. SS#7 will be the glue that unites ISDN networks. The third phase will involve providing Broadband capability for the efficient transport of video, data and other high bandwidth services.

These initial ISDN islands will be of great benefit in data applications. They should reduce the cost of any information transport to the user and enable the telco to participate in the data communications transport business to a far greater extent. ISDN islands will provide a good reason for telco Account Executives to make friends with the EDP manager while cementing the relationship with his traditional customer, the voice telecom manager.

Telcos will build on the ISDN islands with Signalling System #7 capabilities. This will allow business and Centrex services to be extended across multiple central office service areas. Before 1990, a geographically dispersed business will be able to receive Centrex services as if they were in a single location. This will enable many new service capabilities including Class, LATA wide business services and private virtual networks. It will create an increased opportunity to interwork between the public network and private branch exchanges. The result will be enormous opportunities for the business subscriber and for the telephone operating companies. Signalling System #7 will include Transaction Capabilities (TCAP) that will allow access to intelligent network databases. Intelligent peripherals and feature nodes will be additional network elements allowing telephone operating companies and non-Telcos to trial new features and services quickly, with new ways to implement these features.

The introduction of Broadband ISDN will occur in the early 90's. The technical feasibility of Broadband has already been demonstrated by a number of pilot projects. As has already been mentioned, Broadband services are targeted for both residential and business customers.

The features in the ISDN should be largely bandwidth independent. That is, forwarding a call should operate in a standard manner whether the call is a voice call, a data call or a video call; whether the call is analog or digital at 16 kbps, 64 kbps, 384 kbps, 1.536 mbps or about 150 mbps; whether the call is placed over fiber or twisted pair. Modular ISDN add-ons to POTS switches allow for this smooth feature growth as the needs of the end user evolve from POTS to Broadband ISDN.

Before we leave the subject of network evolution, we should review the forces that will be driven by potential competition to the telephone company within the context of ISDN. The loop competition is cable TV. The proper competitive response is to be the first with fiber to the home. The cost of the fiber and the CPE electronics could be financed by a small tax deductible mortgage. The tougher competition is from both toll network providers and PBX suppliers. The common carriers can directly connect to PBX's and other private networks. The proper competitive response from the local exchange company is to offer enhanced services and capacity to PBX's by interworking with ISDN central offices. The proposal is to obtain account control by utilizing the power of ISDN/Centrex and the public network to add value to existing PBX's. ISDN has the power to increase the variety and quantity of information that can be efficiently carried over the public switched network.

CONCLUSION

In summary, ISDN is a technology that will encourage increased utilization of the public switched network and increased opportunities for telcos to define and sell new services. The future is bright for those who can properly define the new services, intelligently test market their concepts, and quickly bring new offerings to end users. Those network providers that invest early in ISDN and that are customer oriented will be winners. We all will painfully learn that end users will not accept inconvenience or the loss of existing features if they choose to try ISDN. Lastly, the early introduction of ISDN to the minority cannot burden the majority who do not yet perceive a need for ISDN services.

The central office switch will become increasingly critical as the hub of these new network services. Switch architectures are already becoming more flexible to allow graceful migration to ISDN capabilities and SS#7. Huge BHCA capacity will become available as tomorrow's ISDN call mix will be much more complex. The OSS framework will be redefined in this new network environment. Lastly, central office interfaces will be opened to interact with a variety of service modules. ISDN will propel the central office switch of the '90s to become the hub of future network services. The ISDN capable central office switch should be more properly called the INFORMATION NETWORK CONTROLLER.

I submit that marketing of ISDN is now equally, if not more, important than technology for a successful launch of ISDN. I am confident that the local telephone companies will also become experts in data and video communications as well as voice. I am certain that the enhanced capabilities of the ISDN capable public network will cause the power of the network to be more fully utilized. Communication among men will be enhanced. Thus, ISDN is indeed a basis for optimism.

A Computer Supplier's View Of ISDN

Bill Johnson, Tom Ermolovich,
Ralph E. DeMent

Digital Equipment Corporation
Maynard, Massachusetts

This paper looks at ISDN from a computer supplier's viewpoint. A methodology for incorporating ISDN into a general distributed computing architecture is developed by looking at future communications requirements. These requirements are developed by taking a historical look at computer networks and projecting future user needs. ISDN is then examined to see how it best fits these needs.

1. INTRODUCTION

Rapid changes in semiconductor technology are driving the computer and the telecommunications industries towards each other. The communications capabilities of computers and the computing capability of switches are constantly improving. This trend has certainly sparked a great deal of controversy in the industry because it is becoming increasingly difficult to partition functions between the two systems. A lot of this controversy is centered around ISDN, since in many aspects, ISDN falls in an area where the two industries meet. Although there has been a great deal of debate between the two industries on this subject, there has not been a great deal of meaningful debate as three of the words in ISDN, integrated services and network share no common meaning across the two industries.

This paper examines these issues from the perspective of the computer supplier. Setting the stage for a discussion of a computer supplier's view of ISDN requires a brief discussion of distributed computing systems and the use of computers in telephony applications. The view of ISDN is followed by a description of ISDN application areas.

2. CHANGES IN DISTRIBUTED COMPUTING SYSTEMS

Distributed Computing Systems, DCS, has evolved from the early days of computer networks. To understand the future communications requirements of these systems and how ISDN may meet the needs of those systems, it is necessary to take a brief look at the changes that these systems have undergone.

The first function provided in computer networks was computer to computer communications. Programs and files on one computer were able to be copied to another. However, one could not build a network because of the lack of routing. The addition of route through capability enabled the building of small computer networks. The size of these networks was limited by the total number of "hops" in the network. Larger networks were enabled by local area networks, LANS, since store and forward was only required from LAN to LAN rather than computer to computer. LANS also provided a platform for distributed computing. The distributed computing type of architecture allows a building block approach to computing. These building blocks may be associated with an organization structure, i.e. what department the person is in, or they maybe associated with a certain function, i.e. payroll. Files from these different computers are exchanged over the LAN. A major change is on the horizon for these networks. The addition of distributed system services will change these computer networks into Distributed Computing Systems. Users will subscribe to services such as data base services, file services, print services and name services.

Coupling with the above changes in user functions have been changes in interoperability. Originally, communication was restricted to computers from the same type of computer with the same operating system. The first stage in this evolution was to allow computers of different types and different operating systems to communicate with each other as long as they were supplied by the same supplier. This restriction was removed by providing intervendor communication through multivendor gateways. True interoperability will come from OSI.

Distributed Computing Systems have also undergone physical changes. In the beginning, all communication capabilities were based in the host. With the advent of LANS these communication capabilities moved out onto the network. The topological flexibility of LANS was greatly increased with the introduction of bridges that would perform level 2 routing. These bridges allowed the creation of subnetworks at the department level connected together by a backbone network. In the future, this backbone network will evolve to a high speed fiber network.

Throughout these changes, there has been great controversy about wiring. Wiring technologies have come in and out of favor: coax, broadband, twisted pair. A number of recent changes has simplified this problem for the user. First, there are a number of low cost flexible wiring schemes in the market that allow the office user to employ a wide variety of technologies. These schemes provide all the physical connectivity that would be required for the office. Secondly, engineering techniques are becoming available such that LANS can become media independent. Lastly, users have come to realize that the real problems in networking are not wiring but are distributed services and multivendor interoperability.

These changes in technology will drive the communications requirements for the desktop in the 1990's. Initially, the computing industry had a two tier structure, terminals and computers. The introduction of the personal computer changed this structure to a three tier structure which required vastly higher communications speeds to the desktop. Similarly, the introduction of windows and pull down menus changed how people used personal computers. The combination of all these trends gives an indication as to how the personal computer of the 1980's will evolve into the workstation of the 1990's. Not surprisingly, this system will be a multi-windowed workstation executing a combination of local and distributed systems services. Such a workstation will require more than the 4 M's we have today: 1 million instructions/second compute engine, 1 megapixel display, 1 million bytes of memory and a 1 million bits/second communications channel. The computer supplier must provide their customers with communication technologies that will meet these needs.

3. COMPUTERS AND TELECOMMUNICATIONS

Due to the vast size of the telecommunications industry, it has always been a leading consumer of computer products. Most of the applications centered around using computers to automate the operations of the phone companies. Computers were used to manage large data bases of customer and equipment information. Computers have also been used in ancillary functions to call processing. Such an application could be managing an 800 number data base or performing a directory assistance function. While the industry has been applying computers in large data base applications, little has been done to use computers to help the office worker use the telephone. This has been due to a lack of integration between the telephone and computer systems. This lack of integration has resulted in a strange dichotomy in the office. Most office workers have either a terminal or a personal computer executin sophisticated programs such as word processing or spreadsheets. The ease of use of these devices has improved greatly over time. However, there has not been an equal improvement in how to use the telephone. Improvements could be provided by using a menu driven computer to control the telephone.

The customer support center application highlights the need for integration. Customer support centers make heavy use of computer data bases and PBX automatic call distribution, ACD. The interesting problem occurs when the operation of the support center requires a multi-tiered approach, that is, when more than one agent is needed to process a call. A disruption in the work flow occurs when an agent has to transfer both the telephone call and the information from the caller that has been typed into a computer data base. Today this is accomplished by either the use of scraps of paper or by the unnecessary phone calls between agents. In an ACD environment this can be a major factor limiting agent productivity. With integration between the telephone system and the computer system, this transfer could be accomplished by hitting a single key. With one key both the telephone call and the computer data base could be transferred to another agent. Furthermore this delivery could be done via the normal ACD delivery mechanism.

In order to provide integration there must be a command link between the telephone system and the computer system. Minimally, this link must have the following capabilities: make a call, call progress, answer and disconnect; monitor a call and call transfer. Once this capability is put into place, a wealth of applications can be provided to the user. This capability has been termed Computer Integrated Telephony, CIT. The amount of application areas is essentially boundless. Some examples include: Horizontal office automation, i.e. message desk and dialing; Telemarketing; Hotel/Motel; Financial and Credit Verification.

4. VIEW OF ISDN

The computer supplier has multiple views of ISDN. By examining each view individually, a more complete view of ISDN can be developed. The first view is that ISDN is the next generation of the the telephone network. The telephone network has evolved from mechanical to analog electronic to digital electronic to finally ISDN. ISDN can be viewed as the next step towards intelligent networks.

ISDN is also viewed as a set of services. The primary role of an ISDN is to provide a set of bearer services that will meet the needs of a broad range of telecommunication services. To support these services, ISDN provides an end-to-end digital network, powerful subscriber signaling supported by the D-channel packet protocol and support for both circuit switched and packet switched connections.

ISDN is also a communications technology representing the lower 3 layers of the ISO reference model. Historically, computer networks have supported a wide range of transmission technologies at the lower levels of the ISO model: RS232, parallel interfaces, local area networks, broadband, X.25, HDLC, etc. ISDN is another lower level technology that computer networks will support.

The ISDN technology has the attractive property of combining voice and data on the same physical pair of wires. This physical integration has created a user demand for functional integration between voice and data. This integration can only be accomplished by the ISDN system and the computer system cooperating at level 7 of the ISO model. To accomplish this, improvements will have to be made to ISDN protocols at the Q931 layer.

ISDN is also viewed as an implementation. As with any implementation there are problems to be resolved. There are many questions to be answered: When will there be a volume deployment of ISDN? When will there be cost effective silicon available? What will the tariff structures be? When will there be common interface standards between the major suppliers? Who will certify ISDN interfaces? Until there are answers to these questions, computer suppliers will be reluctant to commit ISDN products to volume production. Just as the computer industry continues to take a leadership role in local area networks and OSI, the telecommunications industry must continue to take a leadership role to make ISDN a reality.

5. ISDN POSITIONING

The positioning of ISDN falls into three areas: Wide Area Networks, Local Area Networks and Computer Integrated Telephony. When ISDN becomes deployed, its use will be best suited to the Wide Area Network. Building LANS will be connected to ISDN via a level 3 router. ISDN will be an enabling technology that will allow the penetration of computers and telecommunication services into businesses that have many locations with a wide geographic dispersion. Such a penetration is already occurring today with many customers in retail. Today, the retail store is usually connected back to headquarters with leased 56 kilobit lines. The provisioning and the cost of this service is a limiting factor for penetration in this business.

ISDN also has the promise of solving the problem of remote terminal and remote personal computer interconnect. This remote interconnect problem occurs when there is a geographic dispersion of computer subscribers. Some examples of this are a sales force working out of their home or college students working out of their dormitory. Today, these users are relegated to slow speed communications, usually 2.4 kilobits/second and must give up the use of their telephone while they are communicating with the main host. ISDN will give them a 64 kilobit transfer rate plus simultaneous use of the telephone.

Inside the building in the LAN environment, ISDN will not be recommended for the transport of data. The 64 kilobit/second transfer rate of ISDN basic rate will be too slow to support distributed system services. This point can be easily illustrated by an example. One service that may be requested is to load a personal computer program for which the smallest size today is 256 kilobytes. Loading this file over ISDN basic rate would probably take about 100 seconds when the protocol overhead is factored in. This same transfer over a LAN will be 100 times faster, having a load time of about 1 second. Users will not tolerate 100 second delays every time a remote file service is requested. Cost is another issue surrounding ISDN use for local area data transport. It appears that for the same product cost as an ISDN basic rate interface, it is possible to have a 10 megabit LAN interface. Additionally, it appears that for the same product cost of an ISDN primary rate interface it is possible to have a connection to a 100 megabit fiber optic LAN. Why is this true? There are two reasons. First, ISDN has a different design center than a LAN. It accommodates much greater transmission distances and has a much richer connectivity. Secondly, ISDN is an attempt to redesign the entire telephone system worldwide. This creates an inertia which slows down ISDN's ability to incorporate the latest technologies available to LAN technologies.

In the last area, Computer Integrated Telephony, ISDN protocols need to be improved to allow computers to signal on behalf of a network subscriber. Once ISDN can deliver this capability, it will replace all the proprietary links that exist in the marketplace today for this purpose and will serve as the standard connection between a switch and a computer. This capability will enable ISDN to deliver functional integration of voice and data and will stimulate application development which in turn will create more pull for ISDN.

6. CONCLUSION

As the computer and telecommunications industries come closer together, new opportunities will be generated centered around the capabilities that can be offered to the customer through close cooperation. A foundation for this cooperation is a positioning of the technologies offered to the customer. From a computer supplier's viewpoint, ISDN will complement the existing network offering by connecting LANS together to form a wide area network and by facilitating the connection of remote terminals and remote personal computers. With this complementary viewpoint, the two industries can work together to find the applications that will fuel ISDN's growth.

Rapid changes in communication technology will produce winners and losers in both industries. The winners will be those firms who look for these new synergistic opportunities. Success will come from new markets as well as new uses for equipment in existing markets. Firms that don't come to this realization will find themselves competing in a zero sum game with an increasing number of competitors. One of the key opportunities in the future for ISDN will be its use for Computer Integrated Telephony. The two industries must work together to incorporate this capability into the ISDN standards process. Additionally, the two industries must work together in order to achieve rapid deployment of this technology.

ISDN: A Strategic Approach To The Market

R. A. Schwantes

Research Triangle Park, NC
Northern Telecom Inc.

In this paper, an approach to introducing ISDN to the market is discussed. The challenge to suppliers of ISDN is to provide pre-ISDN and ISDN solutions in a time frame consistent with end-user needs. Strategic activities to address the market are outlined; they include: offering of pre-ISDN solutions, ISDN market trial activity, and development of an ISDN product portfolio.

1. INTRODUCTION

ISDN will bring many benefits to the network and the end user. One of the characteristics of ISDN from which many of these benefits are derived is the reduction of many network interfaces to initially two standards: basic rate and primary rate followed by broadband ISDN. It is this standardization that gives ISDN its "plug-in-the-wall" convenience: the ability to continue to use embedded investment in terminals, PCs, printers, hosts, and telephones in an integrated, transparent way. However, fully defined, standardized ISDN is a product of the 1990's. Yet, those who would be the initial users of ISDN have network needs that must be met today, and will require the ability to interwork with ISDN islands as they evolve. This paper addresses this discontinuity in the marketplace and outlines one supplier's strategy to overcome it.

2. ISDN ADDRESSES END-USER REQUIREMENTS

In the business environment, the increasing importance of data communications is driving new network demands. Many measures point to the increasing volume of data communications:

1) Purchase of T-1 multiplexers is growing at 51% per year;

2) Data traffic is growing at all speeds: from a 50% increase at lower speeds, to a 700% increase at 1.5mb and higher by 1990;

3) By 1990, 39 PCs will be installed for every 100 white collar workers;

4) The number of PC LANs in use will increase 5 times by 1990, while

5) Use of modems will double.

With the emphasis in the market toward data communications, end-user needs have arisen which can well be satisfied with ISDN on the public switched network:

A) The end user needs reliability and quality.

ISDN provides the end user a ubiquitous, seamless connection to a multitude of services, data bases and controls that private networks, given their specific, defined nature, cannot satisfy.

With end-to-end digital connectivity and the increased functionality of ISDN, the ISDN nodes in the network are intelligent surveillance units which can identify, route around, and initiate corrective action when problems occur. Additionally, the end user can capitalize on the around-the-clock, automatic maintenance and administration systems that telephone operating companies use to maintain public data and voice networks.

An end-to-end digital network offers simpler equipment configurations and lower error rates than analog and hybrid networks. Degradation is dealt with on a link-by-link basis. Regeneration of the digital signal occurs at the NT1, the line card, the outgoing channel, and all the way across the network (see Figure 1).

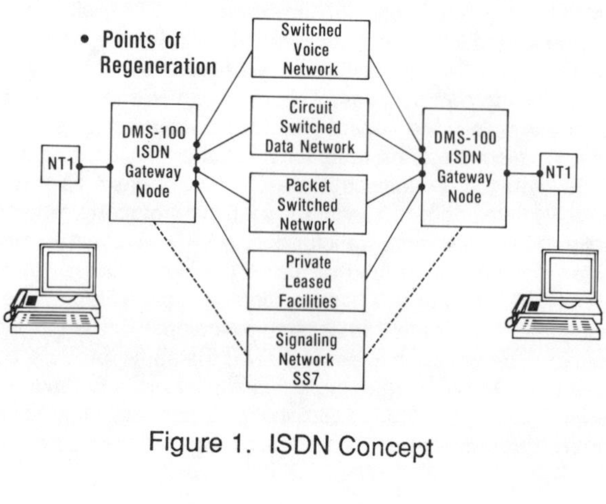

Figure 1. ISDN Concept

B) The end user needs PC LANs and interconnection of PC LANs.

End-user applications frequently require rapid, reliable data transfer between PCs. Existing PC LANs, however, impose geographic distance constraints which hinder the application of LAN capabilities. ISDN offers a means of providing increased LAN functionality by allowing remote PCs to perform as if in a LAN without the distance constraints (wide area networking), and by providing metro area connectivity of high speed LANs (see Figure 2).

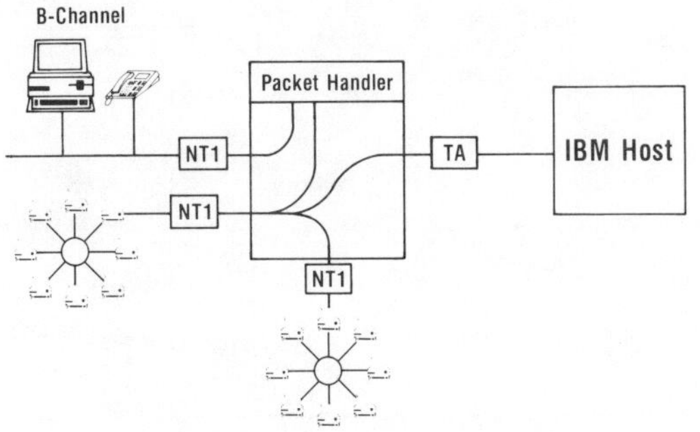

Figure 2. ISDN as LAN Gateway

Because ISDN access rates are worldwide, open standards, any terminal with an appropriate terminal adapter can connect with any other terminal or PC independent of manufacturer, protecting existing terminal investments.

In a central office LAN application, the B and D channels of the basic access loops are switched at the ISDN access node to allow PC to PC communication for a single locale. This type of application can be economical for the small business user, needing connectivity for a small number of PCs, yet which cannot justify the purchase and wiring requirements of a LAN.

A LAN application can be easily expanded to a wide area network, by switching B and D channels from locale to locale. ISDN, then, allows remote PCs to participate in a LAN with near equivalent performance.

PC networking does not have to be limited to the office. The remote branch office may also be the employee's home. With standard 2B+D voice/data basic rate connectivity to the residence, ISDN can support the cottage society evolution driven by the desire of the work force to spend less time traveling and commuting, and more time at home.

Use of ISDN with the appropriate network terminating equipment, makes existing LAN to LAN communication a reality by providing initially a 64 kb transport between LAN gateways, and ultimately, with broadband ISDN, multimegabit transport.

To satisfy the high speed (greater than 1.544 mbps) voice, data, and video needs of campus and metro area network environments, plans are already being formulated in the communications industry to define a third, multimegabit standard referred to as the H channel or broadband ISDN. The same ISDN-like structure, an access pipe with out-of-band signaling, would be used.

C) The end user needs management and control of the network.

Given that large end users have purchased DS-1 facilities for voice and data networking, the 24 channel characteristic of ISDN primary loops conveniently allows the use of this existing DS-1 backbone for primary rate access (see Figure 3). Use of primary rate loops for PBX to PBX, PBX to central office, or LAN to central office communication allows various service facility types to be consolidated to one access pipe. With call-by-call allocation of B channels on primary access through D channel control, the end user has the ability to allocate bandwidth as it is needed, rather than be confined to single purpose pipes which may be under-utilized.

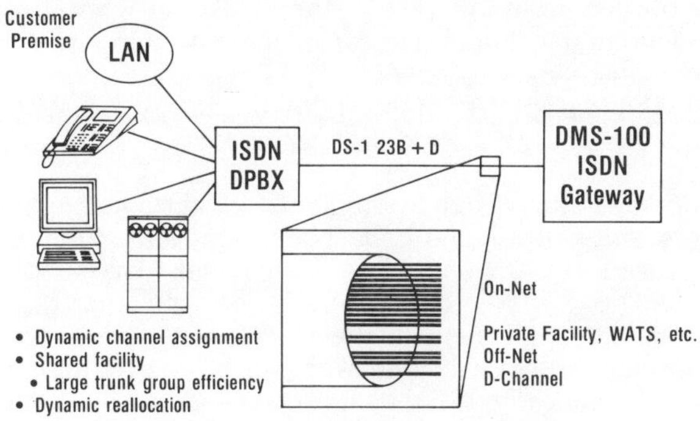

Figure 3. Primary Access

D) The end user needs reduced cost.

The above points that demonstrate the advantages of ISDN can also be translated to cost reduction. However, it should be highlighted that use of ISDN on the public switched network for data communications can be lower in cost to the end user than the dedicated facility, full-time connectivity approach often used in point-to-point and multipoint private data networks. This is especially true when cost of resources in the public network can be shared by many users that do not require full-time connectivity.

Data service can be provided to isolated users without the dedication of under-utilized cluster controllers, high speed lines, modems, and front end processor ports. Locally attached controller response characteristics are available to the remote user; and the need for specialized coax wiring at new sites or areas is eliminated (see Figure 4).

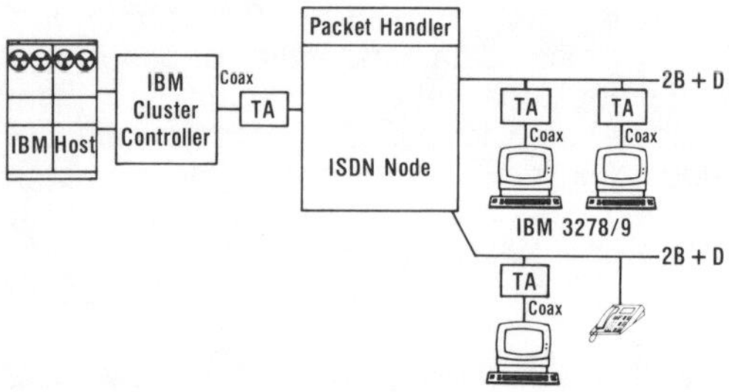

Figure 4. Coax Elimination

3. A STRATEGIC APPROACH TO THE MARKET

Given the above discussion, it is apparent that ISDN provides substantial, tangible benefits to the end user. However, there is a major discontinuity in the marketplace. End-user demand for reliability and quality, PC LAN interconnectivity, management and control, and reduced cost are needs of today. "Pure" standardized ISDN, because of the standards negotiating process and lead time needed for standards verification and volume production, will not be available until the early 1990's. End users that must solve today's problems cannot wait for ISDN to mature. Consequently, alternative solutions are being sought. This has critical implications for ISDN because end user demand is pivotal in financing the necessary investment required to bring ISDN to the market.

To address this discontinuity from a vendor's standpoint, three strategic thrusts have been identified: First, the approach to ISDN should be incremental. The supplier should provide products and services that have immediate application in the market yet, at the same time, can evolve with the evolution of ISDN standards and interwork with ISDN as it is deployed.

Second, suppliers should lead the industry in ISDN field trials and applications to learn the true interests of the marketplace and to assist the telephone operating companies in introducing these services to their customers.

Third, to truly understand the advantages and implications of ISDN, the supplier must be more than a niche player, but rather an end-to-end network product supplier.

A discussion of each of these strategic activities follows.

3.1 Incremental Introduction

At Northern Telecom, we produce a wide range of offerings that meet the business users' current voice and data requirements, yet are designed to interwork gracefully in the ISDN environment as it matures.

The products and services available today which will seamlessly interwork with ISDN include:

A) Digital centrex service - central office based business services, for integrated networking and virtual private networks, are offered and maintained by the telephone operating company. Since 1983, Northern Telecom's centrex software has been accessible via out-of-band signaling techniques. For the ISDN environment, we have extended this capability to allow the out-of-band signaling messages from the ISDN channel to access that same centrex software giving non-ISDN and ISDN business users access to the same centrex features (see Figure 5).

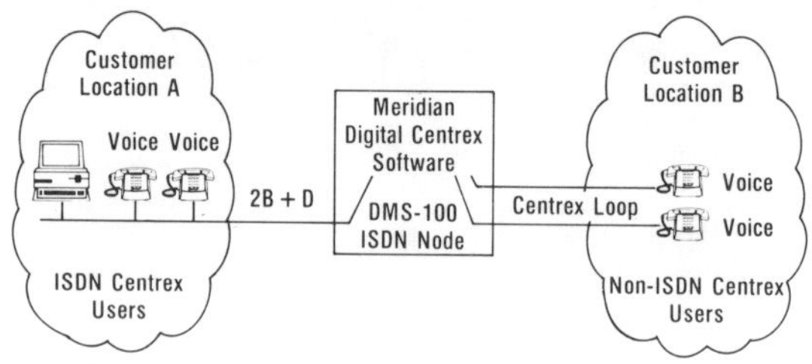

Figure 5. ISDN: Centrex Interworking

B) Switched digital data capability - this service is provided via a high speed, circuit switched data transmission system. Asynchronous speeds to 19.2 kbps and synchronous speeds to 64 kbps can be accommodated, providing digital connectivity from end user to the central office. As a result of two years of discussion, negotiations and trials involving Northern Telecom, Bellcore, the Bell Operating Companies, and AT&T, nationwide Public Switched Digital Service is now being introduced. As ISDN matures, ISDN data users accessing the network will be able to exchange information with non-ISDN circuit switched data users via established data rate adaptation techniques (see Figure 6).

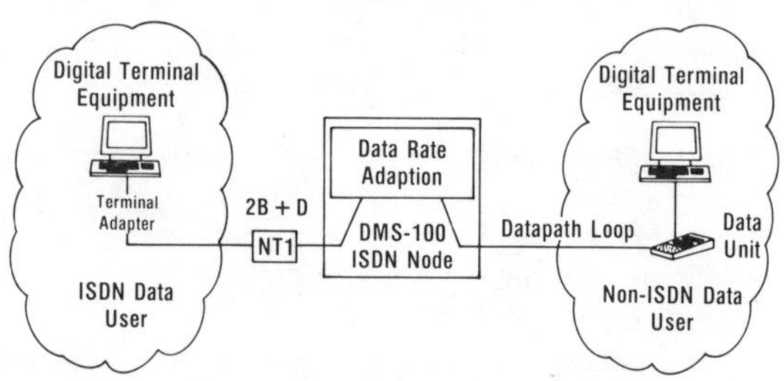

Figure 6. ISDN: Circuit Switched Data Interworking

C) Local area networking capability - PCs and other peripherals can communicate using the protocol conversion capability of a public network gateway while circuit switched voice and data are routed through the central office. With the introduction of ISDN, all terminals still have the same connectivity to the gateway, but facilities for analog and digital circuit switched messages, as well as packet data, can be consolidated onto primary rate access links (see Figure 7).

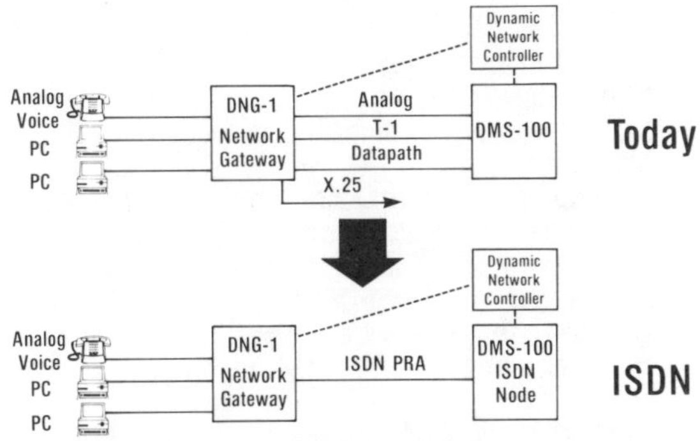

Figure 7. ISDN: Network Gateway Interworking

D) Dynamic network management capability - allows control of station moves and changes, control of network routing, end-user testing of trunks, real time access to network usage, and ability to control usage of available bandwidth. This capability becomes even more valuable as a wider array of features becomes available to the end users through ISDN.

In addition to providing products that meet today's needs, equipment suppliers must adopt the standards defined by the standards bodies (ECSA/T1D1, CCITT)to ensure that ISDN is an open, multi-vendor environment. Where there is an absence of existing standards, Northern Telecom is defining and implementing logical interfaces to allow market trials and applications of ISDN to proceed, while at the same time maintaining the flexibility to integrate standards as they are defined.

A good example is the decision last year of the Exchange Carrier Standards Association (ECSA) to use 2B1Q as the line code for the U-interface. Prior to this decision, Northern Telecom selected an AMI based interface in order to proceed with our ISDN introduction program. We are now in the process of converting to the 2B1Q line card. The necessary new chips have been designed, the interface has been tested, and test data has been sent to the standards committees.

3.2 Trial Activity

To begin fulfillment of the second strategic objective, Mountain Bell and Northern Telecom made communications history when the first ISDN call in North America was made November 18, 1986 in Phoenix, Arizona. The Arizona State Government and Honeywell are the trial customers. The trial includes one ISDN switch equipped with packet handling, three remotes, and many vendors' voice and data terminal equipment (see Figure 8).

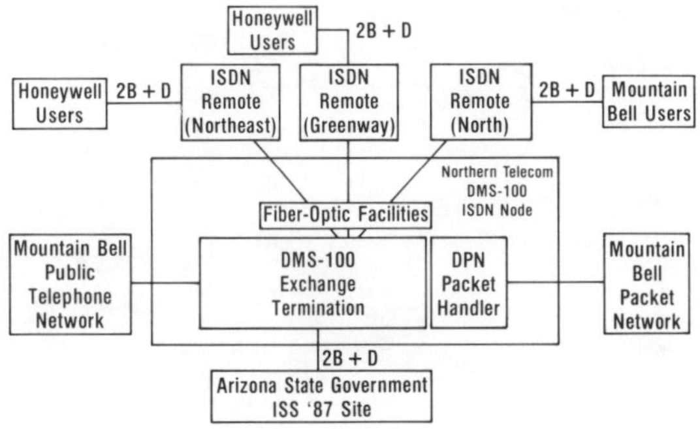

Figure 8. Mountain Bell/Northern Telecom Phoenix ISDN Trial Configuration

The following applications have been successfully demonstrated:

A) Example: Arizona State Government Department of Administration employees use asynchronous terminals to access, in a packet switched mode, a Honeywell host to retrieve the status of job applicants and a list of job seekers (see Figure 9).

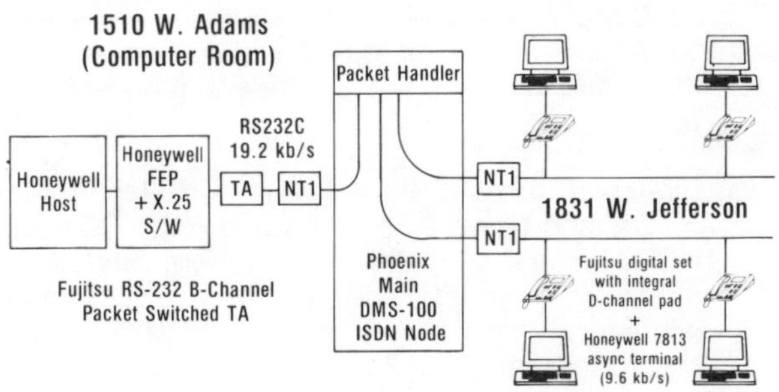

Figure 9. Mountain Bell/Department of Administration: Applicant Tracking

Value: Asyncronous data and voice are combined onto one channel; terminal access is now at a higher transmission rate; the terminals have access to other data bases because they are now in a dial-up mode and not a private line connection; ports can be shared at the host site; modems are eliminated; and the transmission is digital.

Prior to the application of ISDN, the voice terminals used centrex extension lines with key system equipment and the data terminals were connected to the host via private lines.

B) Example: Arizona State Government Purchasing Office employees use PCs and terminal adapters running SNA/3270 emulation to access an IBM host through an IBM front end processor that has been equipped with Network Packet Switched Interface (NPSI) software (see Figure 10).

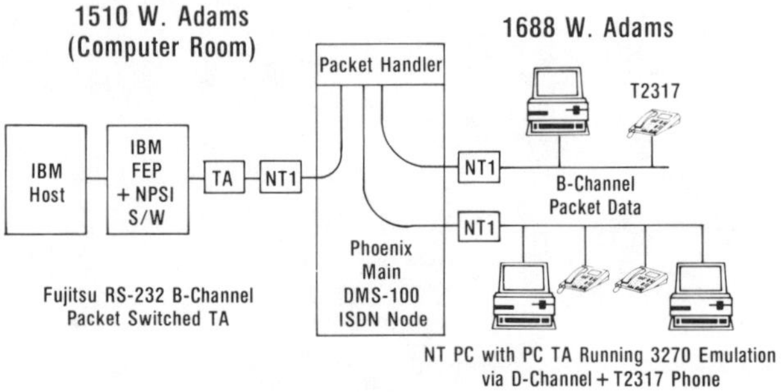

Figure 10. Mountain Bell/Department of Administration: State Purchasing Office

Value: Cluster controllers, coax cable, private lines, and modems have been eliminated; terminals can now communicate with other data bases because they are in a dial-up mode; ports are shared on the front end; voice is now provided over the B channel, eliminating key equipment used previously.

Prior to ISDN, voice service was provided by centrex extension lines. Data transmission occurred in a normal 3270 SNA environment with 3270 terminals linked to a controller via coax, with the controller linked to the front end processor via private lines.

C) Example: Arizona State Government Department of Transportation employees use NCR PCs and NCR terminal adapters which perform SNA/3270 emulation. The terminals, through an ISDN node, access an Amdahl host with an NCR front end processor (see Figure 11).

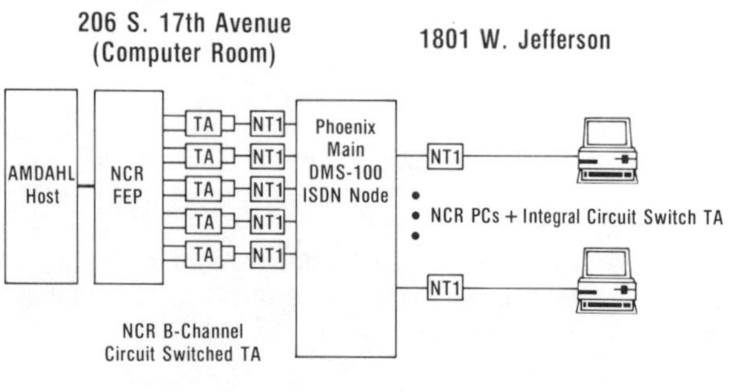

Figure 11. Mountain Bell/Department of Transportation NCR-COMTEN Application

Value: An ISDN compatible terminal can now be placed anywhere on the ISDN network and access the host without a private line, controllers, or coax links.

Prior to ISDN, the 3270 terminals were coax linked to a local controller that was in turn dedicated to the host via a private line. Voice was provided by a PBX off premise extension line which was an additional private line dedicated to the user's desktop.

D) Example: Ten Honeywell employees have an ISDN basic access line into their home served from ISDN remotes. Two of these homes also have a security monitoring system installed. The equipment used in the homes is the Fujitsu digital telephone set with D channel Packet Assembler and Disassembler/Terminal Adapter. The employees use their standard home PC that is connected to the Fujitsu set. The terminals access the Honeywell data bases via the D channel. In those homes that have the security monitoring system installed, the security information is transmitted over the D channel to the Honeywell security monitoring station. This security status information is displayed on a PC (see Figure 12).

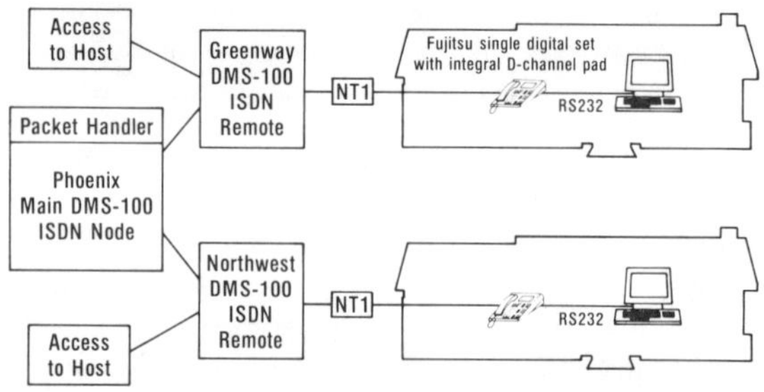

Figure 12. Mountain Bell/Honeywell: Residential Access

Value: This application demonstrates the ability to establish a work at home environment for voice and data without establishing a secondary line to the home, and the ability to have security monitoring over the D channel using the same facility.

Prior to this application with ISDN, work at home would require a secondary telephone line with a modem for data access, or a modem on the existing line, in which case a voice call could not be placed or received.

E) Example: Mountain Bell users share files from a central server in a wide area networking application and access an IBM host. IBM PCs use PC terminal adapters to achieve 3270 emulation. Using ISDN basic access lines, the PCs can access the shared files from the server over the public switched network. In this example, access is provided via an ISDN remote. The PCs, emulating 3270 terminals, can also access an IBM host over a public packet switched network (see Figure 13).

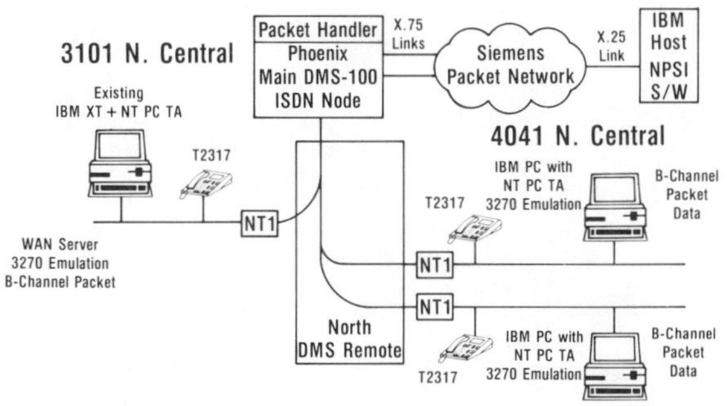

Figure 13. Mountain Bell Wide Area Network Application

Value: This application demonstrates PC to PC communication without a LAN; high speed data base access without private lines, and wide area networking of PCs. Geographic coverage is the range of the ISDN node and this can be extended further by connection to Public Packet Switched Networks or to other ISDN nodes via Signaling System 7.

U.S. West, supported by Northern Telecom, is now introducing ISDN to end users currently served by DMS-100 Meridian Digital Centrex. Several applications are targeted for completion during 1987. The first of these is taking place in Portland, Oregon where ISDN access is being added to an existing DMS-100 central office that currently serves approximately 3000 Meridian Digital Centrex subscribers.

More trial activities are scheduled with all of the Regional Bell Operating Companies and major Independents.

3.3 End-to-End Network Knowledge

The third strategic thrust is to be an end-to-end network provider of ISDN components. By being a provider of the five key product groups that support the ISDN network, a supplier can assess market interest and develop new network applications (see Figure 14).

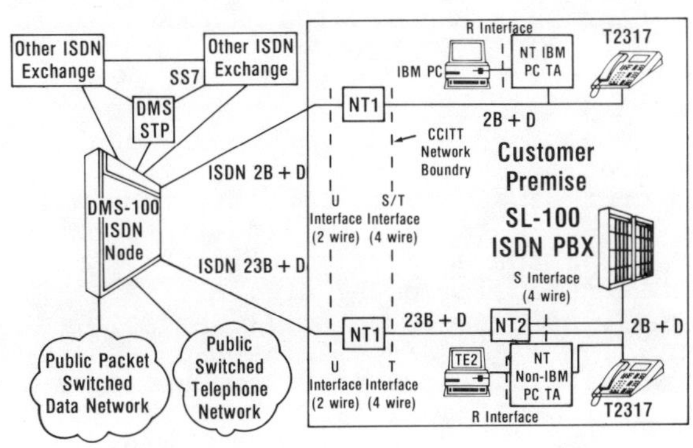

Figure 14. ISDN: The Environment

The ISDN building blocks are characterized as follows:

A) The ISDN access switch must be flexible to change and designed to expand its processing capacity as users needs increase. B and D channel packet handling capability must be an integral part of the system to allow access to X.25 host interfaces and X.75 public and private packet networks.

B) A Signal Transfer Point (STP) is required in the network to switch Signaling System 7 (SS7) messages. SS7 is integral to ISDN as it provides the nationwide signaling transport capability to allow the entire network to communicate, provide data base services, and to provide dynamic network management and control beyond a local calling area, across service network boundaries.

C) Network Termination (NT) equipment terminates both basic and primary rate loops and provides a standard interface for access to the ISDN switch.

D) Terminal Adapters (TA) must be available to allow non-ISDN terminals to access ISDN. To be truly integrated, ISDN must be a multi-vendor environment in which there is no need or tolerance for proprietary standards.

E) ISDN Terminal Equipment (TE) includes digital, ISDN compatible telephone sets such as the T2317, ISDN PBXs such as Meridian SL-100 equipped with primary rate access capability, and other terminals/peripherals.

In summary, our message to computer communicators is that ISDN is real and will evolve along with your needs. There are many public network services and products available now that solve today's communications problems, that are cost effective and economically justifiable in their own right, and that are fully evolvable to ISDN. By adopting these pre-ISDN solutions, the end user is also adopting a no-risk path to ISDN.

The Introduction of ISDN in the F.R.G.

Heinz Kunze

Fernmeldetechnisches Zentralamt Darmstadt, F.R.G.

ABSTRACT

The Deutsche Bundespost (DBP) has set up a framework of basic principles for the coordinated development of the telecommunication infrastructure. The medium- and long-term network development concept is normally reffered to as the "ISDN network development strategy".
The technical features of the German ISDN approach consists on several steps. After digitization of the telephone network the 64 kbps ISDN will be realized now. Many voice and non voice services will be provided. Before the final introduction of the broadband ISDN the DBP will offer a provisional broadband network with some facilities for video and high speed data transmission. Prerequisite is the installation of a glas fibre network. Within the next ten years the main parts of this ISDN will be realized.

INTRODUCTION

Network planning in the Federal Republic of Germany is based on the firm intention to introduce ISDN in 1988. The paper deals with the philosophy how the existing services will suitable migrate into the ISDN and presents the plans to integrate the data and telematic services which are currently handled in specialized networks.
Finally the broadband communication in ISDN will be discussed.

The paper treats the following topics:
- Voice services in ISDN 64 k
- Circuit switched data communication in ISDN
- Packet switched data communication
- Leased lines
- Text communication in ISDN 64 k
  - . Teletex
  - . Telefax (Facsimile)
  - . mixed mode
- Image communication in ISDN 64 k
  - . Videotex interactive
  - . slow motion pictures
- Terminal adaptors
- Network and Service interworking unit
- Broadband communication
  - . Optical fibre lines as a presupposition
  - . Provisional broadband network
  - . Broadband ISDN (ISDN-B)

General Situation

The ISDN implementation plan is largely dependent on the development plans for the digitization of the telephone network. Trunk digitization started in the FRG in the early seventies. Digital trunks have been implemented in the regional network since seven years and in the national backbone network since two years for network expansion on a regular basis. Major concern has been devoted to the coordinated introduction of digital transmission and switching systems as all digital switches, in principle, have to be interconnected by digital transmission systems. The ISDN backbone network in its initial phase will go into operation in 1988 and consists of 8 ISDN-mode local exchanges in Hamburg, Hanover, Berlin, Düsseldorf, Frankfurt, Stuttgart, Nuremberg and Munich. Each of these will be connected to an ISDN transit exchange. The transit exchanges are fully meshed from the routing point of view. Signalling System No 7 will be employed. The local exchanges will handle to a major part analogue telephone accesses but will be dimensioned to switch 1,000 ISDN subscribers each as well. From 1989 onwards, the Deutsche Bundespost will only purchase switching systems with ISDN capabilities. Already in 1989, 100 new local exchanges will be installed providing an ISDN capacity of about 100,000 ports. Full nationwide coverage will be achieved by 1993.

Prior to the introduction of series-produced equipment, an ISDN pilot project took place at Mannheim and Stuttgart in the 1986/87 time period.

The digital subscriber access line will allow two 64 kbit/s channels and 16 kbit/s for signalling (B+B+D) on the symmetrical copper pair. An important principle is the clear delimination of the scope of responsibility of the DBP and that of the subscriber. The DBP's competence will end at the CCITT-defined interface S_0 or S_{2M}, respectively. This permits subscribers and manufacturers on the terminal equipment sector and the administration on the network side free development in their domain. The S_0 interface allows up to 8 terminals to be connected on a (passive) local bus. The essential feature of ISDN, namely the integration of different services for simultaneous communication on only one network access line, is therefore inherently linked to the digitization of the subscriber access line.

INTEGRATION OF EXISTING NETWORKS/SERVICES

Digitization of the telecommunication network is obviously not only an innovative process from a purely technical point of view. From a vendor's point of view, integration of different services leads primarily to a new product. It improves existing services considerably and makes new services possible. Prime candidates for integration are certainly the existing specialized networks and services. The DBP operates for text and data transmission a leased line network, a circuit switched network and a packet switched network within the "Integrated Text and Data Network" (IDN). The Table shows the number of data terminal equipment connected to these networks.

The survey reveals that data and text communication terminals in the FRG amount to more than 600 000 ports at the end of 1987. This number is not to impressive compared to the bulk of 27 million voice terminals. However, voice communication has a growth rate of only 4 % compared to more than 25 % with non-voice terminals. The tenfold higher traffic load of non-voice terminals compared to average voice terminals throws a new light on these relations. The integration of speech, picture, text, facsimile will very probably stimulate the growth rate of multifunctional terminals. In the mid-nineties with about 1 million terminals, the ratio between speech and non-voice will have reached 60 % to 40 % also in Europe.

Integration of different services on one digital subscriber line provides not only a special comfort for the subscriber but leads also to considerable savings for the network provider. Presently, the 23 exchanges of the CS network and the 50 nodes of the PS network are concentrated on only 17 sites. As a consequence, subscriber access lines average to a length of 40 km compared to only 2.3 km access line in the telephone network with its 6 000 switches. Also the 53 Btx-nodes are situated in only 39 sites.

Network/Service	ports
PSTN (modems)	105 000
CDSN	20 000
PSDN (including access with NUI)*	27 000
Leased Lines	150 000
Teletex	16 000
Telex	170 000
Btx (interactive videotex)	78 000
Facsimile	45 000
Sum	611 00

Table Number of ports for text and data transmission in existing networks (9.87)
* NUI network user identification

ISDN will not only terminate the unfavourable splitting in specialized networks but will offer tremendous chances in network optimization: The twofold usage of (already existing) copper pairs which represent investments in the order of 30 billion DM will practically double the value of these investments. Even when only 10 % of the telephone subscribers will use the new ISDN offerings, the savings will still be 3 billion DM.

What are the services which will be integrated in the digital network (see figure 1)?

VOICE SERVICES IN THE ISDN

Analogue audio signals with a bandwith of 3.1 kHz will be coded in the digital telephone set to the standard transmission rate of 64 kbit/s. Application of lower bit rates, e.g. 32 or 16 kbit/s, is not intended for general use. Apart from improved service qualitity through digitization, the customer will be offered a wide spectrum of new telephone services. Speech transmission is the fundamental service in the ISDN and will consequently be offered from the very beginning in 1988. ISDN offers the opportunity to transmit audio a bandwith of 7 kHz instead of 3.1 kHz but retaining the standard transmission rate of 64 kbit/s. In this case, the PCM coding will be different to CCITT Rec. G.711. The specification for this new service is available. This new feature is intended to be introduced in 1990.

PABXs offering voice and data services will be connected to the ISDN via the S_0 or S_{2M} interface (figure 2). Extensions of the PABX will be connected via the S_0 interface. Many of the manufacturers of PABXs will provide in addition a special S_0' interface. The main reason for this solution is the restriction to a maximal loop length of 200 meters with the S_0 interface for PABX-extensions. Some manufacturers have therefore developed this solution which permits extension loops of several kms at lower costs. A great number of the telephone sets are connected via PABXs to the PSTN. It is apparent that active participation of PABX manufactures is therefore crucial for the quick adoption of ISDN. This is given as most of the PABX manufacturers participate in the ISDN pilot project at Stuttgart/Mannheim.

DATA SERVICES IN THE ISDN

- Circuit Switching

In relation to the lower signalling rates in existing networks, 64 kbit/s in ISDN offer new dimensions for text and data transmission. The DBP had already provided a 64 + 2,4 kbit/s "model" network with interface X.21 in the CSPDN as a forerunner of the ISDN. Certain subscribers and manufacturers interested in early application of 64 kbit/s terminal equipment use this network quite intensively. With the advent of ISDN, this network will merge into the new digital environment.

64 kbit/s data transmission will be a transparent end-to-end connection which will give the subscriber freedom to use his proper protocols in all protocol levels above layer 1. This service is inherent in the ISDN and needs no further specification. It will be provided in 1988 from the very beginning.

- Packet Switching

It is not intended to provide a packet switching service in the ISDN in the first phase of implementation. Packet switching in full range will very probably be provided in a later phase from 1994 onwards. In the meantime, subscribers will have access to the PSPDN via the ISDN. Two different procedures will be offered to the subscriber.
In the introducatary phase terminals connected to the ISDN will reach the PSPDN via the B-channel according to the procedure defined in CCITT Rec. X.31. CCITT has characterized this method as the minimum integration scenario (figure 3).

Special emphasis has to be given to the claim that terminals of the ISDN should experience no operational or technical disadvantages compared to those connected directly to DATEX-P. DBP has carried out thorough investigations to minimize impairments to the packet terminal on the ISDN side. In a first approach only the different numbering scheme gives an indication to the ISDN subscriber that packet switching is not yet fully integrated in the ISDN.
Subscriber equipment is connected to the S_0 interface through a terminal adapter TA X.25 which has the following characteristics:

- automatic call set-up through ISDN to a dedicated port of the interface processor IP when a packet switching call is initiated

- rate adaption according to Rec. X.30

- layer 1 interface according to Rec. X.21 and X.21 bis

In a later phase (abt 1991) packet terminals will access the PSPDN also via the D-channel. As the D-channel is largely unused after the ISDN call has been established it provides an economic vehicle to carry lower speed packet information. With this version both B-channels remain available for ISDN traffic and the IP can be reached by multiplex transmission.

- Leased Lines

Leased line connections through the ISDN can be established in different ways. After thorough scrutiny of the various possibilities, the DBP will introduce the following implementation: 64 kbit/s leased line connections will be realized within the ISDN as semipermanent connections 'permanently' switched through the switching matrix. Two variants are envisaged to be offered:

- connections between terminal equipments are established by the user himself.

- connection between terminal equipments are established by the network provider.

In both cases a 64 kbit/s transparent connection is at the subscribers disposal. In both cases unrestricted end-to-end-signalling on the D-channel (e.g. between PABXs) can be provided at the subscriber's discretion.

TEXT COMMUNICATION IN THE ISDN

- Teletex

Lack of sufficiently precise standards may possibly hamper introduction of the 64 kbit/s Teletex service. Presently the CCITT Recommendations for the Teletex service cover only existing networks.

Standards for Teletex procedures on network layers 1-3 in an ISDN environment are not yet fully agreed. In principle, it would be possible as well to start the service with the CSPDN protocols as employed in the present 2.4 kbit/s class of the CSPDN. Changeover to the final ISDN protocols would occur when the appropriate recommendations are finalized (figure 4).

- Facsimile 64 kbit/s (Group 4)

This service will be based on the already existing recommendations, mainly Recs. T 5, T 6 and T 73

With Group 4 facsimile equipment, quality of the copies and the transmission time will improve dramatically. Opening of this service is still on unsafe ground as it is dependent on the availability of a sufficient selection of affordable facsimile equipment. Visible endeavours of certain manufacturers give us good hopes that the service can be introduced in 1988. An another important fact is the availability of transition methods between terminals group 4 and group 3 in existing networks.

- Mixed Mode

Rec. T 73 is not yet defined well enough to allow an early start of a mixed mode service - a service based on facsimile communication equipment with text capacity. It is expected that this service can be offered from 1989.

INTERACTIVE VIDEOTEX -

The existing interactive videotex service called "Bildschirmtext" (Btx) started in 1983 in the FRG. The present 78.000 subscribers access the nodes generally via 1200/75 bit/s modems. Picture set-up time is accordingly slow, and this is very probably one of the reasons for the low growth rate of this service. With 64 kbit/s on the Bildschirmtext node access line, picture set-up time will be reduced to a degree that even highest demands will be satisfied. Moreover, the high signalling rate permits extensive use of geometric and photographic screen representation according to the CEPT standard. The videotex nodes are capable of processing the new signalling rate, but the network with its communication protocol has to be upgraded as a whole. Consequently, integration of the Btx service into the ISDN will take place in two development steps: In a first phase from 1988 onwards, only the access line and some peripheral parts of the local Btx-change will be operated with 64 kbit/s. In a second phase, all components of the Btx service will be able to exploit the higher signalling rate, thus providing the subscriber with enhanced service quality.

PICTURE SERVICES AND MIXED SERVICES

With simultaneous use of two services like audio and text, text and picture or audio and picture, the ISDN offers new exciting applications. Video equipment on the market designed for the extremely low transmission rate of 64 kbit/s show stupendous quality specifically when rapid movements of the objects represented are avoided. Studies have commenced whether services of this type are feasible and can be offered in the 1990 time frame.

TERMINAL ADAPTORS

In addition to the above services, which will all rely on the capacity of the 64-kbit/s user-information channel, it will be necessary to connect existing terminal equipment used in conjunction with today's telephone network or in conjunction with the CSPDN and PSPDN. While ordinary analogue telephone sets will be replaced by digital versions, other equipment used in the telephone network like modems, facsimile equipment (Groups 2 and 3), answering machines, automatic dialling equipment, etc. can be connected to the ISDN via analogue/digital terminal adaptors.

Terminals in the 2.4/4.8/9.6 kbit/s user classes will find a terminal adaptor X.21 employing the transversion protocol laid down in Rec. X.30. Teletex terminals which are shifted from the CSPDN to the ISDN can be connected via terminal adaptors. If the ISDN Teletex protocol version (layers 2 and 3) is not yet internationally agreed in time Teletex terminals in the ISDN will continue to work still for some time with the existing protocols. In many cases, however, terminal adaptors will not be necessary because the modular software structure of the terminals allows replacement of the interface according to Rec. X.21 by an S_0 interface (layer 1).

Packet switching terminal equipment will be able to access the PSPDN which will continue to be operated, via terminal adaptors X.25.

NETWORK/SERVICE INTERWORKING UNIT (VU-S, figure 5)

With the advent of ISDN it is not expected that the existing text and data networks will disappear abruptly. Data network users insist that the CSPDN remains in operation because of certain technical advantages and a tariff structure which fosters connections with very short duration time. Nevertheless, it is not intended to provide interworking between ISDN and CSPDN for data traffic. Hosts wishing access to the ISDN can easily use ISDN access links, possibly with terminal adaptors, which make the provision of interworking units superfluous.

In the first step the transition Ttx 64 ISDN --- Ttx 2.4 IDN (Dx-L) will be realized. Fig. 5 describes the transition between the terminals in different networks. Beside the speed conversion the interworking between the CCITT signalling systems NO.7 and X.71 must be provided. This will be realized with an independent transition device VU-S (not included in any exchange device) from the beginning of the ISDN service in 1988.

For facsimile subscribers, service interworking will be supplied to make sure that Group 4 equipment can communicate with Group 3 equipment.

In the very last step, TELEBOX (electronic mail service) will be included too.

BROADBAND COMMUNICATION IN ISDN

The strategy in the F.R.G. consists of the following major steps:

- 1985 digital telephone network
- 1988 64 kbit/s ISDN for 64 kbit/s services
- 1988 provisional broadband network
- 1992/93 Broadband ISDN (ISDN-B) for 64 kbit/s and broadband two-way communication services
- later on Integrated Broadband Communication Network (IBCN) for 64 kbit/s, broadband two-way communication and distribution services

The dates mentioned above indicate the years in which the introduction of network components based on a fully standardized technology is expected. All standardization work relies on the results and agreements which can be achieved in international standardization bodies. Only where international standards are not available will national amendments complete the required detailed specifications.

In particular for broadband applications, several trial networks will be realized by the DBP prior to the introduction of the full standardized technique. These activities are also necessary to test technical components in a live environment and to stimulate the design of broadband applications and related customer equipment (project BERKOM, Berlin-West).

OPTICAL FIBRE LINES

For some years, the DBP has been installing optical fibre lines in its trunk network on a regular basis, thereby deploying an economical replacement of the conventional facilities on copper cables. Today, the implementation of the optical fibre lines operating with 140 Mbit/s transmission systems has reached a considerable scale. Furthermore, the DBP intends to install 565 Mbit/s systems on monomode fibres from the end of 1987 onwards.

In addition to the provision of optical fibre lines in the trunk network, the DBP has started to implement optical fibres in the subscriber network. Hence 14 local network have been equipped with optical fibres in the year 1986, and another 15 local networks will be retrofitted with optical fibres in the year 1987. By these means the DBP is creating the necessary prerequisites for the eventual provision of broadband cable systems permitting broadband communication over the subscriber line.

Within the scope of this programme the fibres to be implemented in 1986 will be graded-index fibres. From 1987 onwards all fibres used will be monomode ones.

PROVISIONAL BROADBAND NETWORK

There are signs that a communication requirement combining videoconferencing, videophony and high-speed data communication is emerging. A key feature of the use of these services is the subscriber dialling capability. The DBP has therefore decided to embark upon the introduction of a provisional broadband network, starting at the end of 1987. This network is eventually intended to replace the existing experimental videophone and videoconference network, and should therefore allow both subscriber dialling and the establishment of reserved connections. The technical concept (figure 6) is based on existing standards. The capacity of this network should be increased to cater for a greater number of subscribers (ca. 1 000 subscribers by 1990).

This provisional broadband network will also make use of 140 Mbit/s information channels. Charge metering will therefore be based on the switching of channels with such a capacity. A network termination will be provided on subscriber premises, which, compared to that for the 64 kbit/s ISDN, will perform a broader range of functions. The dialling facilities for the activation and establishment of connections will be incorporated in the network termination. Two groups of user-network interfaces may be made available with subscribers: the analogue video interface with the composite colour video signal for video communication, and the analogue sound interface for accompanying speech. Speech and video communication may be supplemented by data communication, for which a 2048 kbit/s channel and two 64 kbit/s channels will be provided. As an alternative it will also be possible to use the 140 Mbit/s information channel exclusively for transparent data transmission. To this end, the subscriber will be able to use a digital interface with an information-carrying capacity of 137216 kbit/s. Both applications can be used for switched and reserved connections.

The provisional broadband network will have a two-level hierachical structure consisting of the subscriber exchanges and the transit exchanges.

It ought to be pointed out, however, that this network will not provide access to other existing networks such as the telephone network, the 64 kbit/s ISDN, and the dedicated data networks.

BROADBAND ISDN (ISDN-B)

The Broadband ISDN will be the logical consequence of the enhancement of the 64 kbit/s ISDN in terms of the principle of upward compatibility. The concept of this network will conform to the basic principles of the 64 kbit/s ISDN. This means that digital user-network interfaces will be placed at the disposal of subscribers at network terminations providing primarily transmission related functions. Apart from the already standardized user-network interfaces for the 64 kbit/s ISDN (S_0 and S_{2M}), another user-network interface (SBB) is being defined for the broadband two-way services. Ongoing discussions in international Study Groups are concentrating first and foremost on the structure and design of this interface, the user-information channels to be made available and the multiplexing of the various information channels. One of the possible SBB interface structures proposed is H4 + 4xH1 + 2B + D16, where H4 = 140 Mbit/s, H1 = 1920 kbit/s, B = 64 kbit/s and D16 = 16 kbit/s. Furthermore, discussions are also centering on the splitting up of an H4 channel into subchannels for which a so-called basic block of approx. 34 Mibt/s is envisaged, which, if used 4 times, would supply the capacity of the H4 channel.

Since the Broadband ISDN conforms to the basic principles of the 64 kbit/s ISDN, its conception will differ considerable from that of the provisional broadband network. One of the differences, for example, is that the coding equipment required for video communication will no longer be integrated in the network termination but in the terminal equipment. Furthermore, the dialling facilities needed for call set-up and for the utilization of the supplementary services will also form part of the terminal equipment.

One of the key features of the Broadband ISDN is that it will allow both narrowband and broadband two-way communication. The exchanges (figure 7) yet to be developed are considered to be 64 kbit/s ISDN exchanges whose range of capabilities is to be extended by the ability to switch broadband dialogue channels. The ensuing local exchanges will have suitable access units for analogue telephone subscribers, for subscribers using exclusively 64 kbit/s applications and for subscribers using 64 kbit/s and broadband applications (figure 8).

Up to now, no corresponding international standards have drawn up. In cooperation with the German telecommunications industry, the DBP will attempt to expedite the adoption of concept-related specifications by CEPT and CCITT. Only when this has been achieved it will be possible to furnish a detailed description of the Broadband ISDN by means of DBP Technical Standards and DBP Equipment Specifications. These time factors are the reason for the relatively late introduction of the Broadband ISDN on a commercial basis and for the intermediate solutions adopted by the DBP in the form of the experimental videoconference service and the provisional broadband network.

CONCLUSION

With this network development concept the DBP has created the prerequisites for the further development of its telecommunications network. Furthermore, the early conception will permit appropriate service concepts to be developed, which in turn will pave the way for the technical development of terminal equipment and subscriber installations.

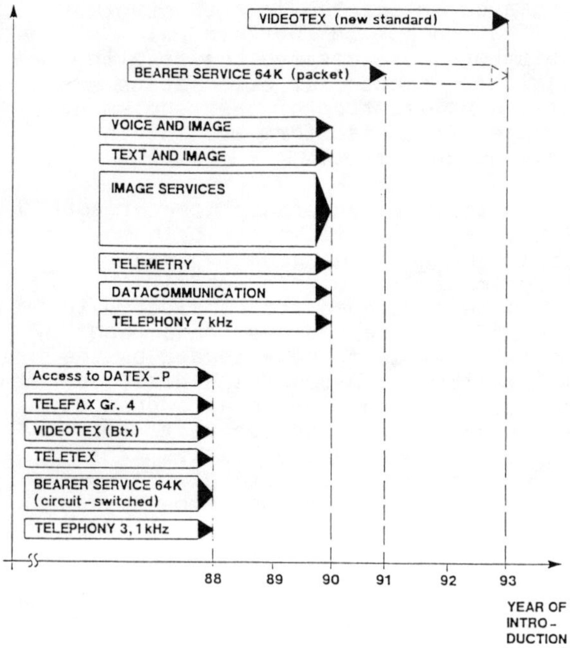

Figure 1: ISDN Services

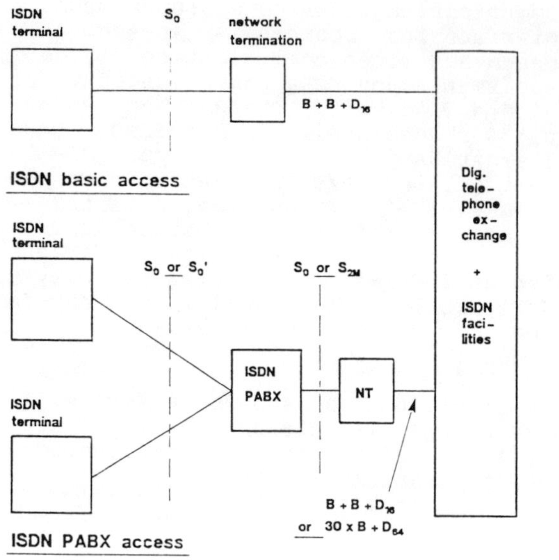

Interfaces

S_0	up to 8 terminals / S_0 bus with simultaneously 2 (3) connections
S_{2M}	up to 30 simultaneous connections
S_0'	for PABX terminals with additional facilities (S_0 interface + options)

Figure 2: Access of ISDN Terminals

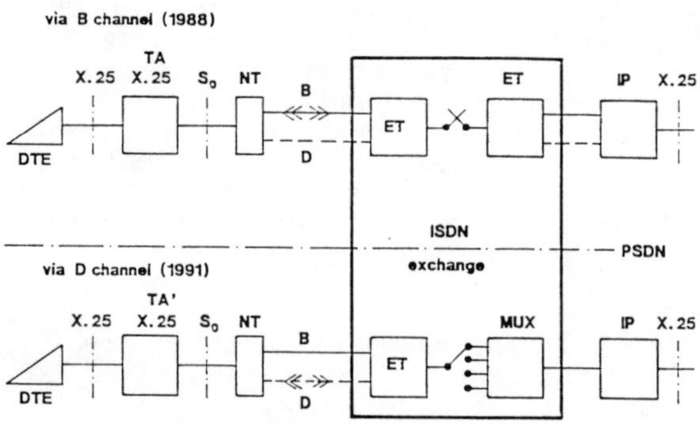

PSDN	packet switched data network
DTE	data terminal equipment
ET	exchange termination
IP	interface processor
MUX	multiplexer for D channels
NT	networktermination
TA	terminal adapter
S_0, X.25	CCITT interfaces
⟵⟶	packet switched data flow
☐	additional equipment

Figure 3: Access packet switching in ISDN

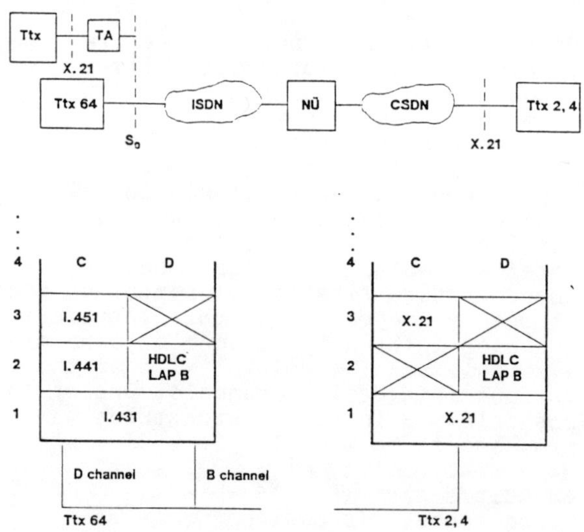

NÜ	Network transition
TA	Teletex adapter
C	Protocols for connection control (terminal-network)
D	Protocols for connection phase (terminal-terminal)
CSDN	Circuit switched data network
S_0, X.21	CCITT interfaces
I4. . .	CCITT recommendations ISDN D channel

Figure 4: Interworking Teletex 64 - Teletex 2,4

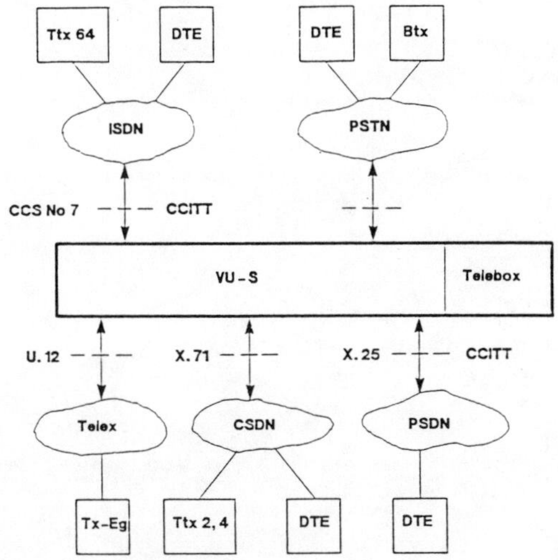

Telebox	electronic mailbox system
DTE	Data terminal equipment
Tx-Eg	Tx - Terminal
VU-S	Interworking system between networks and for service transitions
Btx	Videotex interactive
PSTN	Telephone network
CSDN	Circuit switched data network
PSDN	Packet switched data network

Figure 5: Network transitions in DBP area

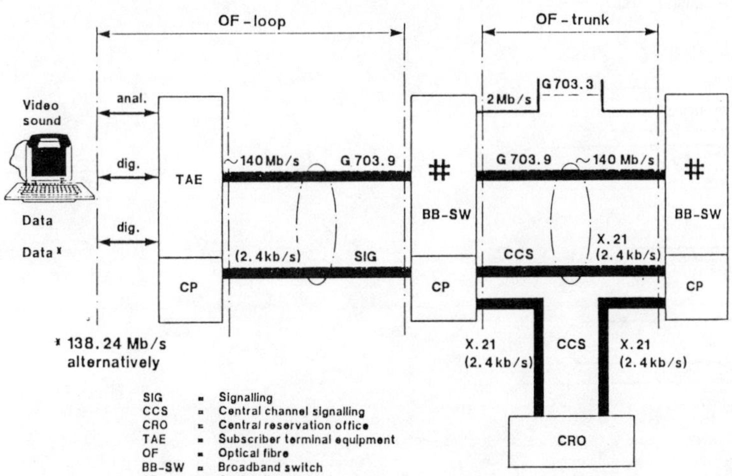

Figure 6: Technical Concept for Provisional Broadband Network

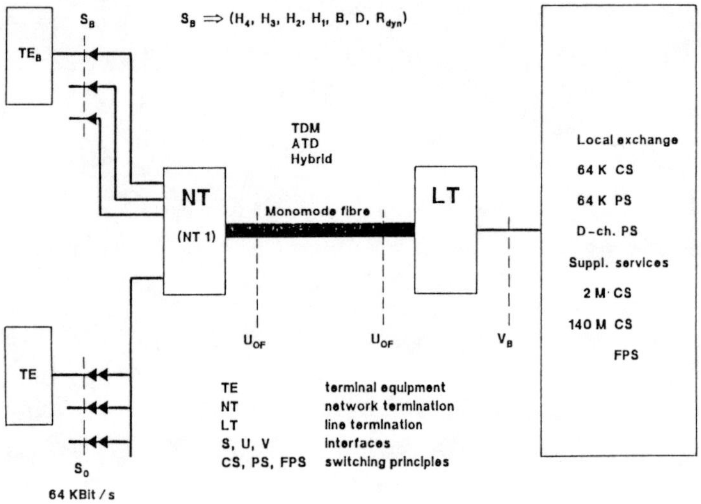

Figure 7: ISDN-B: Loop Concept for Basic Access

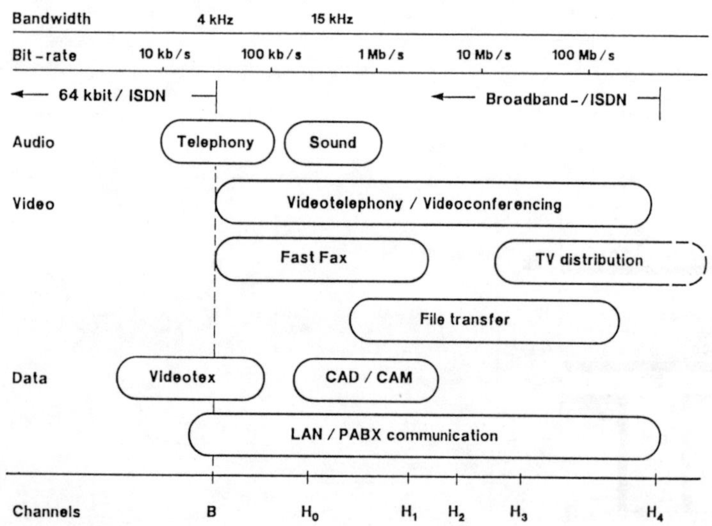

Figure 8: Broadband Services

NTT's Commercial ISDN Services

M. Kuwabara

Japan
Nippon Telegraph and Telephone Corporation

In the first quarter of 1988, NTT plans to start INS commercial services. This service will provide standard user-network interfaces such as Basic access and Primary rate interface in accordance with CCITT "I" series recommendations.

To ensure smooth establishment of ISDN, it is important to confirm the required technologies and study the social effects of ISDN. Accordingly, NTT carried out INS field trial over a two and a half year period.

This paper presents the experiment of INS Model System at Tokyo and examines several ISDN technologies to realize INS including the user-network interface, the addition of an "I" series interface subscriber module to the existing digital switching system, subscriber loop architecture, synchronous digital networks and the No.7 common channel signaling system.

1. INTRODUCTION

Recent developments in digital technology have enabled construction of more economic and efficient telecommunication networks. Moreover, these networks can be greatly improved in function through digitization. As its major effort in this direction, NTT has been promoting construction of an advanced digital network called Information Network System (INS)[(1)].

Many countries are presently carrying out ISDN field trial. NTT started constructing its INS Model System in 1982 and carried out a field trial from September 1984 to March 1987[(2)]. While , CCITT has been actively studying ISDN specifications. Basing on the result of INS Model System and CCITT recommendations, NTT has been planning commercial ISDN service in 1988, which is the practical innovation of the INS. The user-network interface supported by NTT's INS commercial system is coincident with CCITT standard.

2. CHANGE IN LEGAL STRUCTURE GOVERNING JAPANESE TELECOMMUNICATIONS

2.1 Privatization of NTT, and Introduction of Competition in Telecommunications Business Field

Since its foundation in 1869, the telecommunications business in Japan was monopolized by a national enterprise or public corporation.

However, by the nineteen-eighties, circumstances surrounding the telecommunications business had greatly changed.

Namely, the need increased for more diversified and enhanced telecommunication services . Thus, a organization to flexibly and effectively respond to these new expectations was required. In order to cope with this requirement, it was thought that the

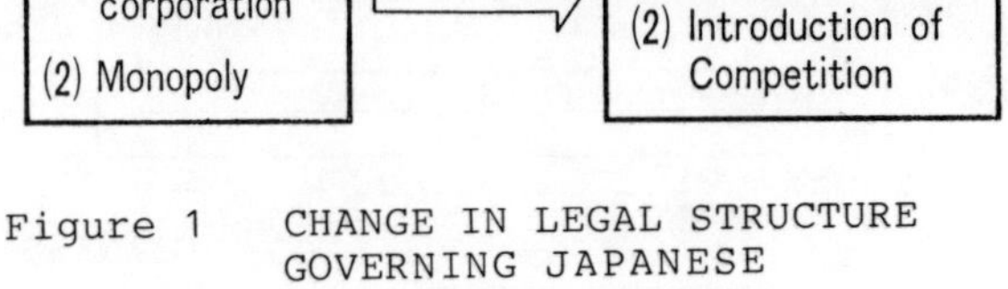

Figure 1 CHANGE IN LEGAL STRUCTURE GOVERNING JAPANESE TELECOMMUNICATIONS

telecommunications business should be opened for plural companies in competition basis. Because of these factors, a revolution in the telecommunications business system was enforced in April of 1985 (see Figure 1).

There were two major aspects to this change.

First, the NTT public corporation was privatized and reformed to NTT as a joint-stock corporation. New NTT is involved in the same business fields as the NTT public corporation. NTT started out with 780000 million yen of capital stock and 300000 employees.

Second, advantages of competition system was introduced. The telecommunications business fields formerly monopolized by the old NTT were opened up to plural companies.

2.2 The New Telecommunications Business Law

In Japan, telecommunications carriers are broadly divided into two categories going by the names of Type I and Type II telecommunications carriers(see Figure 2). Type I carriers have their own transmission facilities to provide their services. On the other hand, Type II carriers rent telecommunication circuits from Type I carriers based on a tariff system to run their business.

The distinction between Type I and Type II does not imply a difference in the telecommunications service provided. Both carriers can supply Basic as well as Enhanced service.

In the U.S., telecommunications services are divided into two categories; regulated basic services and non-regulated enhanced services. The distinction between these two, that is, where to drew the line, is presently under discussion. In Japan, regardless of local or long line services and basic or enhanced services, all telecommunications business fields are open for competition.

NTT is a Type I carrier, and 9 other companies have also permitted as Type I carrier. With regard to Type II carriers, more than 300 companies have begun conducting business.

The structure of the new Telecommunications Business Law in Japan is noteworthy in that it does not prevent carriers from providing integrated services. That is, there is no restriction on providing ISDN services combined with enhanced services.

3. Existing NTT Telecommunication Networks

NTT has three switched networks for public use: the telephone network, packet switched data network and circuit switched data network , these are illustrated in Figure 3.

The telephone network is largest, consisting of 46 million subscribers including mobile.

With regard to non-telephone services, circuit switched data network with X.20 and X.21 interfaces switched network called DDX-C has been in service since 1979. For computer communications, packet switched data network with X.25 and X.28 protocols called DDX-P have also been in service since 1980. In addition, NTT has a facsimile communication network and a videotex communication network as VAN having communications processing functions.

Facsimile network services have been available for registered tele-facsimile terminals connected to the telephone network since 1981, and make it possible to effectively and inexpensively transmit facsimile information. Also,

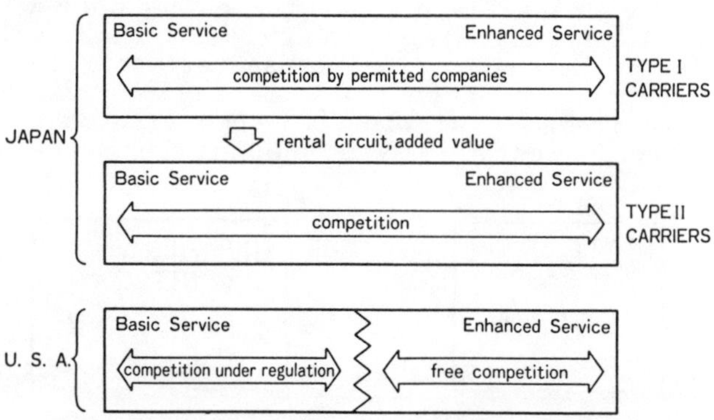

Figure 2 COMPARISON OF TELECOM.BUSINESS BETWEEN JAPAN AND U.S.A.

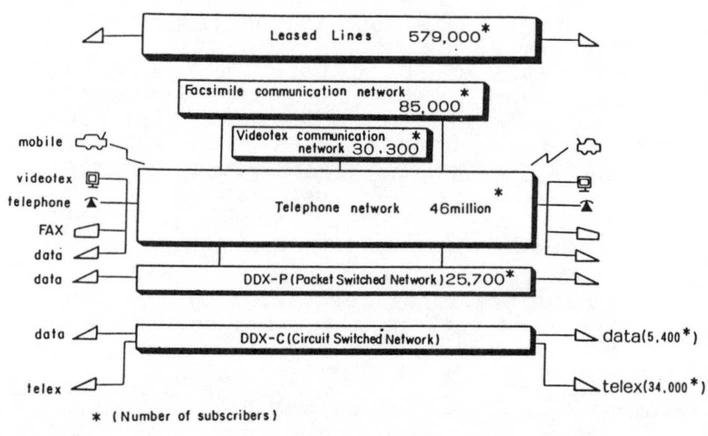

Figure3 EXISTING TELECOMMUNICATION NETWORKS IN NTT

the network offers various services such as multi-delivery or confidential communication.

Videotex communication network services connecting videotex terminals and information centers have been available for videotex terminals connected to the telephone network since 1984. This services offers still pictures containing various kinds of information.

4. ISDN

4.1 Features of ISDN

In existing telecommunication networks, both analog and digital information is transmitted as analog signals. However, advances in semiconductor devices and optical fiber transmission technologies have now made it possible to economically send not only digital signals but also analog signals.

The features of the ISDN can be broken down for discussion into network digitization and integration of services (see Figure 4).

With digitization, network cost can be reduced and a high quality end-to-end digital path can be established. By using this digital path, non-telephone information up to 64kb/s can be transmitted.

NTT is presently aiming at structuring of the INS , which will be able to comprehensively support all types of telecommunication services.

NTT has been planning commercial ISDN services , which is the first step to the INS. The INS evolves to provide the following attributes.

(1) High-speed access such as H2 channel
(2) Wideband or high-speed subscriber loops using optical fiber
(3) Unifying the tariff system based on transmitted information.

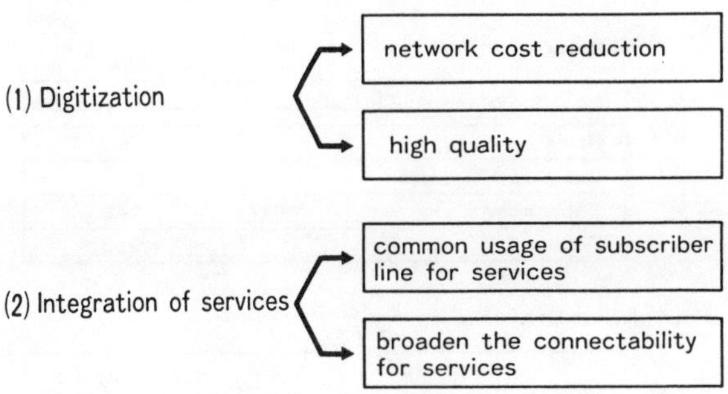

Figure 4 FEATURES OF ISDN

4.2 NTT's Network digitization Activities

NTT has been promoting digitization of transmission and switching systems.

The optical fiber transmission system, named F-400M, transmits the signals of 5760 64kb/s channels at a 400Mb/s bit rate for each core fiber. This system was installed from Kyushu Island to Hokkaido Island, as a Japan Traversing Route in February 1985, which will act as the backbone of NTT's future digital networks. NTT has also developed radio system. The 2S-P1 system was introduced for short-distance section in 1969. The 20L-P1 system, which also transmits at a 400Mb/s bit rate for each system, was introduced in 1977. Moreover, 4/5/6G-200M system with 50km repeater spacing was introduced in 1981. By 1987 NTT will have introduced digital transmission systems to the major cities shown in Figure 5.

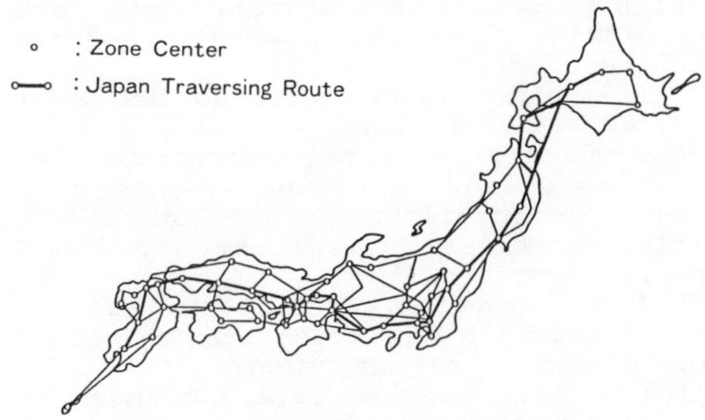

Figure 5 DIGITAL TRANSMISSION SYSTEM(1987)

Figure 6 LOCAL DIGITAL SWITCHING SYSTEM(1987)

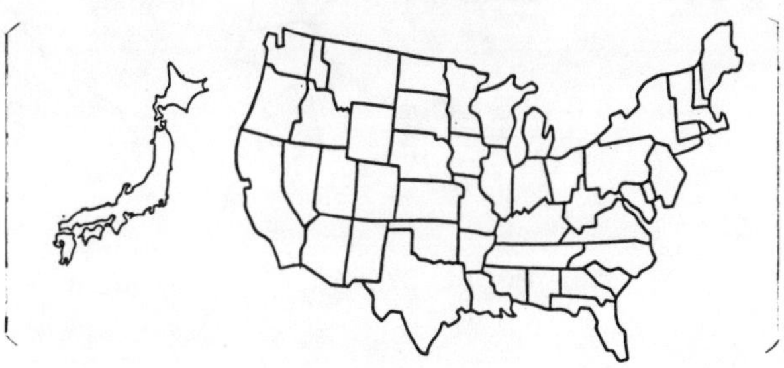

Large capacity transit digital switching systems, called D60, were introduced in 1981. They have been installed to zone centers as illustrated in Figure 5. On the other hand, local digital switching systems, D70, were introduced in 1982. The installation areas for the D70 at the end of 1987 are illustrated in Figure 6. Over 500 D70 units will have been introduced by the end of 1987.

5. INS MODEL SYSTEM

To smoothly introduce the ISDN, NTT has established the INS Model System project and carried out the field trial in advance of commercial service. Construction of this Model System started in September, 1982 in Mitaka-city, Musashino-city, and a section of the Tokyo metropolitan area (see Figure 7). Various private businesses, public offices, department stores, factories, hospitals, shops and households coexist in these areas.

The main objectives of the INS Model System are as follows.

(a) To obtain a better understanding of user needs and investigate the social and economic effects of INS on private life, business activities, and local public services.

(b) To confirm effectiveness of digitization on operation and management of telecommunication networks.

(c) To determine necessary changes in training programs that must be made in accordance with these results.

This Model System was over at the end of March 1987. However, part of this system continues to be utilized for regional administration at the request of Mitaka-city office.

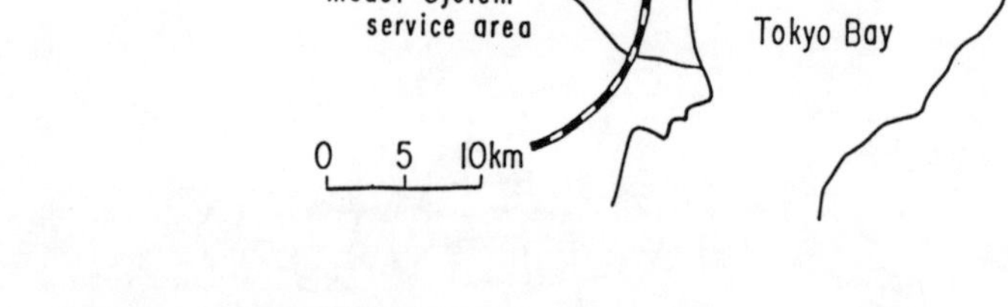

Figure 7 INS MODEL SYSTEM SERVICE AREA

5.1 Scale of the INS Model System

In the Model System, many kinds of terminals were offered. A list of terminals is shown in Table 1. This system uses 1,500 terminals, including 450 digital telephones, 310 digital facsimile units, 310 digital interactive videotex terminals and 310 high-speed or broadband terminals.

Furthermore, about 10,000 existing analog telephones were taken in the digital local switch of the Model System. Moreover, subscriber optical fiber cables were installed to transmit information from broadband or ultra-high-speed terminals. The total length of subscriber optical fibers were approximately 4,000km.

To fulfill the objectives of this Model System, a number of participants were asked to join as monitors and Information Providers (IPs) (see Table 1). Since the number of applications exceed the facilities, monitors were rotated every six months, and as a result, a total of approximately 1990 monitors were able to experience ISDN

Table 1 THE SCALE OF THE INS MODEL SYSTEM

(1) Total length of subscriber optical fibers 4,000km

(2) Terminals

Terminal		Number of terminals
64kb/s	Digital telephone	450
	Digital facsimile equipment	310
	Digital interactive videotex terminal	310
	Digital telewriting terminal	70
	Data terminal	50
Broadband	Video terminal	80
	Video telephone terminal	160
	Video conference terminal	30
	Video camera	10
	Ultra-high-speed facsimile equipment	20
	Color facsimile equipment	10
	High-definition TV	3
Total		1,503

(3) Number of monitors

Location of terminal installations	Number of monitors
Home	1,266
Business office, etc.	575
Showroom	149
Total	1,990

(4) Number of IPs

Media	Number of IPs
Video Response System	167
Voice storage service	66
Digital interactive videotex service	120
Total	353

services.

The major feature of this Model System is that it not only controlled end-to-end connections between monitors but also provided center-to-end communications. In order to popularize center-to-end communications, a variety of useful information must be provided from the centers. In the Model System, IPs provide video information retrieval service including moving pictures, voice storage service(pre-recorded information) and digital interactive videotex service (see Table 1).

5.2 System Configuration

The basic configuration of the INS Model System is shown in Figure 8. The system consisted of a 64kb/s network, which provides telephone and 64kb/s digital path, and a broadband network for such services as video conferencing and ultra-high-speed facsimile transmission.

To provide INS services for users, ISDN capabilities such as digital subscriber interface and advanced signalling need to be added to local switching systems.

NTT's D70 digital local switching system has been widely introduced in the existing telephone network. Thus, specific functions for INS services were added to the D70 digital switching system.

Furthermore, to provide center-to-end services, many computer centers were provided. To ensure mutual inter-connection capability between terminals and computer centers with different protocols or with different media, communications processing system was implemented in this Model System.

For broadband subscriber transmission, a 4MHz analog system is employed. Because of the high cost of CODEC, it is not economical to digitize moving picture information signals on the subscriber's line.

Multiple services including a 64kb/s and video services have been provided through this broadband subscriber line transmission system by means of a wavelength division multiplexing transmission technique.

Moreover, broadband system employs the space division switching system differed from D70. This is because that transmission speed of the broadband signal is too fast to constitute an economical time division switching system.

(1) User network interface for 64kb/s digital network

The user network interface provided in the INS Model System is a simple structure interface called "Y". The "Y" interface structural and signaling system are different from those of CCITT "I" series interface as shown in Figure 9.

The Y interface structure is B1(64kb/s) + D1(4kb/s) / B2(16kb/s) + D2(4kb/s), where the symbol "/" means the different physical interfaces appearing on both sides of the symbol are simultaneously provided for the user. Moreover, its signaling system is channel-associated. That is, signal channel(D1 or D2) are each paired with an information channel(B1 or B2). Because signal processing can be simplified with this simple signalling approach, a cost-effective user-network interface can be attained for both terminals and switching systems. Moreover,the "I" series interface had not yet been standardized by CCITT at that time.

For these reasons, NTT experimentally provided NTT's original interface.

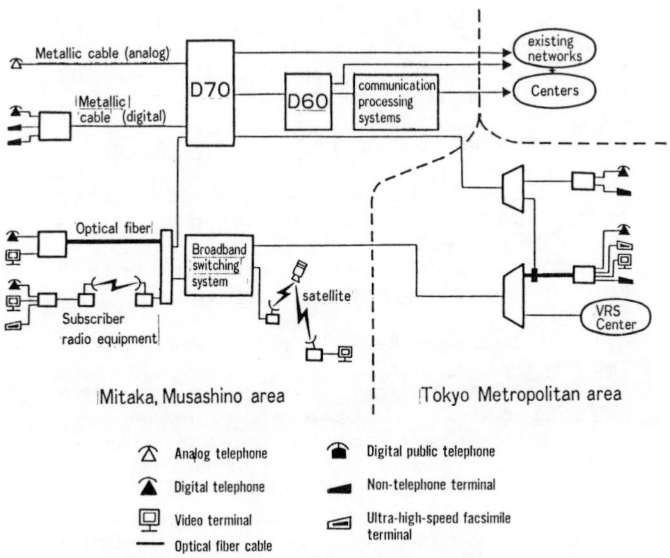

Figure 8 BASIC CONFIGURATION OF THE INS MODEL SYSTEM

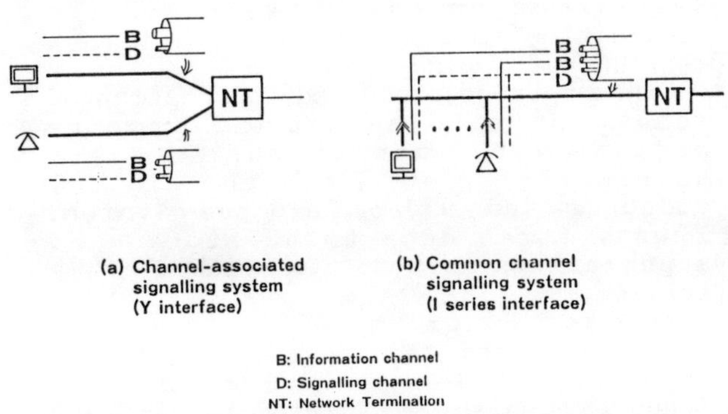

Figure 9 CHANNEL-ASSOCIATED SIGNALLING SYSTEM AND COMMON CHANNEL SIGNALLING SYSTEM

(2) Subscriber loop transmission system

In order to make INS construction feasible, it is necessary to offer digital end-to-end connection to subscribers using the existing metallic wires.

To achieve this, NTT has developed a time compression multiplex (TCM) digital subscriber transmission system, which can utilize the existing pair of copper wires (see Figure 10). In this system, burst mode digital signals are reciprocated at more than twice the original signal bit rates between a telephone office and customer's premises according to a centralized timing clock. This enables bi-directional transmission over a metallic pair. The system is expected to be the most appropriate one because of the long transmission spans even in a high cable accommodation condition, due to the freedom from near-end crosstalk and good compatibility with highly integrated digital LSIs. The system can transmit full duplex 88kb/s digital signals: 64kb/s channel and 16kb/s channel for information and two 4kb/s channel for signaling over an existing metallic pair.

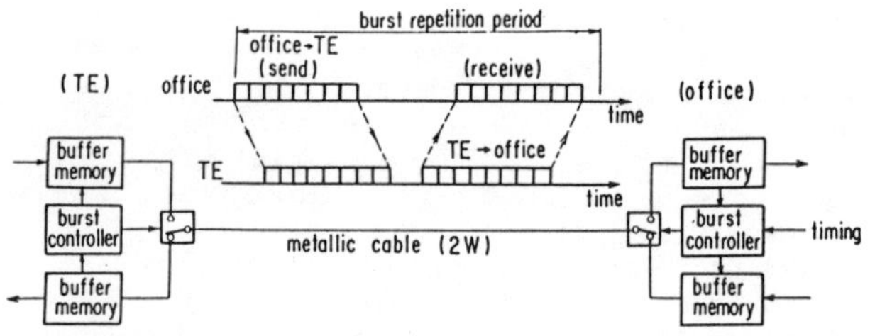

Figure 10   TIME COMPRESSION MULTIPLEX TRANSMISSION SYSTEM

5.3 Service utilization

A variety of services provided by the INS Model System were found to be adaptable to diverse social activities. Several major examples of service utilization follow.

(1) Home-shopping

This service presented information on goods for sale and special events provided by department stores, supermarkets, etc., via the digital videotex and Video Response System. Twenty-three department stores and supermarkets joined in providing this service.

(2) Home-banking

Home-banking made possible automatic deposits, withdrawals, and transfers of funds. It also facilitated balance and transaction inquiries, and information concerning interest payments. Thirty-three different banks joined.

(3) Telecommuting satellite offices

To save mental and physical energy, as well as the time wasted in daily commuting, an individual could work in his home or at a satellite office in his neighborhood using INS services. Some enterprises opened experimental satellite offices. Terminals used include Japanese teletex units, which incorporate word processors, and high-speed facsimile units.

(4) Municipal service

The City of Mitaka has actively adopted new telecommunication services in carrying out its public service functions. It has provided the following four service.

(a) Local administrative information service

The videotex system is used to transmit municipal public notices, information concerning various events sponsored by the city or the community center, and information concerning public facilities.

(b) Quick issuing of residence certification documents

It was found that various certificates could be quickly issued, such as residence certification documents and tax payment vouchers, utilizing ultra-high-speed facsimiles connecting city offices and local neighborhood offices. Records stored in city offices can thus be easily accessed without the need for citizens to travel to the central records office.

(c) Motion picture service providing information of major events

Information of major municipal events can be provided by means of the Video Response System.

(d) Manuscript communication service for the deaf

Communications can be performed by hand-writing on a digital telewriting terminal installed in the city hall or the individual's household.

5.4 Confirmation of ISDN technologies

The INS Model System was operated stably without interruption over a two and a half year period.

ISDN technologies confirmed through the INS Model System operation were as follows.

(1) 64kb/s system Digital switching technology

Confirmation was made of the switching control technology necessary when both analog telephones and digital customer premises equipment are simultaneously accommodated in the same digital switching system. We also were able to confirm our remote line concentrator technology which enables early, easy and economical provision of ISDN services to areas where digital switching systems are not installed in the initial stage of ISDN introduction.

(2) Communications processing facilities

Five kinds of communication processing facilities were installed,and each function in these facilities was confirmed.

(3) Digital subscriber loop transmission technology

A TCM system was employed for digital subscriber loop transmission, we were able to reliably operate at transmission distances of up to 7Km.

The NT (Network Termination) installed at customers' premises has been confirmed to consume less electricity (750mW, including power consumption for one digital telephone terminal) due to the use of LSIs. This enables the supply of electricity from the telephone office for each call. The technology for this was confirmed, paving the way for cost-efficient ISDN services.

(4) 64kb/s customers' premise equipment

Tests were conducted for 21 kinds of customers' premise equipment. We confirmed control technologies to select the appointed equipment and complete the communication. We confirmed interfacing between customers' premise equipment and the NT, and the basic requirements for the user-network interface. To find a trouble at subscriber line and customers' premise equipment, a turn-back test function at ET (Exchange Terminal; a local function capabilities) and NT was prepared. However, this step turned out to be insufficient. On the basis of these experiments, NTT has proposed the improved test function to the CCITT SG18. NTT's contributions were reflected in the new loop test specification.

5.5 User Opinions and Requests

Human monitors participating in the trial were able not only to gain an understanding of the digital network, its equipment, and available services and functions, but also to see many future possibilities. Since corporate, local community and household needs differ, suitable ways will have to be considered to meet each need. It is necessary to coordinate customer needs for new equipment and services. Advanced information needs cannot be met by equipment alone. "Software" aspects, and how they are matched to corporate or home activities, are vital. With regard to issues related to development of an information society, a variety of useful opinions were received regarding rules for new media communications, the way in which information should be provided, and tariff system.

6. NTT's COMMERCIAL ISDN SERVICES

6.1 Schedule for NTT's Commercial ISDN Services

NTT plans to start ISDN services on the basis of CCITT standardized user-network interface in March, 1988.

As described previously, NTT originally developed the "Y" interface for its Model System and planed to use

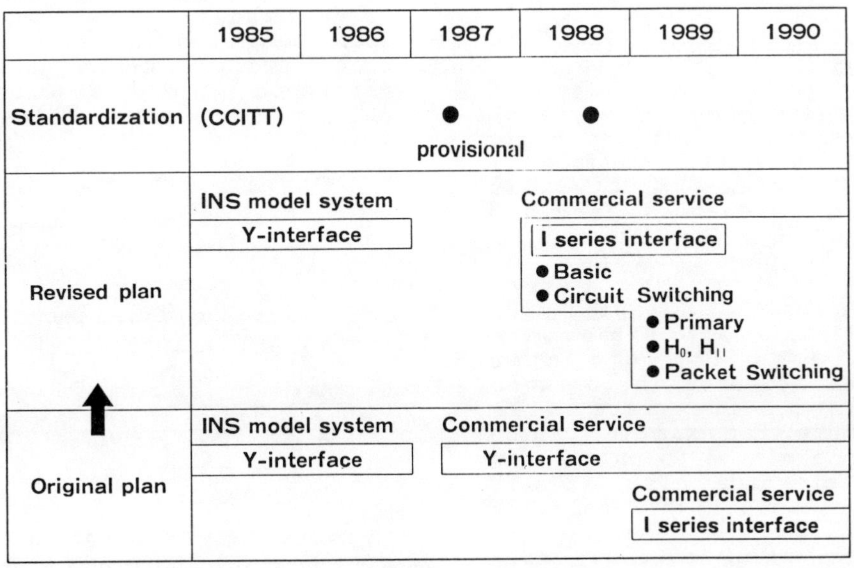

Figure 11 SCHEDULE FOR NTT'S COMMERCIAL ISDN SERVICES

it at least early stage. However, it now changes a mind to adopt the "I" series interface in accordance with CCITT recommendations from the beginning of the INS provision. NTT plans to offer ISDN services beginning in March of 1988 as shown in Figure 11.

NTT has advanced the schedule for introducing the "I" series interface from 1989 to 1988. In order to establish standards for connection between terminal equipment (users) and network, etc., within Japan, the Telecommunications Technology Committee (TTC) was founded in October 1985. TTC has been promoting standards linked with CCITT study group activity. CCITT issued the first version of its ISDN recommendations in 1984. Since then these recommendations have been brushed up. and CCITT approved recommendations related to layer 1 (BA) and layer 2 in 1987.

What is more, expectations are that CCITT will finalize a complete set of recommendations including layer 1(PRI) and layer 3 in 1988. In Japan, May 1987, TTC issued standards based on the 1987 Recommendations (provisional) and draft recommendations. As according to TTC standards, NTT is now preparing the "I" series interface for commercial ISDN services.

In March 1988, the basic interface (2B+D) is scheduled to be introduced. After one year, in March 1989, the primary rate interface will be added. At first, the commercial service will first be started in Tokyo, Nagoya and Osaka. It will then be expanded nationwide.

6.2 Introduction of CCITT Standard Interface

The CCITT has recommended two major interfaces (see Figure 5): the basic interface and primary rate interface.

The basic interface structure is composed of two B channels and a D channel, that is 2B+D. The bit rate for the D channel in this interface structure is 16kb/s. The 64kb/s information channel both enables mass data transmission. It also expands the application area for computer communications. Moreover, by using a 64kb/s channel, the facsimile transmission time, for example, can be reduced from one minute to only a few seconds. This interface also makes it possible to transmit high-resolution video signals.

For computer communications, many interactive applications are available. However, these 64kb/s communication circuits are not used efficiently due to the time operators spend in thinking and manipulating equipment. That is to say, there are two application areas types of data communication networks, one is packet switching type and the other is circuit switching type. The "I" series interface provides the packet switching mode, which makes more efficient use of these circuit mode applications, along with the circuit switching one.

On the other hand, the primary rate interface is based on a primary PCM group recommended in G.703. This primary rate interface can be used to provide plural multiplexed 64kb/s and H0 (384kb/s),or H11 (1,536kb/s) channels (see Figure 12).

NTT will introduce both the basic and the primary rate interface.

Prior to discussing user-network interface, it is important to clarify the demarcation point between user equipment and network facilities. In Japan, the NT1 function, called the DSU, is offered by carriers. Namely, a user-network interface is prescribed at reference point T described in CCITT "I" series recommendation (see Figure 13).

Items		Structures	
Basic		2B+D	B = 64 kb/s D = 16 kb/s
Primary Rate	multiplexed	23B+D	B = 64 kb/s D = 64 kb/s
Primary Rate	high-speed	mHo+nB+D H11 / D	Ho=384 kb/s H11=1536 kb/s D = 64 kb/s D = 16 kb/s

Figure 12 USER-NETWORK INTERFACES

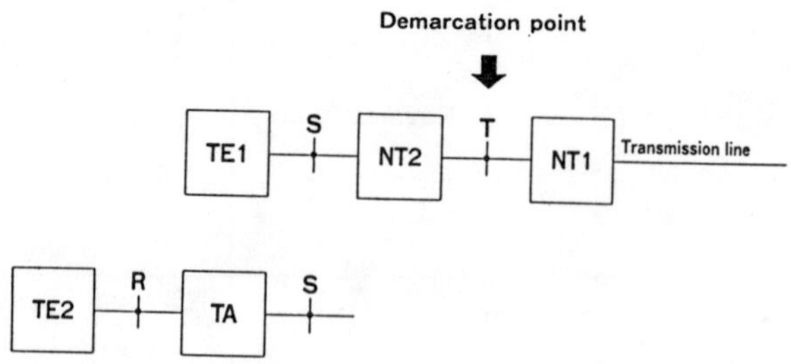

Figure 13 DEMARCATION POINT BETWEEN USER AND NETWORK

6.3 D70 Digital Switch ISDN Capabilities

To supply ISDN services, digital local switching system (D70) must employ both CCITT "I" series interfaces; the basic interface (2B+D) and the primary rate interface. These ISDN capabilities should be added to the existing switching system in order to provide the "I" series interface in a rapid, flexible and economical manner. Figure 14 shows how the "I" series interface is introduced into the existing switching system (D70). To add ISDN capability to the existing system, NTT chose the method of ISDN module addition and has been developing it. This module is called the ISM ("I" series interface Subscriber Module). The ISM provides the interface for subscriber access and interconnection with transit networks.

6.4 Subscriber loop architecture

NTT has developed a 320 kb/s line bit rate TCM system for the basic access interface (2B+D) using a pair cable, on the basis of the confirmation in the INS Model System.

As two-wire digital subscriber transmission system, NTT has studied two systems, one is TCM and the other is ECM (Echo Cancellor Method). Unfortunately an ECM system circuit suffers from near-end crosstalk (NEXT) caused by other ECM circuits. Therefore, limited use of cable pairs is required to extend ECM system coverage. On the contrary, a TCM system does not suffer from NEXT because bursts sent from a central office are synchronized with the office clock. In the case of a 320 kb/s TCM

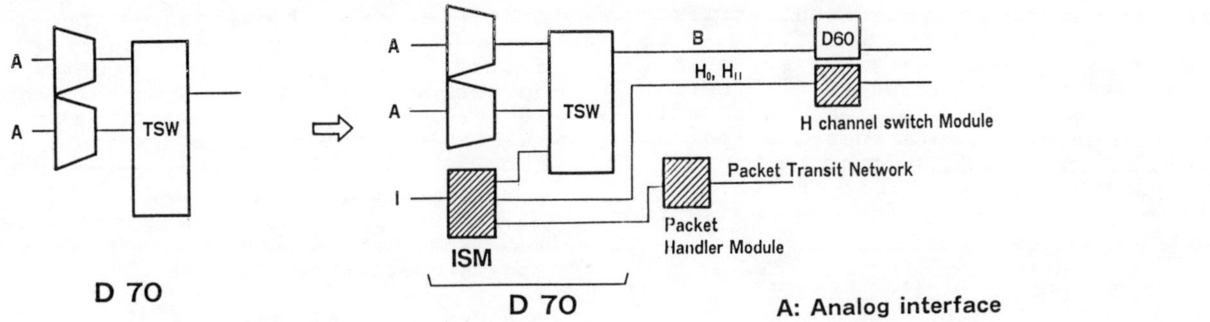

Figure 14 INTRODUCTION OF "I" SERIES INTERFACE

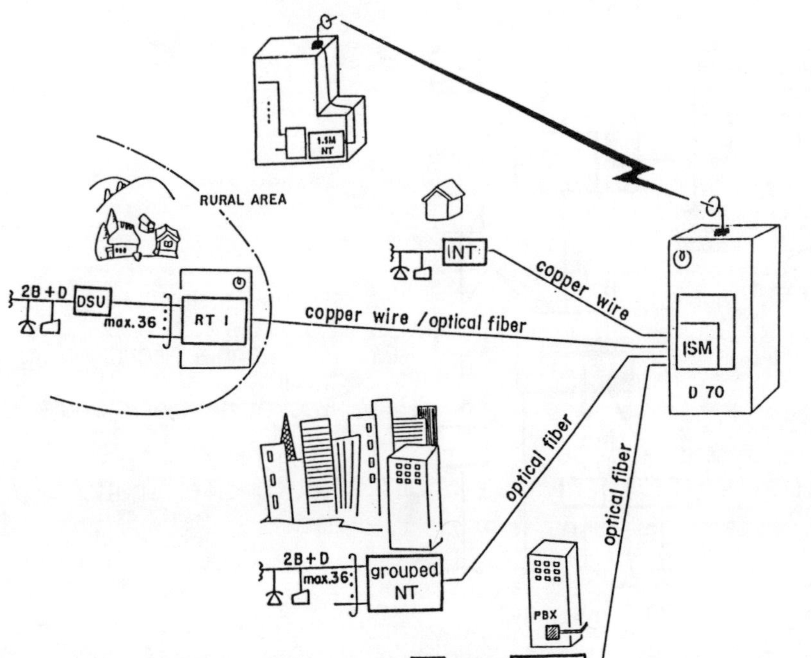

Figure 15 SUBSCRIBER LOOP ARCHITECTURE

system, although impulse noise is assumed to be a major range limiting factor, the 320kb/s line bit rate TCM is more appropriate in NTT's area coverage than the 160kb/s ECM for the Basic access (2B+D) transmission. Due to line transmission attenuation restriction, the TCM system can be utilized by subscribers within a 7 km radius maximum from the digital local exchange under NTT's subscriber cable condition.

For subscribers outside the 7km radius, NTT has developed a digital subscriber loop multiplex transmission system called RT1 (Remote Terminal Type1). Moreover, NTT has been developing Grouped NT to provide plural basic interface. 1.5MNT is also under development to provide the primary rate interface on the customer's premises. This is because TCM cannot be utilized for high-speed transmission.

NTT has also developed digital subscriber radio system that utilizes the 26GHz band. By modifying this existing system, a digital subscriber radio system providing an "I" series interface can be realized. This radio system ensures rapid installation as well as economical performance for high-speed digital subscriber transmission line, especially when demand is scattered. (see Figure 15)

6.5 Synchronized Digital Network

In the ISDN era, various bearer service are needed in the transit network. Thus, synchronous multiplexing,

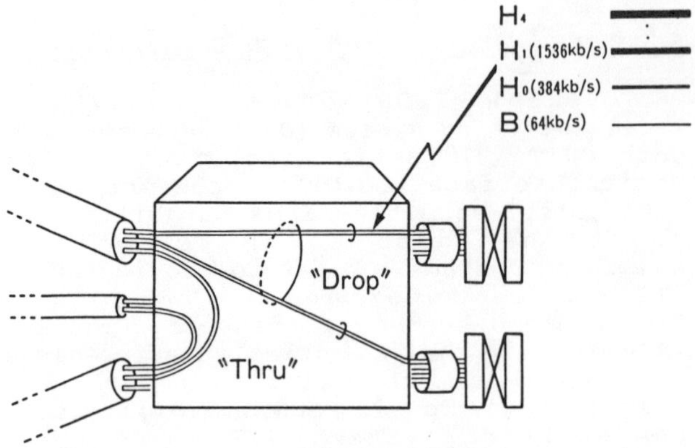

Figure 16 SYNCHRONIZED DIGITAL MULTIPLEX

which can effectively deal with connection of different bearer services, is a key technology.

NTT has developed synchronized multiplexers that operate at speeds of up to 6.3Mb/s. They play a key role in assuring flexible and economical operation of digital networks. Figure 16 images 12 synchronized digital multiplexers. In order to realize a more economical and flexible synchronous digital network, NTT is now studying on our next generation of synchronized multiplexers which will operate at speeds up to the DS-4 level.

6.6 Network Structure

As shown in Figure 17, the INS is

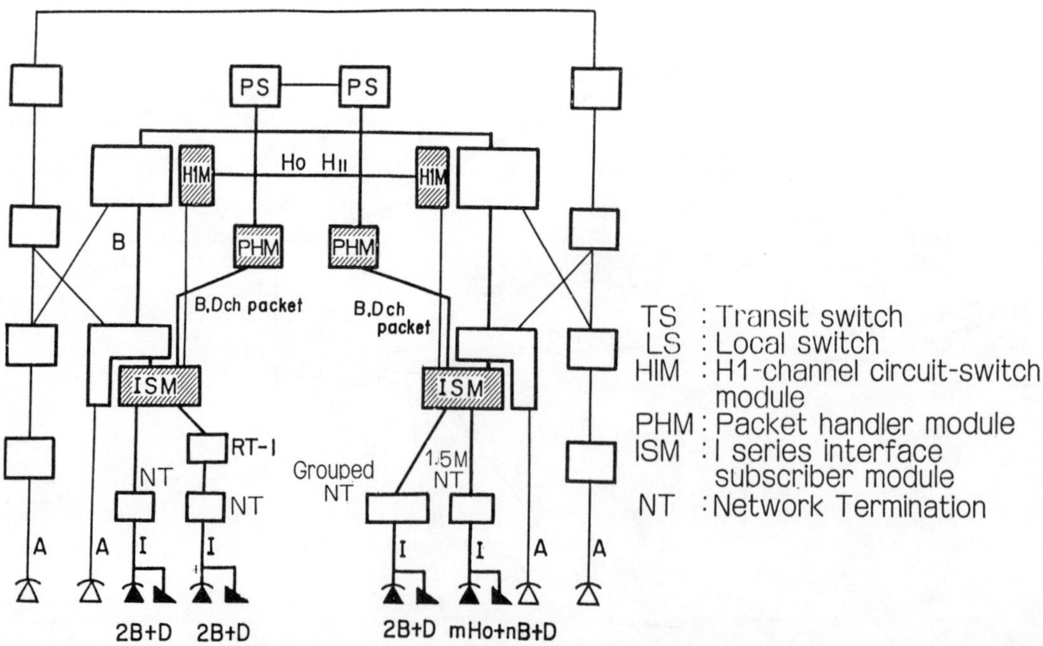

Figure 17 ISDN CONFIGURATION BY OVERLAYING ON EXISTING TELEPHONE NETWORK

constructed by overlaying the existing telephone network. That is to say, INS subscribers are accommodated to the ISM added to the existing D70 switching system.

Moreover, transit networks are constructed for each circuit switched B channel, packet switched B and D channel, and circuit switched H0, H11 channel. That is, circuit switched B channel calls are connected from the ISM to the transit network through the transit switch (D60). Packet switched calls are connected from the ISM to the PHM, and then connected to the Packet transit Switching (PS) network. Circuit switched H0, and H11 channel network calls are connected from ISM to the high-speed digital transit network.

6.7 Common Channel Signalling System (No.7)

Purposes for introducing common channel signaling system can be broken down into two types:

(a) Transmission of various signaling information for ISDN; and

(b) Efficient provision of supplementary services

To utilize the D channel signaling provided in the user-network interface, a common channel signaling system has been adopted for inter-office signaling large quantity of signals and transmit signals while the communications path is in use. It was also adopted because of its various signalling functions. For ISDN usage, a newly defined ISDN user part of common channel signalling system No.7 -ISUP(ISDN User Part)- will be standardized by the CCITT. ISUP expands the functions of TUP(Telephone User Part) and will make possible the transmission of information essential to ISDN. It can convey, for example, information of bearer capability, the calling identification number, a sub-address, and user-user information. On the other hand, common channel signalling system is also essential to provide new services and control network efficiently. A database system has also been prepared in the common channel signalling network for new services and network control. This database is called Network Services Point (NSP).

The network configuration for common channel signalling is shown in Figure 18. To ensure high reliability, it has duplicated the A and B plane structures.

6.8 Supplementary Services

For various supplementary services over the "I" series interface, a layer 3 protocol, which makes it possible to realize supplementary services, needs to be standardized. At present, the layer 3 protocol concerned has been studying. NTT intends to keep pace with study at CCITT. That is to say, provision of supplementary services will be added as shown in Figure 19.

6.9 Terminals

For ISDN services to become popular, a variety of terminals need to be made available to users. Examples of "I" series interface terminals under study are listed in Table 2. During the initial phase, NTT plans to provide a

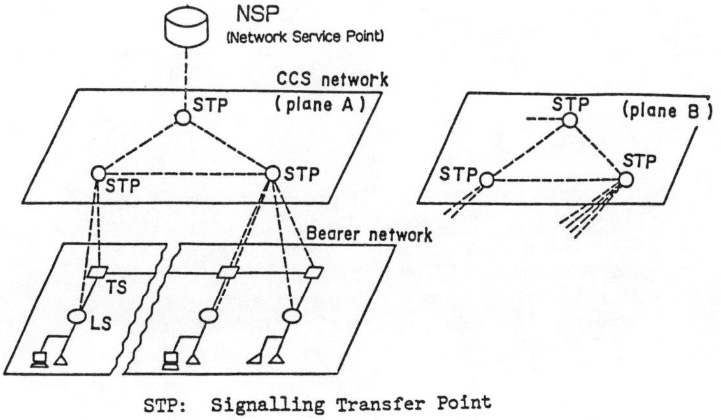

Figure 18 NETWORK CONFIGURATION FOR COMMON CHANNEL SIGNALLING

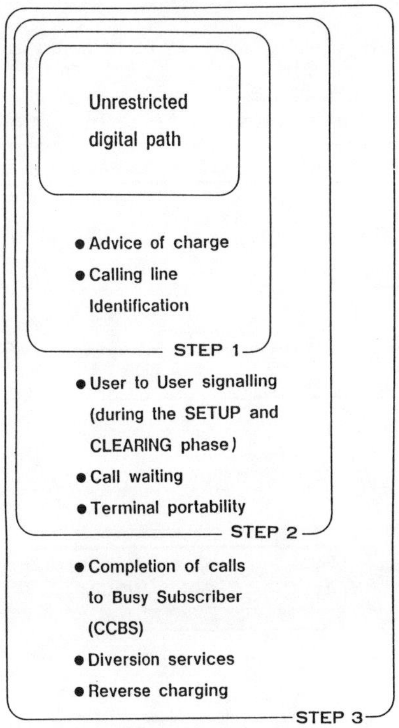

Figure 19 SCHEDULE FOR IMPLEMENT OF SUPPLEMENTARY SERVICES

Table 2 EXAMPLE OF I SERIES INTERFACE TERMINALS

Digital telephone (including PC interface)
Terminal adapter
Voice conference terminal
Integrated key telephone system
Integrated PABX
LAN
Digital facsimile (high definition and monochrome half-tone)
64kb/s Video conference terminal, 384kb/s Video conference terminal
Still picture terminal
Slow-scan TV
CAD/CAM

digital telephone including a PC interface, digital facsimiles , voice conference terminals, and others.

The terminal market has been completely opened up to companies. It is thus expected that companies besides NTT will provide a wide variety of terminals.

7. FUTURE ARCHITECTURES OF THE INS

While the ISDN concept allows for more economical and efficient construction of digital networks and diverse services, this is not the final goal of the INS. In order to completely meet user needs including those for video services and in order to reflect back hardware and software progress into the network architecture , NTT is carrying out R&D on the following subjects.

(1) Access network using optical fiber

Optical fiber makes possible a common access network for voice , data transmission and video services, including CATV. So, users will be able to access the network easily.

(2) New transfer network

Potential demand is great for high speed, broadband services. To cope with these demands, new transfer network based on such technologies as high speed transmission(H2) and high speed packets needs to be realized.

(3) Telecommunications management network

In order to simplify a network structure, management functions are separated from such basic networks facilities as the transfer and access network, in accordance with a hierarchical concept.

(4) Bit-based tariff system

By integrating of circuit switching and packet switching systems, it will be able to introduce a new bit-based tariff system.

Figure 20 is an image of the future network architecture of the INS.

To realize the INS, much study and development must continue to be carried out. Moreover, for INS to gain its due popularity, various new utilizations must be further developed in partnership with users.

From March 1988, NTT will start commercial ISDN services.

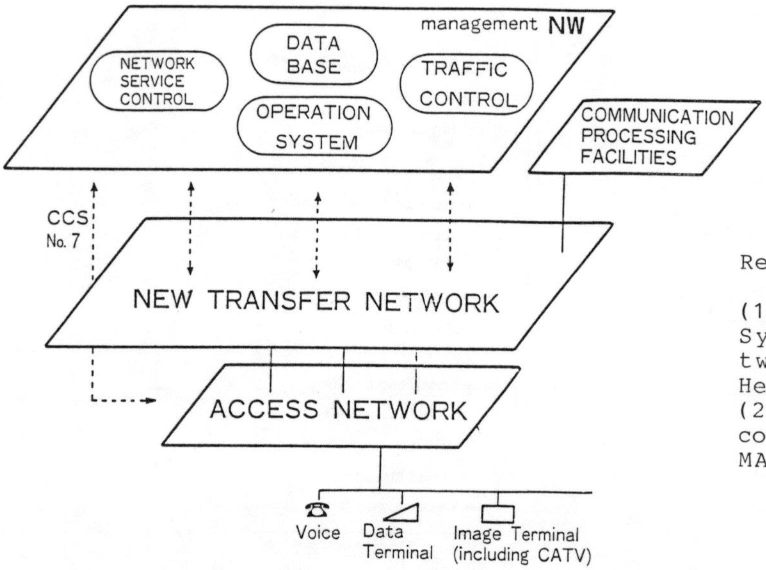

Figure 20 FUTURE NETWORK ARCHITECTURE

References

(1) Y.Kitahara, "Information Network System - Telecommunications in the twenty-first century -",London,England Heinemann,1983.

(2) M.Kuwabara, "Japan tests its model communications system" IEEE SPECTRUM, MAY 1984.

ISDN IN EUROPE: CHALLENGES AND OPPORTUNITIES

Claudio Carrelli

SIP Headquarters
Rome - Italy

Maurizio Decina

Polytechnic of Milan
Milan - Italy

This paper gives a snapshot of current challenges and opportunities for the evolution of ISDN. The main challenge lies in the crucial matchup of technology and market demand, as it is needed for successful ISDN evolution. This matchup depends on the development not only of cost-effective technologies, but also of adequate marketing skills and strategies. A further challenge is given by the regulatory environment that makes even more critical the technology-market intersection of "feasible" information services and applications. The ISDN opportunity is twofold: a growing role for the "common" telecommunications network in the business and residential markets, and a promise for the desired worldwide connectivity of information services (voice, data and image), today limited, expecially in the case of data communications, by proprietary lines of products.

1. INTRODUCTION

The ISDN (Integrated Services Digital Network) represents the evolution of the common carrier telecommunications network **[1]**.Today we characterize the ISDN by means of the following five attributes:

1) **end-to-end digital connectivity***;*

2) **integrated services access**, *for multimedia (voice data and image) applications;*

3) a reduced set of **world-wide standard interfaces;**

4) **intelligent networking services:** *these are the special services to meet the need of business users and to allow customer control on the network resources;*

5) **enhanced networking services:** *these are the network services beyond information transport, such as protocol mediation and data base access for videotex applications.*

With reference to attributes 1 and 2, in the early stages **the narrow-band ISDN** is being implemented using the present copper wire distribution plant, while in later stages the **broad-band ISDN** is going to be realized using new fiber cable installation up to the customer premises. **Narrow** and **broad** refer to the information capacity offered to users that ranges from the million of bit per second (narrow) to the billion of bit per second (broad) and hence to an increasing multiservice provisioning capability including up to **high definition** motion **television** (HDTV).

While the first two attributes are only of **technical** relevance, the remaining three are key factors for the successful **market** development of such ISDN networks. This paper focuses on these latter factors and gives a snapshot of current challenges and opportunities for the evolution of the ISDN, with particular emphasis on the European scenario.

2. COMPETITION FOR INFORMATION SERVICES

The ISDN concept arose more than ten years ago, as the **network** response to the challenge of the Information Age for the provision of information services. Information **transport** and **processing** are the two originating **core businesses** in this competition scenario, as schematically depicted in Figure 1.
The three building blocks of network evolution can be found in this picture:

- *the transport block (network)*
- *the transport-associated processing block (network data base)*
- *the operations support block (network operator)*

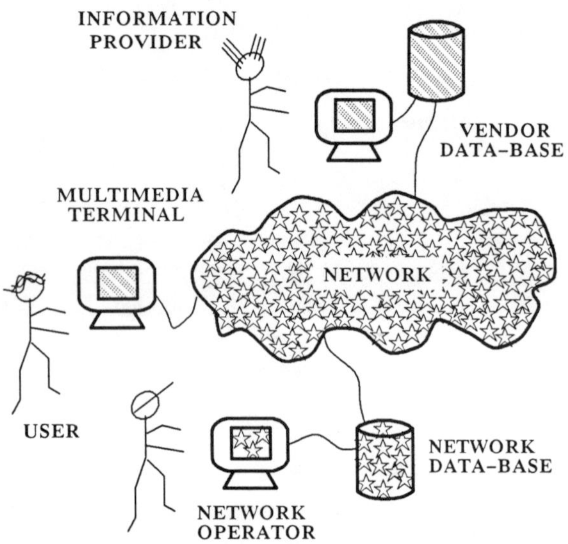

Figure 1 - Competition for the Information Services

These three building blocks are (in generalized applications) used to interconnect user terminals with vendor data bases and information providers. In this scenario the network makes use of **intelligence** to compete for the business information market. Service applications are oriented both to the residence market (consumer services) and to professionals (business services).

Figure 2 displays this intelligent network concept in form of the **onion scheme** commonly used by computer people. At the core of the onion there is the transport mechanism. Technologies of **transport** are rapidly changing, and in this evolution, ISDN represents the **digital** and **high-speed** (**fiber transmission** and **fast packet switching**) stage. The first inner onion skin layer is the **network operating system** implemented by the routing and addressing mechanism. The remaining upper onion skin layers deal with network applications and they constitute the market appeal of ISDN.

The title intelligent network services refer to customer control capabilities over network resources and features. These are obtained using the two blocks: network service data bases and operation support (this latter to cater for customer interactions). Intelligent network services include:

- *the Private Virtual Network (PVN) service*
- *the Enhanced 800 service*
- *the Network-wide Centrex.*

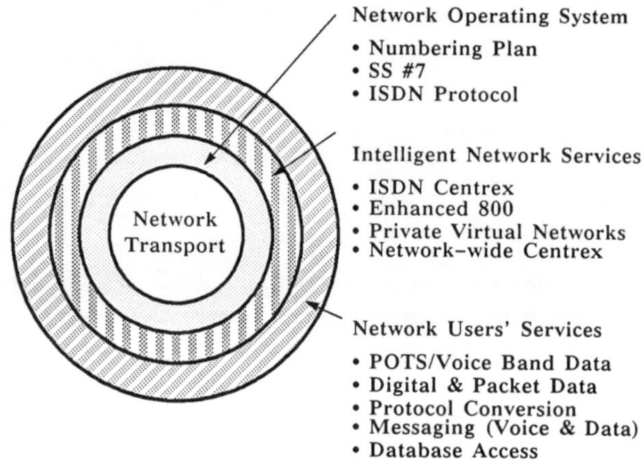

Figure 2 - The logical Architecture of the Intelligent Network

The PVN service, for example, implements multiple private virtual networks on a call-by-call basis via common carrier network infrastructures. It is worth while to be noted that **network intelligence** can be implemented regardless of the transport capability, i.e. the information characteristics involved in the service (voice, data, and image).

The last layer deals with application services. These were traditionally confined to POTS (Plain Old Telephone Service) and voiceband data. Today a number of **enhanced networking services**, making use of network database and application software, are emerging. These services include: protocol mediation, gateway access to a variety of vendor data bases, and messaging.

Of course, it is understood that both intelligent and enhanced networking services can also be provided by outside vendors **in competition** with network operators.

The onion scheme of Figure 2 makes definitely clear that the ISDN Reference Protocol Model is multidimensional as shown by the model of Figure 3, that represents an evolution of the OSI (Open Systems Interconnection) model **[2]**. There are three concurrent processes: user data, signalling data and management data. Signalling is a full protocol layered process with specific application processes (e.g. the enhanced 800 service).

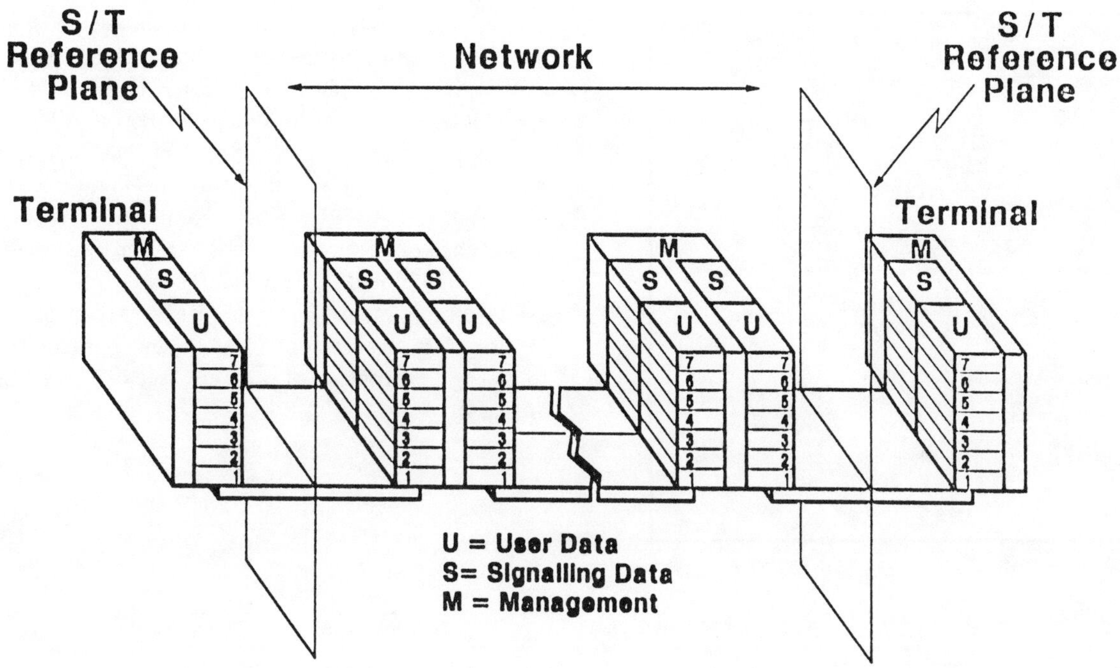

Figure 3 - OSI Extension to ISDN: the ISDN Protocol Reference Model

Management is ubiquitous to allow operation support and customer control; it can use both signalling and information transport protocols and facilities. Moreover, users' information can be processed within the network up to Layer 7, e.g. for protocol mediation, data base access and messaging.

Specific applications of protocol mediation and data base gateways are today implemented in European networks, for example, to provide the **Teletel** videotex service in France. In fact the successful implementation (today more than 2.5 million subscribers) of the French Teletel is largely due in our opinion to three synergistic factors:

- *the creation of a kick-off installed base for the electronic directory service (0.5 million Minitel terminals)*
- *the low cost and simplicity of the Minitels*
- *the provision within the French telephone network of several* **Videotex Gateways** *embedded into digital central offices.*

Of course there are other factors that positively contribute to this success, namely the tariff structure, the anonymity of subscribers and the **centralized billing.**

3. THE CHALLENGES

The main challenge for successful ISDN development lies in the crucial **matchup of technology and market demand.** A suitable matchup depends on the development not only of cost-effective technologies, but also of adequate marketing skills and strategies. Figure 4 shows the strategic directions of ISDN by using the Harvard Map of the information business. MacLaughlin's Map [3] is characterized by two dimensions of information: the services/products dimension and the form/substance dimension. We have only shown the information business at the extreme corners of the Map (mail, paper, books, professional services), while we remember that at the center of the Map sets the "computer" (hidden by the ISDN spot sketched in Figure 4). ISDN comes from the telephone services at the upper left of the Map, encompasses broadcast services and is strategically oriented to both the residential and the professional markets.

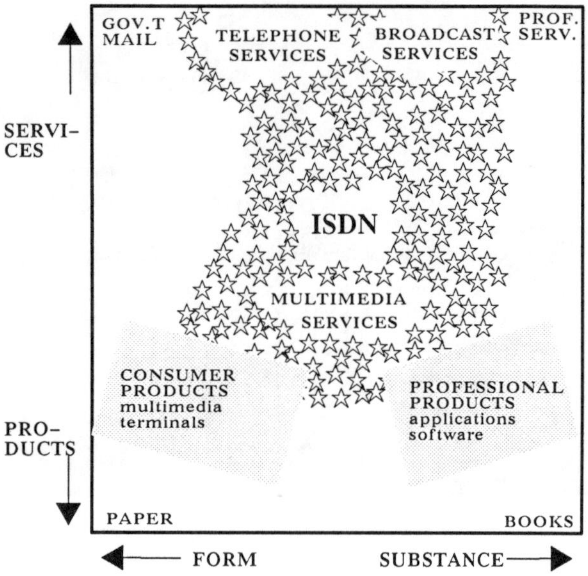

Figure 4 – Harvard Map of the Information Business: where does the ISDN go?

While the residence is the golden goal of ISDN at later stages, it is today understood that the ISDN **top priority** are the large, medium and small business markets. Building the suitable knowhow to sell professional specific service applications is the true challenge for network operators, in particular for the European PTTs. A number of key strategies are needed to adequately face this challenge that is emerging in concurrence with the trends toward privatization and liberalization of telecommunications in Europe.

The main strategic directions should include:

1) **restructure of the PTT organizations** *to serve business user needs;*

2) establish **partnership between large** *(business and institutional)* **customers and PTTs** *for shaping future services offerings including tariff structures; an example of such successful alliances may be the fiber optic cable facilities provided by PTTs for shared use with large customers;*

3) develop **consultative selling** *capabilities to understand* **medium business** *user needs and to customize* **cost-competitive** *and* **open** *networking solutions;*

4) **establish partnerships between PTTs and professional applications software vendors** *to conceive and sell specific service applications* to **small business** *customers.*

We have to point out that unfortunately the existing and planned ISDN field trials in Europe have given more focus on technical issues, interworking issues and exploration of applications, rather than on proving application economics and target market acceptance of such applications. On the other hand we believe that there has been in the last two years an increasing understanding by European Administrations of the importance of application economics for the successful ISDN development. In fact there is today a joint strong effort of the major European PTTs to approach the commercial aspects of ISDN. In particular the following examples of market sectors are being addressed:

- *retailing;*
- *banking;*
- *brokerage;*
- *travel;*
- *health;*
- *press, editing and advertising;*
- *security.*

Businesses in Europe are in a state of change, and PTTs are revising their marketing techniques; methodologies from abroad are being evaluated and amalgamated with local practices. Emphasis is given to more efficient operating practices, and separate profit centers are created to allow the rationalizations and cost reductions needed to keep abreast in an eventual competitive environment.

With particular reference to the North American environment we can also identify a further challenge to the ISDN evolution as given by the local regulatory situation that makes more critical the technology/market intersection domain of "feasible" information services and applications **[4]**. We refer here to some limitations on the network providers to allow competition with other service vendors on enhanced networking applications such as the one mentioned for the French videotex service.

4. THE OPPORTUNITIES

The key-word for the ISDN opportunity is simple:

- *world-wide connectivity for information services (voice, data and image).*

The aim of ISDN is indeed to provide a common infrastructure for open communications, incorporating and extending the concept of **Open Systems Interconnection** to multimedia (voice, data and image) communications. Customers will be able to use the same terminals and standardized service applications across international frontiers. Today such communications transparency is seriously hindered by the proliferation of proprietary lines of products, especially in the case of data communication and processing services.

5. THE TRANSNATIONAL EUROPEAN ISDN

In Europe the major PTTs started a collaboration to jointly promote ISDN and in particular the transnational aspects of ISDN. The informal activity of these PTTs will be reported at TELECOM '87. The main objective of the collaboration is to undertake every effort to provide international ISDN interconnection as soon as 1989-90.
The common activity covers harmonization work on:

1) **user-network interfaces:** *including terminal adapters, private network signalling, testing and maintenance;*

2) **inter-networking:** *including SS no.7 signalling (ISDN User Part), numbering and routing for the services specified below;*

3) **engineering and planning for interconnection;**

4) **services:** *including bearer services, supplementary services (e.g. calling line identification, freephone, subaddressing, terminal portability) and teleservices (e.g. photographic videotex, teleaction, picture transmission, file transfer and Personal Computer Services);*

5) **terminals:** *including telephony, telefax,teletex and Personal Computers.*

Moreover, the cooperation covers the key-item:

6) **commercial aspects:** *including review of national plans for ISDN introduction, identification of target customers and marketing strategies, tariffing principles (although different tariffs will be retained at early stages, tariffs will be progressively harmonized and will eventually converge) and promotional activities (e.g. by common logo and brochures).*

The work sketched above stresses the real opportunity of ISDN in terms of benefits to network users resulting from the synergy of international standard interface/applications and of transnational availability of common interfaces/applications.
We believe that the current effort of the European PTTs in:

- *producing detailed specifications to promote development of compatible services and equipment;*
- *planning the interconnection of their networks;*

constitutes a significant step for the successful deployment of ISDN.

Further international collaboration is also needed on a world-wide basis, including Europe, North America and Japan, to realize the promise for the much desired world-wide connectivity of information services while entering the Information Age.

5. CONCLUSION

We have given a short review on the main challenges and opportunities for the evolution of the ISDN. Main reference was made to the ISDN in Europe, but similar considerations do apply to other parts of the world as well. The main ISDN challenge is given by the crucial intersection of technology and market applications. Adequate marketing skills and strategies should be developed to make successful the ISDN deployment

The main ISDN opportunity lies in the realization of the greatly desired world-wide connectivity of information services. To reach this goal, an higher level of international cooperation is needed among network operators, beyond the present standards activities, to promote **the ISDN concept of open multimedia communications,** and to plan both **international network interconnections** and **transnational service offerings.**

REFERENCES

[1] **M. Decina, W.S. Gifford, R. Potter, A.A. Robrock:**
IEEE Journal on Selected Areas in Communications,
special issues on: *"ISDN–Recommendations and Field Trials"* May 1986, and *"ISDN Technology and Implementations"*
November 1986, Vol.Sac–4 nos. 3 & 8

[2] **M. Decina, A. Roveri:**
"ISDN Architecture and Protocols"
Chapter 1 of K. Feher's book
"Advanced Digital Communications, Systems and Signal Processing Techniques"
Prentice Hall, N.J., 1987

[3] **S.F. McLaughlin, A.L. Antonoff:**
"Mapping The Information Business"
Special Report for the **Program on Information Resources Policy,** Harvard University, Oct. 1985.

[4] **I. Dorros:**
"The New Future – Back to Technology"
IEEE Communications Magazine, Vol. 25,
No. 1, January 1987.

ISDN
An Investor's View

Richard C. Toole

Merrill Lynch Capital Markets
New York, N.Y.

The pragmatic investor wants to know one thing about ISDN: What does it mean to the bottom line?

1. Overview

In the investment field, the decision maker is the portfolio manager—whether individual or institutional. And while the advisors, fact gatherers and security analysts busy themselves with the mean little details, the portfolio manager's approach is much simpler.

Instead of getting into all of the whys and wherefores of ISDN, for example, the portfolio manager has two basic questions, "Is the concept viable?" and "How do I play it?"

ISDN must be for real. If not why would so many important people gather at the ICC-ISDN '87 Conference in Dallas, Texas in mid-September? Unlike the much publicized SDI (an acronym with remarkably similar components to ISDN), ISDN is now in the testing stage. By comparison, SDI seems to be a need (real or imagined) looking for a conveyance while ISDN is a conveyance looking for a need. Superconductivity—the state in which electric current passes through a material without resistance—also is analogous to ISDN. Breakthroughs in superconductivity have piqued investor. The possibilities of 100% efficient power conduction are enormous. However, one venture capitalist quoted in the New York Times had this to say: applications of new advances in superconductivity "are all too far off to be of real interest to us right now." [1]

Similarly for ISDN—the superconductor of telecommunications if you will—the real bet is whether or not it will fizzle or prosper depending on development of a significant market within a reasonable time frame.

A new book on telecommunications had this to say: "ISDN is suitable for all communication services, the customer can choose what it should do, and the network can evolve as service offerings develop. Uniformity of standards and procedures and a digital system are keys to its success." [2]

The other knotty question for portfolio managers is how to invest in ISDN. It's a little early but there appears to be no such thing as a "pure play". As with so many technological breakthroughs it can be frustrating to find that the major players often are so big that the contribution of something new may be lost in the shuffle or so tiny that the risks of survival are enormous. Timing, too, is all important. For example, Comsat was once a pure play on a new technology. When the stock began trading at $20 in the early 1960's, the concept was so appealing that a mystique developed which pushed the stock into the $70 range.

This move took place before the company even had a satellite system or the hint of profits. When reality set in and Comsat's potential could be defined and its regulatory scheme became apparent, the stock plummented.

Initially, the interest in ISDN is high, similar to Comsat. The concept is bright but at this point not clearly definable from an investment standpoint. No doubt distortions will develop. Some guru will have a sure fire way to make money on ISDN, and, at the outset, it will be hard to argue against the pure play. Distortions will occur but eventually truth will out. There is money to be made investing in ISDN but the crystal ball lacks clear definition at this point.

References

[1] *New York Times* June 17, 1987 page D6.

[2] Hyman, L., Toole, R., and Avellis, R., *the New Telecommunications Industry: Evolution and Organization* (Public Utilities Reports, Arlington, Va., 1987), p. 58.

The information set forth herein was obtained from sources which we believe reliable, but we do not guarantee its accuracy.

I STILL DON'T KNOW

ANNA-MARIA KOVACS

EATON VANCE MANAGEMENT
BOSTON, MASSACHUSETTS

Like most attendees of ISDN conferences, I have come away with a few ISDN jokes. My favorite is that ISDN stands for "I still don't know". There is a great deal I still don't know and, as far as I can see, even the experts have more questions than answers. I certainly do. It is unclear what all the standards will be, in what sort of equipment they will be embodied, how they will be implemented, and at what cost. As you know, standards are only partly set. CPE standards, in particular, lag. Equipment is being designed and to some extent manufactured in the hope that standards that have been defined will stick, or that equipment in the field may set "de facto" standards. Trials are being run, but they represent only islands of equipment, not end-to-end network solutions. And each manufacturer's solution is different. I have asked both AT & T and Northern Telecom, but I still do not know how you would get an ISDN #5ESS to talk to an ISDN SL1, or even to a DMS. Yet late 1987 - early 1988 is the ISDN rollout target I was given by a ROC. Needless to say, that rollout can only be partial, gradual, fragmentary. Will ISDN ever be anything more?

As an investor, of course, I have to funnel all these technical issues and general anxieties down to the specific question of "whose stock will go up?" In answering that question, I focus precisely on the vagueness of the situation. The winners will be those who can act before knowing. The uncertainty around ISDN gives a great edge to those who can plan and hedge R & D bets well. The survivors will be those who have the muscle to set de facto standards, bear great financial risk, and dodge nimbly from one product to another as standards are set.

There is, fortunately, one thing I do know. There will be a market for ISDN. I know that because the ISDN functions are in great demand already. Voice-data integration, connectivity, high-speed data transmission, terminal portability, dynamic bandwidth allocation, and centralized network management all exist, even if in somewhat rudimentary and isolated form, in today's CPE. And there are real customers buying them. PBXs, LANs, T1 networks, packet networks, protocol converters and Centrex are all attempting to fill in fragmentary ways the needs ISDN promises to fill with more elegance and completeness.

This should all mean an enormous market someday. New and upgraded equipment and software at the central office and on customers' premises will be needed and more digital transmission will follow. Even if all central office switches and PBXs are not fully replaced, most of their parts eventually will be. Switch upgrades will require, at the very least, new software, additional processor capacity, new line cards, and a packet switching capacity. Network test and management will also be needed. Those are great opportunities which I predict with some confidence.

The timing is much harder to predict and is critical to the industry's development. I do not know whether it's a three or ten-year process overall, and how long the different phases will take. For example, how long will the politics of the standards negotiations give ISDN central-office equipment a lead over ISDN CPE? Will the phase-in of ISDN be gradual enough for existing small vendors to keep up with product development, or is a big R & D lead critical? Will it be fast enough to put the less nimble major vendors at a disadvantage in product development? Those are the key questions. Above all, they will determine whether the current small telecom equipment vendors have any chance of survival.

It is fair to assume that the big players, AT & T Technologies, Northern Telecom, IBM/Rolm, Siemens, etc. will dominate and survive this process, because of their R & D strengths, their market dominance, and financial reserves. That is not to say that the road will be smooth. Their R & D spending may far exceed budget, they may miss a market or, alternately, leapfrog competitors by a year or two. Undoubtedly they will show uneven earnings for a while, if only because they will be facing another generation of confused and shell-shocked customers, who just cannot sign that purchase order. So the major vendors will probably focus on the sophisticated central office customers first, but will be ready with at least some CPE products when the corporate customer is finally ready to spend. We should note that ISDN may provide a particular opportunity to the foreign vendors who now have a chance to catch up by leapfrogging. With ROCs anxious to add sources of supply, Siemens, LM Ericsson and NEC have a real shot at the U.S. market. Their past history in the U.S., however, does not give much ground for optimism. The conflict between U.S. - European standards committees may disadvantage the foreigners, as well. Nevertheless they will all survive.

It is the small vendors whose life is at risk during this time. They have little R & D capability, little money in the bank, and no central-office equipment sales to help get through the paralyzed-customer phase. So they have to make the right product bet the first time, at the right time.

It is fair to assume that the majors will leave lots of niches open to these little guys, as they have in the past. In fact, if ISDN really brings full connectivity, the role of the niche supplier should be easier than in the past. So it is a safe bet that there will be lots of small vendors in five years, some doing very well. The question for an investor is whether they are likely to be new, or survivors from the current crop.

When I invest today in a vendor of LANs, PBXs, packet or T1 networks or any other "ISDN-like" equipment, I do not know how soon real ISDN equipment will obsolete my vendor's products. I know it will obsolete it, but the timing is critical, as are the company's skills. Will my vendor be able to step up to an "ISDN generation" product? If not, is his bankruptcy inevitable? Even if he can ultimately develop products, can he have a cost-effective product evolution? The major vendors can probably manage their customers' migration path. The majors can absorb the cost of upgrading their customers' interim purchases to minimize the customers' risk in buying pre-ISDN solutions. For small players, it is much more difficult to come up with the intermediate product generations and to finance the migration. It is hard enough just to finance development costs. It is also more difficult for the small niche player to survive "frozen" markets, when uncertainty kills demand. Revenue stumbles are so very obvious in undiversified little companies. Thus, it is likely that most of the small equipment vendors of today will not survive. Some will be absorbed by larger vendors who need their specific skills. A new generation of niche players will rise.

Despite all the uncertainties, I think a reasonable scenario for ISDN rollout can be drawn. I am making some assumptions here about customer behavior. In particular, I believe that confused customers are risk averse and are unlikely to invest as long as they fear their equipment may prove to be of "the wrong standard." Because ISDN CPE standards have not been set yet, that is likely to drive customers to Centrex, as the low-risk way to try out ISDN. The ROCs will encourage that trend, because they see ISDN as the resurrection of Centrex. The major vendors, in response, will focus on central office-based development. There will be little room for small vendors in this phase. However, when CPE standards emerge and some products follow, it is quite likely that the desire of corporations for self-controlled networks will re-emerge. The ROCs may face another wave of stranded Centrex investment. The big vendors will have to refocus their attention on CPE. This will be the time when small vendors, mostly new ones, will get their chance to innovate, to fill in around the edges. As a cynic, I predict that by this point, we will be close to the next wave of technology-breakthrough, probably broadband, and will go through a similar cycle again.

I entitled this speech "I Still Don't Know" because to me as an investor, it is the uncertainty surrounding ISDN that is most important. Not only will uncertainty affect Wall Street psychology, making the stocks very volatile, but, more importantly, it will affect the industry players' positions. It is the confusion around ISDN that is providing the opportunity to excel, to get an edge. Understanding the timing of generations of technology, navigating the conflict between CPE and central office-based systems, playing on the psychology of shell-shocked communications managers, are the key skills that will determine which telecom vendors will succeed. Deciphering these trends correctly will provide the payoff for the investors.

As I try to invest today, I feel safe putting my money on the big guys. For them, ISDN is a big opportunity and they can handle it. I also, of course, try to differentiate among them, looking particularly for R & D and marketing skills. I can also begin to pick out the losers among today's small manufacturers. In fact, that is the easiest job: find the vendors whose products ISDN will obsolete, and who are also too poor or inflexible to move on with ISDN. It is the small-company winners I cannot pick yet, because they have not been born yet.

An Investment Banker's View of ISDN

William R. Becklean

Kidder, Peabody & Co., Incorporated
Boston, MA

This presentation looks at ISDN from the point of view of an investment banker. If the financial community is to contribute the capital to deploy ISDN, it would like to be assured that ISDN represents a sound business proposition. Given investors' disenchantment with the telecommunications sector and the appearance of confusion on the part of prospective users, the telephone industry needs to reassure the investment community that there is a real market for ISDN services.

My assignment today is to present a point of view from the investment banking community. Our role, as you probably know, is to provide access to the capital which will be required to manufacture the equipment and build the physical plant necessary to deliver ISDN services. We do this by finding investors who are willing to invest their money in your enterprises, creating securities which represent a legal and financial interest in your business and selling them in the public or private debt or equity markets. Our value added to the user of capital is knowing who these investors are and therefore being able to tap the best (which translates into the least expensive) sources of capital. Our value added to the providers of capital is our ability to provide some judgement with regard to what constitues a good investment opportunity. That judgement process, which we call due-diligence, involves not only looking at the financial aspects of a transaction but also trying to understand, to the best of our ability, the business underlying the investment. Obviously, if our judgement is good and the deals we have sold to the investors have worked out well, they will be more receptive to our deals in the future. On the other side of the coin, sell them several deals that don't work out and the market tends to become skeptical. Frankly, investments in the leading edges of the telecommunications industry have, with certain exceptions, been generally disappointing over the last five years and we are facing a generally skeptical investment community. For that reason it behooves us to try to understand ISDN and its potential risks and rewards.

ISDN is certainly not a new concept that has just burst upon the world. My first introduction was almost ten years ago at a presentation on the ITT System Twelve. Since that time I have tried to do my own homework on the subject and I have also organized several panels using experts like yourselves to explain ISDN to the broader community of institutional investors. But I'll have to confess I'm not sure exactly how far I or the investment community have gotten in our understanding of what ISDN really is. But maybe we shouldn't feel so badly about that. It seems that we're not the only ones who are confused. Few of you could have missed seeing the results of a recent survey by a major trade publication which indicated that 56% of the communications managers they polled were confused about what ISDN really is and what it is good for. And despite all the publicity ISDN has had over the last year, that figure is up from a 53% confusion factor in last year's survey.

Now that has to give an investment banker reason to be concerned. If the prospective users of the service don't know what it is and what it is good for, who does? And on what basis can we make a case that there will be a market for it when it is ready to be delivered?

Maybe this thing is so big we have a classic example of the "Elephant Problem". You know, one person looking at the trunk, another looking at a leg and another looking at the tail. They are all looking at the same thing but seeing something different. If that is the case, at least we might ask who are the parties that are looking at the ISDN Elephant and what is it they are seeing.

It seems to me there are at least four parties involved - the service providers, the network equipment suppliers, the suppliers of customer premise equipment and the end users.

The service providers, both the interexchange carriers and the local exchange carriers, are surely looking hard because ISDN will affect their lives over the next ten or twenty years. They are, after all, the organizations that will have to deliver ISDN services to the end users and they are the organizations that will have to raise the capital to build the networks. Their analysis of what the customer wants will be critical because it is their bottom line that will be affected by his acceptance or rejection of their offerings. For the interexchange carriers, ISDN appears to offer an opportunity to meet the needs of large multi-location organizations operating extensive private networks by providing flexible dynamic bandwidth in place of the inflexible leased circuits in use today. Value added services such as the turnkey provision and operation of complete networks represents an attractive revenue opportunity for the interexchange carriers. However, we think this opportunity will be limited to smaller users by the desire of the large network operators to maintain control over their own network operations. The local exchange carriers face the threat of large users bypassing local distribution facilities and going direct to an ISDN-capable IXC point-of-presence in their service area. We think these carriers will have to focus their marketing effort on serving up ISDN functions to meet the requirements of those users who are too small to operate large private networks, to the occasional off-net requirements of the large network operators and to the requirements for local-area networking with a "next generation Centrex" so to speak. Several ISDN trials of the latter sort are just now getting underway and these will hopefully generate some useful user inputs.

ISDN represents both a risk and potential reward for the service providers. We expect market share shifts among the interexchange and local exchange carriers as new services are introduced. And, if ISDN is going to offer more efficient use of bandwidth, the carriers must offer value-added services or stimulate increased traffic to realize revenue gains. It seems pretty clear that if ISDN is ever going to provide its promised payoff it will have to demonstrate the capability to build traffic volume in what we all hope and believe is an elastic market.

The network equipment suppliers are, of course, critical to the whole ISDN revolution because it is only through the features and functions of the installed hardware that ISDN services can be provided. These manufacturers are in the difficult position of trying to respond to the presumed needs of their interexchange and local exchange customers when these customers are uncertain themselves of the market requirements. One outcome which seems certain is the move to more flexible network and switch architectures which will allow the carriers to develop their own ISDN applications on standard platforms as opposed to the historic provision of monolithic switch designs and the seemingly interminable wait for software generics incorporating new and specialized features. Thus, it seems, the manufacturers will also have to adopt a more flexible and market-driven approach to their business.

For the suppliers of customer premise equipment, both voice and data, the arrival of ISDN presents a whole host of new strategic issues. How will IBM and DEC, both of whom have invested heavily in proprietary systems to provide users of their equipment networking capability, adapt to an external system which purports to provide this as a standardized capability. Will these vendors sacrifice "account control" in order to access the presumed benefits of ISDN? My initial reaction is "not likely". So, how can ISDN accommodate the reality of the dominant market position maintained by these large vendors. And will ISDN be the catalyst which finally provides the incentive to combine voice and data functions in CPE or will the economics still encourage users to maintain separate systems. In my mind, the economics will always prevail in the real marketplace. The economic benefits of ISDN-compatible terminals will have to be demonstrated in order to overcome the failure, to date, of an integrated voice/data approach to the world.

And finally, the end users - the poor guys, 56% of whom are confused about ISDN -they're surely looking at this thing and wondering what's in it for them. For sure, it had better simplify their life or provide economic benefits or it seems highly unlikely that it can be successful.

So, who is looking at the whole elephant? Is anybody out there in a position to integrate these different points of view into a whole that is consistent with what the market wants as well as being in a position to influence deployment of the system. My great fear is that right now, the only constituency in a position to do so are the technicians, the engineers who are focused not on the issue of what the customer wants but on what the technology can do. Forgive me for being hard on the engineers but I was educated as an electrical engineer and I know how, in our infatuation with the possible, we can lose sight of the real mission. Something about the forest and the trees, I think.

So, tell me, are the technicians in control of this thing? Do we have an elegant technical solution looking for a problem? Is ISDN destined to be a repeat performance of Integrated Voice Data Terminals, Voice Data PBX's, Digital Termination Systems, Picture Phone, Video Conferencing and Net 1000? All elegant applications of technology but ultimately all market failures. When you stop to think about it the telecommunications industry has not lacked for promising new technology. The problem has been to deliver it in a form the user is willing to buy. Can you deliver on ISDN?

I think it is fair to say that the investment community is disenchanted with its investments in the telecommunications industry - some sectors more than others. As you can see in the accompanying chart, following an initial run-up in anticipation of

STANDARD & POOR'S 500 INDEX
VERSUS
SELECTED TELECOMMUNICATIONS COMPANIES

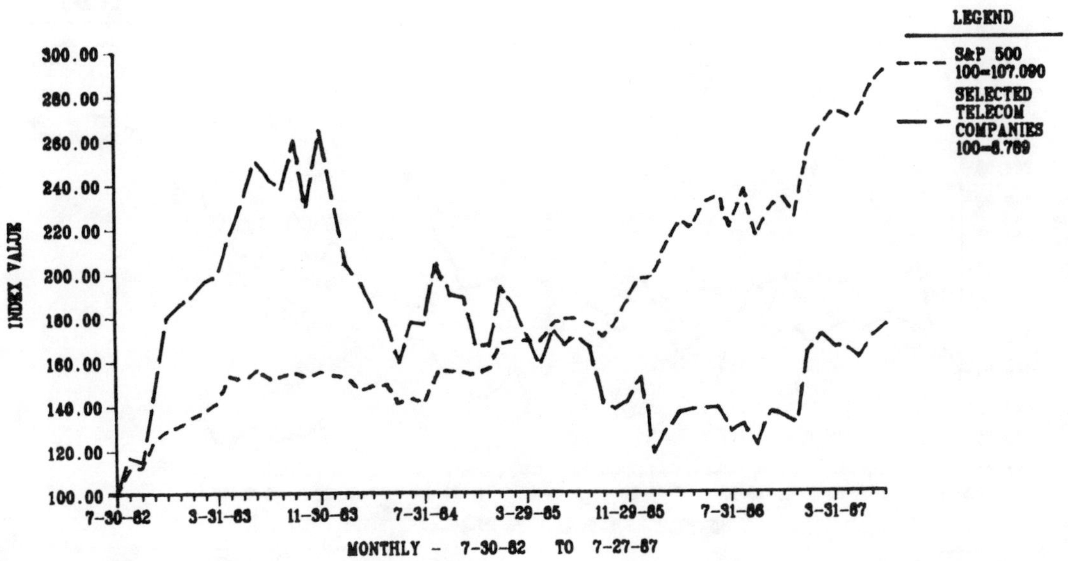

the assumed benefits of divestiture, a representative index of telecommunications equipment stocks significantly underperformed the S&P 500 over the last five years. Using the S&P 500 as the standard against which portfolio performance is measured, you can see why a money manager might be skeptical about a telecommunications equipment stock story today. Even the Regional Bell Operating Companies, the superstars of our industry have provided only an average return over the full period since divestiture although a money manager could have outperformed the averages if he timed his sale to coincide with the

STANDARD & POOR'S 500 INDEX
VERSUS
THE REGIONAL BELL OPERATING COMPANIES

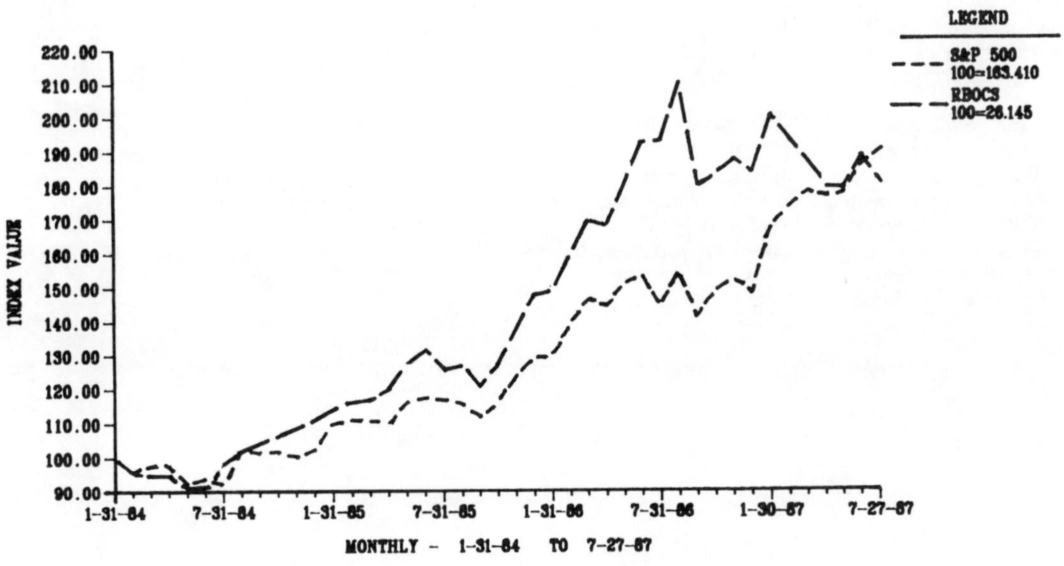

STANDARD & POOR'S 500 INDEX

VERSUS

AMERICAN TEL. & TEL. AND MCI COMM.

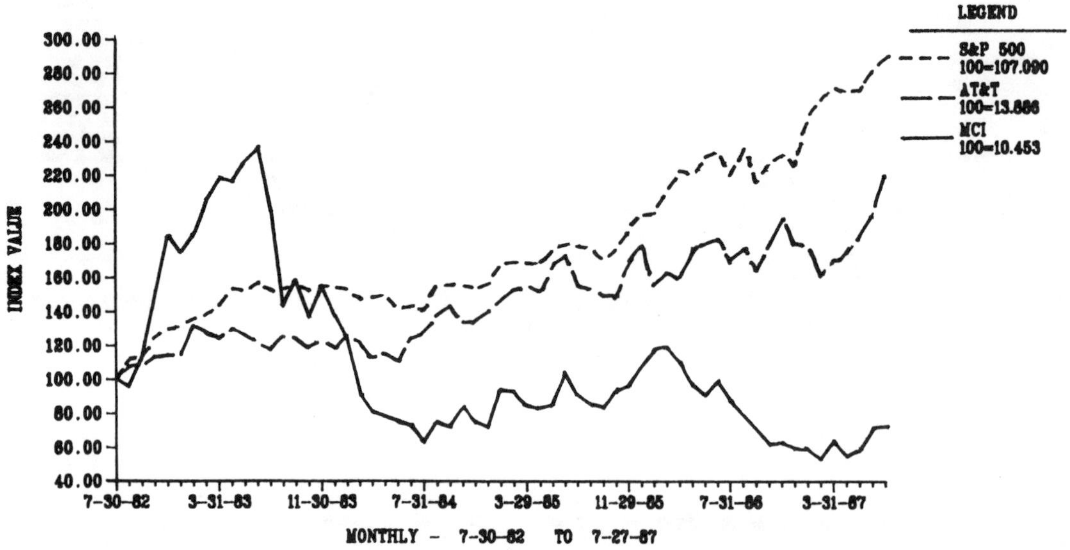

end of the RBOC honeymoon last year. The interexchange carriers, too, have underperformed the averages.

So, why these disappointing results for the investors? Largely because of decreasing investor expectations that the telecommunications industry can deliver better than average earnings growth. Fundamentally, stock prices are a measure of investor expectations concerning the future increase in value of the underlying company's equity. And this future increase in equity value is directly related to the ability of a company to earn a return on invested capital. Stock prices generally rise as expectations for future earnings gains improve and fall as expectations deteriorate. Its simple enough - above average earnings growth will generally yield above average stock performance.

So, what is the relevance of all this to ISDN? In the simplest terms the answer is....will ISDN justify the investment? Can the telecommunications industry generate above average earnings growth through the deployment of this new technology? If you can convince the investment community that you can, expectations for future earnings gains will rise and you will be able to fund the necessary capital investments on more favorable terms i.e. at higher stock prices.

That is clearly a desirable result and it leads to the obvious question of what can be done to achieve it. In my mind, you have a selling job to do on the investment community. You need to convince us that ISDN is not a solution in search of a problem. You need to make the case that ISDN is a market driven response to what you perceive to be the future needs of your customers. And in making your argument I suggest that you focus on answering what seem to me to be three critical questions:

1) Does it meet a user need?
2) Is it cost effective?
3) Are there sufficent incentives for standards to prevail?

If you can deliver ISDN technology that provides an affirmative and believable answer for each of these questions I think the investment community will feel more comfortable that ISDN can provide the basis for improved growth in earnings for both the equipment manufacturers and the service suppliers.

We're always eager to finance a good business proposition. But please - don't give us another turkey.

ISDN - PICTUREPHONE® REVISITED?

Joan E. Sinclair

The Adams Express Company
Baltimore

This presentation gives an overview of how investors weigh new technological developments in reaching investment decisions. In this presentation particular attention is given to how ISDN will impact revenues and earnings of the carriers.

At the outset, I'd like to point out that I don't pretend to be an expert on the technology of ISDN, or, for that matter, any other technologies. Fortunately, I don't have to be. Rather, as an analyst for an investment company, I try to grasp the underlying trends and future prospects for the widely-diverse segments of American industry for which I am responsible. Hopefully, I relate these factors correctly to the price action of the stocks involved.

Speaking about different technologies, let me share with you an incident which took place recently in my firm. The chairman of my company dropped in to ask me the meaning of re-babbitting of bearings. It seems one of the broadly-diversified companies in which we hold stock had acquired a company which performs that service. His query was met by my blank stare. I hastened to assure him, however, that the management of this company had made very few mistakes in acquiring some forty-odd companies during the past decade or so and I was sure this acquisition would enhance the company's prospects. I mention this incident because it brings up a crucial tool of my trade, i.e., subjective judgment of a company's proven track record over time.

When I look at a new concept to determine its investment implications, I normally try to gauge market potential, costs, future profitability, and a host of other factors. In thinking about the concept of ISDN, however, what comes to the forefront of my mind is a very lengthy series of questions, not the least of which is just what is ISDN? I recently attended a conference sponsored by the Information and Telecommunications Technologies Group of the Electronic Industries Assocation and listened to one speaker after another describe new products, all of which can be adapted to ISDN, "whatever that is". These last three words, or similar ones, were used almost as an integral part of each presentation on ISDN.

In looking at ISDN's impact on the telephone industry, therefore, we really have to define what we're talking about. ISDN's potential impact on the Regional Holding Companies, which are restricted from offering information services over the network, is one aspect. Another is ISDN's possible effect on non-Bell telephone companies which are not so restricted. Are we talking about the local loop or the long-distance link? Where will the ISDN intelligence lie - in the equipment on the customer's premises, in the local loop, or in the long distance connection? And, a key question - are we talking about a rate-base, strict rate-of-return regulated service?

I'd like to concentrate on this last question since so much of the industry is still rate-base, rate-of-return regulated and ISDN is being developed within the framework of the regulated, core network. In this environment, who will benefit - the

Picturephone is a registered trademark of AT&T.

investor or the rate payer? Regardless of the capability of a given management, will shareholders benefit from ISDN --- and when?

There are some encouraging signs of relaxation of tight rate-of-return regulation on the part of regulatory authorities. For example, the FCC is currently looking at alternatives to rate-of-return regulation for AT&T. A proposal to allow AT&T to price individual services with a cap linked to the inflation rate is being considered. Precedent for this type of regulation already exists in the United Kingdom as well as elsewhere in Europe.

As has been well publicized, the Justice Department early this year recommended that the MFJ ban on offering information services be lifted from the regional Bell operating companies (BOCs) but this issue seems far from being resolved. Regardless of Judge Greene's decision, appeals seem certain. Bell Atlantic's request to become the first BOC to win FCC approval for a plan to provide network-based enhanced services without establishing a separate subsidiary under Computer Inquiry III's comparably efficient interconnection (CEI) requirement has run into snags. So the FCC may not allow the Bell operating companies to make enhanced services available as part of a network offering until they implement permanent open network architecture (ONA) interconnection capabilities. When that happens, true competition for the local loop will be a reality.

No one, of course, can predict the timing although the Bell operating companies as well as AT&T will be filing ONA plans by February 1988. How Congress and a new Administration in Washington will react are also unknown. And then there is the question of how state regulators will view competition at the local level. Regulatory obstacles currently protect the local franchise. State regulators want low rates and universal service but at the same time want to protect the local exchange carriers' customer base. Balancing cost-based pricing with the needs of the small telephone user is tricky. Still, quite a few state public utility commissions are approving some very interesting telephone company proposals designed to move the companies at least partially away from rate-base, rate-of-return regulation. Also, in a number of states regulators are beginning to allow flexible pricing for competitive services.

Setting aside the issue of regulation of enhanced services for the moment, let's look specifically at what's happening with ISDN in the local exchange network. A number of trials, rollouts, etc. are underway between local Bell companies and large customers. In addition, non-Bell companies are beginning to provide digital hook-ups to major accounts. Centel, for example, faced major problems in Las Vegas some years ago from high growth demands and loss of customers who were moving to private networks. Centel's response was to install digital switches and fiber optic trunking. While Centel did not employ ISDN technology, the company certainly achieved its principal objectives -- It kept its customer base intact. Furthermore, the low cost and maintenance-free characteristics of the installations have enabled the company to maintain stable rates in Las Vegas. Rates have not been increased since 1965, and probably will not change for the foreseeable future.

If all the ISDN trials and rollouts now underway do no more than stem the tide of bypass, they will have served an essential purpose. These trials have brought the industry into much closer contact with its customers. Moreover, the list of companies involved with the local exchange carriers in these trials reads like a who's who of American industry. I also suggest this is ample proof the industry is not implementing ISDN even in the absence of user demand for it, as some critics have suggested.

Time does not permit a lengthy description of the various trials and rollouts currently underway. Approaches vary among the BOCs but most are committing very substantial amounts of money to ISDN development. In talking with various BOC managements, the same underlying themes emerge. In addition to protection of the customer base, there is a sense of urgency to generate new revenues through utilization of the existing investment. One of ISDN's major attributes is its adaptability to today's embedded base. Rate base growth has slowed dramatically and diversified activities have yet to contribute meaningfully to the bottom line. The information industry market in the United States has been estimated by Ameritech at $425 billion and the BOCs do not want to be left at the gate.

If we let our imaginations run wild, there is a multitude of services possible in an ISDN age. With a regulatory okay, the telephone companies may be offering electronic mail, voice storage and retrieval, pay-per-view TV, video conferencing and a host of other services.

Of course, we've been hearing about these new services for a number of years. But the ISDN world is evolving very slowly. ISDN will be evolutionary, not revolutionary. This has been true throughout the entire history of the telephone industry. Technology has always played a leading role. In this evolution, electronic analog central office (CO) switches brought us new services such as call waiting and call forwarding. Now, Custom Local-Area Signaling Services (CLASS) such as call blocking, call tracing, and distinctive ringing are making their debut, thanks to the introduction of Common Channel Signaling #7 (CCS#7). CCS#7 is one of the key network requirements for ISDN. Its introduction is being spurred dramatically by the FCC requirement that the local operating companies provide equal access for 800 service to the other common carriers (OCCs).

Other pre-ISDN offerings proliferate which, from a user's standpoint, can resemble virtual private networks. What I find most interesting about these service offerings is that they follow the Centrex concept in providing businesses specialized communications functions as well as private data networks. We all remember the much-publicized demise of Centrex a few years ago. Centrex didn't die, however. Just the opposite happened. Usage soared as the BOCs upgraded CO-based analog switches through software. And now we see these companies following the same principle to spur usage further. Bell Atlantic and BellSouth have introduced CO-based local area networks (LANs). NYNEX has developed its Pathways integrated digital services which exchange voice, data and images on a single high-speed digital line. Ameritech has developed both the Feature Node/ Service Interface and a CO-based Public Switched Digital Service. And many other Centrex configurations are available.

The BOCs are not alone in offering virtual private network services. Non-Bell companies and long-distance carriers are also aggressively attacking this market. Rochester Telephone calls its network ViPS, MCI's is Vnet, AT&T has its Software Defined Network (SDN) service and so on. All of these companies are using the resources of the public switched network to provide pre-ISDN, customized services to private users. These services appear to be dedicated private networks to the user. The customer enjoys the same privacy and flexibility of a private network with end-to-end capabilities tailored to user needs. Moreover, the user doesn't really care whether its virtual or integrated or ISDN as long as the service provides an economic solution to the user's network problems and someone is available to repair it.

Today's new telephone services are only the beginning of what we, as customers, will enjoy in the future. Once standards have been agreed upon, as customers we'll be able to pick and choose among a plethora of new offerings. And both we and the industry will enjoy enormous efficiencies and cost savings.

But how will we benefit as investors? When I was first approached by Len and Dick to participate in today's panel, I must confess I was very skeptical about the potential impact of ISDN on future sales and earnings of the telephone companies. As I talked to and visited with many industry representatives, however, I found my skepticism fading rapidly. The threat of bypass is very real but the extent to which the industry is fighting this threat had eluded me. Is this beneficial to me as a shareholder today? Without question it is. I can't quantify the benefits in dollars and cents but if customers such as McDonalds, the Commonwealth of Virginia and other state governments, Trust Company of Georgia, and too many others to mention had deserted the BOCs for private networks, the stock market would have been quick to react.

Another development that surprised me is the projected growth in shipments of ISDN lines. Dataquest estimates that, for the years 1988 through 1991, some 2 million ISDN lines may be sold. BellSouth alone expects to ship 1,500 lines between now and 1990 but, by 1990, the company anticipates its shipments will reach 20,000 lines and increase to 200,000 lines in 1992. And BellSouth's customers are already beginning to realize cost savings, as these initial lines have been priced at about 12-1/2% under the previous cost of an access line. By 1990, when BellSouth tariffs ISDN offerings, customers will save even more. In the meantime, the company has begun to realize some profits - not significant profits in terms of total earnings but the direction is right. Portfolio managers are not likely to lose sleep over developments such as these. That's not something that can be said about certain moves some of the telephone companies have made in diversifying, for example.

The goal of revenue stimulation would be furthered substantially if the BOCs were free to offer enhanced ISDN services over the core network. Ameritech estimates revenues from new offerings, including information services, at $100 to $150 million by 1990 and $750 million annually by the mid-1990s were the company free of regulatory restraint. If, as many experts believe, equal access to the local network is inevitable, it seems to me that the BOCs will be freed of these information service restrictions. Meanwhile, the trend towards relaxation of strict rate-of-return regulation appears to be accelerating.

Let's return now to my original question, i.e. what is ISDN? I think I can answer that question now. ISDN to me is not a technology, an architecture or even a concept. To me, as an investor, it is a powerful weapon the telephone companies hold, a weapon to be used in the competitive arena that is rapidly emerging.

When will the benefits of ISDN be realized? Investors are already enjoying the indirect benefits. The direct ones - substantially higher revenues and even greater increases in telephone company profits, the result of the tremendous efficiencies and cost savings associated with ISDN - are a few years down the road.

Which telephone stocks should you buy and when? I'll leave the answers to those questions to my esteemed colleagues from Wall Street.

Thank you very much.

STANDARDS POLICY — BARRIER OR ENABLER?

Dr. John S. Mayo

Network Systems
AT&T Bell Laboratories

Standards have never been more important or more needed than today. They affect vendors, service providers and customers. They affect a nation's global competitiveness. And they affect the prosperity of the world economy itself in profound ways. Why are they so necessary? The simple answer is "complexity" — the dramatically growing complexity of our technologies, the dramatically growing complexity of user needs, the dramatically growing complexity of a world economy driven by the rising expectations of more than five billion people.

Overall, the standards issue comes close to what Ben Franklin once described in another context: if we do not find a way to hang together, then we shall hang separately. That, certainly, is what standards mean to telecommunications. Hanging together. Without standards, and lots of them, it is inconceivable that we would have a national network through which a caller can immediately reach more than 180 million other telephones in the U.S. and more than 300 million others outside the country. That situation, of course, is not nearly so good with respect to data communications, which is why our industry has been working on the international set of standards known as the Integrated Services Digital Network (ISDN).

ISDN will enable voice and data to hang together, as well as the multitude of incompatible data networks that exist today. By providing a means for communications and computer systems from many vendors to play as a team, ISDN promises a vast range of efficiencies and other benefits to users. In fact, with the first ISDN field applications in Illinois and Arizona this year, those gains are already on the way. ISDN trials have been scheduled for all other major local exchange holding companies, and there are plans to introduce ISDN in nineteen countries in 1989. So my immediate answer to the question posed at this session is, "Yes!" Our standards policy is an enabler. ISDN is becoming a reality, and our standards policies have enabled it to happen.

But, of course, our work is not finished. As we go forward, we want to improve the standards process. Whatever we do, I think, we must focus on these fundamental requirements:

- Order, not chaos. Standards must be reasonably easy to use and apply. And in our industry, they must promote interconnectability, compatibility, and open networking of communications and computer systems from multiple vendors.

- Encouragement for innovation. Standards must encourage technological innovation by leaving ample room for the application of innovative technology to meet the needs of the marketplace.

- Economic and technical drivers. Politics, while inevitable and necessary, must not drive the process.

- Timeliness. Standards must not come so early as to stifle innovation, nor so late as to be irrelevant or to necessitate drastic re-design of existing products.

- Voluntary consensus among all stakeholders. Imposing standards, through whatever agency, poses the serious danger of strengthening one party at the expense of others, and so depriving users of benefits that flow from healthy competition.

Some of these requirements are more obvious than others. Rather than explaining each in turn, let me look a little more closely at why we want standards, and at the myths that may hinder us from achieving good ones.

STANDARDS GOALS

It bears repeating that the broad goal of standards policy for our industry should be connectivity, compatibility, and open networking of communications and computer systems from multiple vendors — whether in common carrier public networks, in private networks, or on the users' premises. Users demand and deserve nothing less.

We at AT&T are unequivocally committed to this goal, and to ISDN as an essential, major step toward achieving it. AT&T gives full support to national and international standards bodies working toward this same goal, including the CCITT, the ISO, T1 and a number of other organizations. What's more, the logical structure of AT&T's Systems Architecture is based on the internationally defined framework called the Open Systems Interconnection (OSI) Reference Model on which ISDN is based.

Now there should be nothing terribly controversial about the goal of open networking. In fact, we've seen broad support and commitment to it throughout the computing and telecommunications industries. Examples since the AT&T divestiture include the creation of the Exchange Carriers Standards Association and Committee T1.

Industry is also broadly committed both to the goal of open systems, and to voluntary consensus as the means for setting and implementing standards. One of the best proofs is that many companies and end-user groups, including AT&T, have formed the Corporation for Open Systems (COS) — which now boasts more than 60 members. The purpose of COS Is to accelerate the introduction of interoperable, multivendor products — based on ISDN and the OSI model.

As I've said, the standards-setting process in telecommunications has been in my view, on balance, a success. It vindicates the voluntary, consensual approach, even though for the volunteers themselves, that process is often slow, tedious and frustrating.

STANDARDS MYTHS

Which brings me to the question of what may or may not be wrong with the standards setting process. In part we hear complaints that stem from believing in several myths.

One myth is that standards, like diamonds, are forever. But the fact is that as standards are defined and implemented, we almost always see the evolution of user demands for new features beyond the scope of the current standards — particularly as users work with, and better understand, the new technical capabilities at their disposal. Concurrently, we also see the continuing advance of information technology to provide new capabilities.

For example, we have already defined the two ISDN interfaces for standard network access — the Basic Rate Interface and the Primary Rate Interface. But users are already requesting, and technical work is proceeding, on advanced ISDN capabilities, such as "broadband" ISDN and "packet mode" ISDN. The key point is that standards cannot be static; they must evolve, and applicable products must also evolve, gracefully and economically. To expect otherwise is to endure certain disappointment in the marketplace.

Naturally, the standards process may not keep up with the market, imposing costs on the vendor, or user, that may seem wasteful. My own company's position, is that where market demand is strong but standards are not yet final, as may be the case with ISDN, we will build products according to our expectation of the evolving standards. Future design changes, or "upgrades", will be made, if needed, to conform to the final standards. After all, the whole process is ultimately market driven, enabled by advancing information technology.

Another, more insidious myth is that when open standards are applied, they reduce the associated product to a mere commodity, diminishing strategic competitive advantage or incentives for innovation to gain such advantage. The truth is that appropriately defined standards can advance the goal of interoperability while leaving more than enough room for innovation.

For example, ISDN specifies globally uniform interfaces to public common carrier networks. But ISDN also empowers equipment vendors, network providers, and end users to partner in defining the ISDN services behind those interfaces — services that are limited in variety only by imagination and creativity. In short, user demand for open standards is bringing new benefits to all players, though also making competition a lot tougher.

Still, the trick and the key is timing! We do not want standards until after industry has defined the technical concepts. If standards come too early, they may needlessly restrict the options for innovation, or fall quick victim to important innovations that rise to challenge them. On the other hand, if standards come too late, non-interworking products may have proliferated to the point of creating a no-win situation, where there are heavy economic penalties to pay whether vendors and users adapt to the new standards or not. Delays may also encourage de facto standards to emerge, standards set by a major player and tending to skew the competition heavily in the player's favor. Hence, good timing in the standards process demands great care about two factors: first, the starting point, and then, once started, the interval to completion.

STANDARDS DELAY: CAUSES AND CURES

Of course, the most commonly heard complaint about the standards process in our industries, including ISDN standards, is not that the process is too fast but that it is too slow. We all know the reasons why it is slow. One cause is the peer-level process itself, especially in deliberations about ISDN. Interest in ISDN has brought new volunteers to the standards bodies. Even more people have come as a result of the decision to include customer equipment as well as network equipment. So we have more proposals to consider, and we take much more time to reach consensus.

Finally, there is always a tendency, if left alone, for the volunteers in the standards bodies to volunteer too much, that is, to broaden their work to set standards that are not really necessary or desirable.

We can't eliminate all these causes for delay. Clearly, all stakeholders have a right, and almost an obligation, to participate. We cannot and should not expect a smaller number of players. Stakeholders can facilitate the process by continuing to send people with strong technical backgrounds, people equipped to address the issues on their technical and economic merits, and in the spirit of compromise.

But even so we cannot escape from the fact that with many stakeholders will come many agendas. Despite many pronouncements about the desirability of global ISDN standards, some vendors have had obvious economic reasons to resist some aspects of the standards. In other cases, domestic organizations have redebated published international standards. Perhaps they wanted to give domestic vendors a marketing advantage; perhaps they decided that a domestic version would fulfill a regional need more effectively. Perhaps their reasons have just been emotional.

Still, if we can't eliminate all these causes for delay, I do see ways to reduce them. For one thing, we can impose more discipline on the process by creating careful long-range timetables that extend, say, over four years. The CCITT already does this routinely. Moreover, at its June plenary meeting this year, our own Committee T1 decided to establish a computer-based tracking system to aid in the planning and scheduling of its work on standards. We need to take more steps like this.

Another useful tactic may be simply to ensure that the process takes place in broad daylight. Every company or country will pursue its interests. But the sunshine works wonders in keeping us all honest and responsive to the interests of others. Especially when as vendors, for example, we know how strongly our customers may feel about the need for a particular set of standards. Thus, we now know, if we did not before, that when we as vendors are clearly slow to act, major users will bypass the traditional standards organizations, just as General Motors and Boeing did when they partnered with vendors to develop the Manufacturing Automation Protocol (MAP) and the Technical and Office Protocol (TOP).

Management can also help the standards process by providing constructive guidance and support. It can make sure that the people it sends have time to do their homework; it can stand behind the principle of compromise and make sure that its representatives act accordingly; and, not least, it can make sure that its representatives follow the right priorities. Priorities are, of course, a shared responsibility. Vendors and users should both have a say about priorities. And so should the administrators of the standards bodies themselves, especially in watching over the day-to-day activities that normally fall outside the purview of top management.

SETTING PRIORITIES

Setting priorities has become so important that the process demands special attention. A first step, certainly, is to classify needs, with careful attention to the type of standard involved. There are at least four classes of standards. They tend to vary according to the state of the underlying technology, and they are not all equally important. The first class is architectural. It is the broadest and deserves the highest priority. Examples range from the ISDN Basic Rate and Primary Rate Interfaces to the OSI Reference Model. Standards of this type are vital enablers benefiting the whole industry. They should have first priority because they are needed first, before implementing a technology.

Standards of the first class have not kept up with rapidly growing user demands and advancing technology. For example, we still lack standards in the important new area of open distributed processing. Fortunately, the International Organization for Standardization (ISO) has begun a new modeling effort, larger in scope than the OSI Reference Model that I mentioned earlier. The new model, termed Open Distributed Processing, takes off from the top of the OSI Reference Model to model how systems interact to achieve distributed processing.

The new model will define a general end-user environment, how that is to be achieved in a programming environment, and how computer subsystems — for example, databases, terminals, and programs — interact to create such an environment. This model will be used to coordinate future standards developments, leading to an integrated set of distributed processing standards.

Again, the need to develop architectural standards, is especially critical, because failure to do so erects the most serious potential barriers to the progress of interoperability. The obvious danger is a growing embedded base of nonstandard products, perhaps with significant cross-vendor incompatibilities.

A second class of standard is technology-driven. That is, the technology typically precedes the standard. One example of technology-driven standards is the CCITT's "ADPCM" algorithm for compressing voice (from 64kb/s to 32kb/s), as in AT&T's Bit Compression Multiplexer-32,000. Packet switching standards provide another example. In the 70's, new packet switching technologies drove the industry to establish the family of X.25 standards. Now, in the 80's, powerful VLSI technologies are driving us to ISDN LAP D packet mode standards.

Major vendors have often created de facto standards of this kind simply by introducing their products, Smaller vendors have fallen into line, marketing both peripherals and compatibles that compete on the basis of feature and price. One of the most well-known examples of a de facto standard is the MS-DOS operating system, which IBM introducted for its line of personal computers. In the future, we will probably see somewhat fewer de facto standards and more use of standards set by the standards bodies for a multi-vendor environment.

In a multi-vendor environment, technology drives standards when there is a technological race among several vendors to provide solutions for user needs. To meet those needs rapidly, the vendors must run ahead of standards. At stake are big losses and prizes — ranging from large R&D expenditures to market share.

When technology drives standards, each vendor eventually submits its version of a standard to a voluntary, consensus standards body — such as Committee T1. Typically, the deliberations result in a standard that attempts to incorporate the best of each specific submission, and that is, therefore, different from each. The goal is to obtain the best possible standard, and to obtain maximum buy-in from all stakeholders by "sharing the pain" associated with conforming to the standard.

In the meantime, however, each vendor usually meets the press of user demand with an initial product based on its own technology. Importantly, the vendors who take these risks of leadership must also make a strong commitment to introduce economical upgrades as needed to make their products conform to the standard when it is defined. I've already illustrated this type of commitment in discussing my company's approach to ISDN developments.

Clearly, the overall standards process is imperfect, particularly in regard to technology-driven standards. The process takes time and creates the need to introduce interim products that must be upgraded to conform to the standards that eventually emerge. Nevertheless, efforts to speed up the process will prove counterproductive if they undermine the process itself, which succeeds through consensus based on technical and economic compromise.

Again, there is a time for everything under the sun. To put it another way, standard-setting is a matter of risk management. For example, if powerful customer organizations get impatient and try to "declare" a standard unilaterally, they may eliminate the need for interim products while standards are being considered. But they will also heighten the risks for all concerned, including themselves. (Incidentally, I am not referring here to MAP and TOP, which were created by consensus among users and vendors, even though the users were pushing hard to achieve it.)

Setting standards without consensus can easily translate into the risk of setting standards before the technology in question has matured. In that case, the standard may not hold, because solutions will soon appear that are so much better than the standard that no one can afford to ignore them. Setting standards without consensus can also translate into weaker competition for the customers' business. The reason is that when standards are declared they usually reflect the technology of a specific vendor. So one clear "winner" emerges along with several "losers", who lose in both R&D investment and market share. It is also possible that the "winner" might not have the best technology, meaning that the users could be the biggest losers of all.

More broadly, this approach would hamper U.S. global competitiveness. Because it would increase the risk of declaring standards that diverge from international directions, and it would weaken the linkages among domestic and international bodies that are vital to the global standards-setting process.

In short, these risks suggest that perhaps we cannot, and should not, look for dramatic change in the process for setting technology-driven standards. The debits of the process are a necessary tradeoff for the benefits of vendor competition and the benefits of the voluntary, consensus approach to standards-setting.

The third kind of standard properly requires less effort, even though it is important. Basic technology is not a deciding factor in setting standards of this kind — for example, standards for things like the encoding of service capabilities. Because entire technologies are not at stake, standards of this kind can be somewhat arbitrary, in the sense that any number of viable solutions are often available. The key here is to ensure that the deliberative process does not get bogged down in minor technicalities. People must not lose sight of the fact that the matter is arbitrary and that it can and should be settled quickly.

UNNECESSARY STANDARDS

The fourth class of standard is the one deserving the lowest priority, except in sorting out which standards fall in this category, because these are the standards that we really do not need at all. I have deep respect for the world's standards bodies and their hard-working members. These organizations are zealously working to fulfill an ever-expanding need for standards. To the extent that standards must lead technology, those activities amount to what is in effect a joint development process. In the case of OSI and ISDN standards, the process surely ranks as one of the biggest and most complex in history. So it is not surprising that these efforts have led an ever-expanding spiral of activity, with all the resulting delays and the need to avoid diverting our energies.

How do we recognize when standards are unnecessary? Recall that our first priority is basic interoperability. In contrast, we now see more and more efforts to standardize the innards of systems, performance levels, and many variants of the same thing. In addition, of course, different countries have insisted on different standards. In earlier times, we found it easier to accept different standards for railroad gauges, television or the North American and European channel structures for digital transmission. But as the global economy grows more and more tightly integrated, the world community will certainly view this diversity, and the wasted resources it entails, with much less patience.

My point is not that any one standards effort is necessarily wrong. My point is that we cannot create all the standards we might like and still meet our first responsibilities.

We should put our energies where they will count the most. For example, in ISDN, we should emphasize work on the few major interfaces for connectivity, and curtail work on defining every potential supplementary service. Instead of attempting to standardize all such services, we should provide enough built-in "hooks" so that vendors can innovate by means of value-added capabilities, performance and price.

If standards bodies rush in where angels fear to tread, we shall all pay a price. A price not just of greater delay but of greater bypass of the normal standards process. Add the delays of overambitious standard setting to the usual ones and customers are even more likely to get disillusioned. They will step in to meet their own needs as they have with MAP and TOP. Or the manufacturers will act to provide proprietary solutions that meet pressing user needs.

STANDARDS AND REALITIES

I have already described several myths about standards. Now let me sum up by offering a list of realities. The first is that market needs and user demands ultimately drive the whole standards process, whether we like it or not. These needs are powerful and growing. Our customers know that information technologies are strategic, that information technologies give a competitive edge, that information technologies can make or break a business. No wonder customers want us vendors and service providers to hurry up.

Another reality is that standards are best set through voluntary consensus. When users, vendors, and service providers negotiate their differences, they create a framework in which all can prosper. The process especially motivates manufacturers and vendors to compete with innovations that better serve the user.

Still another reality is the one I began with. The complexity of our world makes standards at once more difficult and more vital than ever. We have little use for standards in building a log cabin, but we do in building a World Trade Center. We don't need standards to balance our personal check books, but we certainly do when reporting corporate results to the Federal Communications Commission, the Securities and Exchange Commission or public shareholders.

To cope with such complexity, we must discipline, streamline and prioritize the standards process. Let us focus on necessities, not on luxuries, on what must be done, not on what might be nice to have. Let us not expand the scope of this joint development process to the point that it falls of its own weight.

Finally, the most powerful reality of all. Whatever our shortcomings, we have proved once again that we can make the process work. ISDN is a reality, because industry created the standards that allowed it to happen. And it is a reality with tremendous promise, a reality that at AT&T we believe will play a major role in fulfilling the vision of what we have called Universal Information Services. These will be powerful services that will let anyone move information in any form — voice, image or data — any time, anywhere. Regardless of the name you give the vision, I believe we all share it. In fact, we recognize that these services are a manifest destiny, a destiny made certain by the onrush of technological progress and user demands. The only question is, "How soon?"

A big part of the answer rests with the volunteers who struggle to make the standards process work — with us, the vendors, service providers and users. By transcending our individual, competitive interests, we can speed the day when our common vision becomes reality.

Dr. John S. Mayo is Executive Vice President, Network Systems at AT&T Bell Laboratories in Murray Hill, New Jersey. In his current position he is responsible for guiding Bell Laboratories thrust in electronics technology, transmission systems, switching systems, and operations systems and network planning. Since joining Bell Laboratories in 1955, Dr. Mayo has been Director of the Oceans Systems Laboratory, Executive Director of the Ocean Systems Division, Executive Director of the Toll Electronic Switching Division and Vice President of Electronics Technology. He assumed his present position in May 1979.

ISDN STANDARDS - AN FCC VIEW

Dr. Thomas P. Stanley

Office of Engineering and Technology
Federal Communications Commission

I. INTRODUCTION

The decade of the eighties has witnessed the appearance of a new and exciting development in communications, that of Integrated Services Digital Networks (ISDNs). It is not so much a new invention as the appropriate merging of products and technologies already available. It is by some definition (and different definitions of ISDN seem easy to come by) an inevitable development, at least technically. In the future, the word "network" will probably refer to what is currently called an ISDN. In ten years or so, it will probably be common practice to refer to the ISDN as being ubiquitous. This development appears quite natural. These future networks should provide increased consumer benefits in types and speed of service, as well as cost reduction.

The FCC has had an ongoing interest in ISDN matters quite simply because such a network will use the facilities of common carriers and the Commission is charged with the regulation of common carriers. The Commission must determine whether ISDN developments pose any threat to current common carrier policies and decisions, and what steps must be taken to assure that the FCC does not intentionally impede ISDN developments. To date industry groups and committees have taken a considerable potential regulatory burden off the agency by their apparently successful activities in the ISDN planning process.

One must be impressed with the extent of ISDN development efforts that have already taken place. We have seen the separate development of products and systems that will eventually be component parts of the ISDN. These include software control of switches (stored-program control, or SPC), digital telephony, and packet-switching. It appears that a number of ISDN services will be commercially available in the early 1990's.

II. ISDN STANDARDS ACTIVITIES

The development of standards for ISDN is an obvious requirement for the successful implementation of the facilities and services that can be offered. Development efforts are of vital concern to carriers, equipment manufacturers, users and government. Much work is being expended both internationally and domestically to devlop standards guidelines that will enhance development, avoid needless risk, and assure compatibility. The FCC's stated concern is that any standards development not frustrate its pro-competitive telecommunications policies.

The International Telegraph and Telephone Consultative Committee (CCITT) has issued a comprehensive series of recommendations in all areas of ISDN. These recommendations, known as the I-series, form the cornerstone of worldwide ISDN development. The next plenary of the CCITT will be held in 1988, at which time a new series of recommendations will be of paramount importance, both to suppliers and users. This is the case because these parties have already taken calculated risks in manufacturing and purchasing ISDN equipment according to the existing recommendations.

The U.S. participates in the CCITT process through representatives from sectors of the communications industry and government. The FCC is an active participant in this process with staff from its Common Carrier Bureau. The FCC staff attends these international meetings so that world-wide ISDN standards developments can be closely monitored to assure that these activities coincide with U.S. interests and policies. The FCC wants to insure that the U.S. position is given full consideration before any recommendations are made. At this time, the Commission believes that there is no reason for concern about the direction of international ISDN standards development. It appears that domestic policies and decisions are being well represented and being given full consideration, and that it will not be necessary for the FCC to consider intervening with any special regulatory process. Study Group XVIII of the CCITT, which is responsible for matters relating to digital networks, including ISDN, met in Hamburg, Germany, in July, 1987. Recognizing the importance of these meetings, the FCC was represented.

Domestically, the primary industry group formulating ISDN standards policy is the T1 Committee of the Exchange Carriers Standards Association (ECSA). This committee submits monthly reports to the Commission. The FCC's interest here is that the ISDN standards work conducted by this group of industry representatives stays within Common Carrier policies and decisions. The FCC staff is available to the T1 Committee in the event that guidance is required.

III. COMMISSION PROCEEDINGS

For over twenty years, the FCC has been continuously involved in the issues raised by the converging of the technologies of communications and data processing. This covergence is a direct result of the development of digital technologies. These issues and subsequent FCC actions should be taken into account when considering development of FCC policy relating to ISDN standards issues.

In the "Computer I" proceeding, which began in November, 1966, computer/communications services were placed in three categories, each of which carried its own regulatory guidelines. The first category contained "data processing" services which, being highly competitive, were not to be regulated. "Communications" services, the second group of services, were to be regulated as common carrier offerings under Title II of the Communications Act. The final group of services was defined as "hybrid" services. "Hybrid" services were described as offerings combining "remote access data processing and message-switching to form a single integrated service". These were to be regulated based on Commission determinations as to which function, either "data processing" or "communications", predominated in the particular service. Under Computer 1, common carriers, other than the American Telephone & Telegraph Company (AT&T) and its Bell Operating Companies (BOCs), were permitted to offer data processing services only through separate subsidiary companies. This "structural separation" was designed to prevent anti-competitive behavior. The Commission assumed that AT&T and the BOCs could not offer any data processing services under the terms of the 1956 antitrust consent decree. One can perceive that these rulings might have had an important bearing on the future development of ISDN. Some ISDN functions, such as stored program control, message switching and signalling are consistent with the description of data processing. However, ISDN technology as known today would have had significant difficulty developing under the Computer 1 guidelines. Some ISDN services would have fallen under the "hybrid" category of service and would have required separate FCC determinations as to their treatment.

In the mid seventies, it was becoming obvious that data processing was expanding to a much wider range of applications. At that time, the Commission perceived three general categories into which most of these data processing applications could be placed. These were arithemtic processing, word processing and process control. However, the Commission also noted two categories of computer use by carriers which did not conform to the definition of data processing. They were network control and routing and input/output processing, which were made possible by advances in computer technology that permitted distributed processing. These more varied applications were making it increasingly difficult and burdensome for the FCC to determine just what was "communications" and what was "data processing" and, therefore, to apply the regulatory guidelines of Computer 1. As a result, in August, 1976, the Commission issued a Notice of Inquiry and Proposed Rule Making for the Second Computer Inquiry (Computer 11).

In the Final Decision in Computer II, released in May, 1980, the Commission concluded that it was no longer possible to draw an enduring line of demarcation between the service offerings of the technologies of communications and data processing. A new regulatory framework was established based on whether a service was judged to be "basic" or "enhanced". Under this scheme, only basic services were to be subject to Title II regulations under the Communications Act while enhanced services were to be deregulated. A basic service was defined as ". . .a pure transmission capability over a communications path that is virtually transparent in terms of its interaction with customer supplied information". Enhanced services were those ". . .offered over common carrier transmission facilities used in interstate communications, which employ computer processing applications that act on the format, content, code, protocol or similar aspects of the subscriber's transmitted information; provide the subscriber additional, different, or restructured information; or, involve subscriber interaction with stored information." The Commission concluded that the public interest would not be served by attempting to distinguish enhanced services as either communications or data processing and regulate accordingly. Based on the changing technological environment, it would be impossible to define a clear and stable point at which communications becomes data processing. By foregoing regulation of enhanced services, the Commission provided carriers with flexibility to provide new and innovative services and to tailor services to individual user needs. Carriers owning underlying transmission facilities and providing enhanced services were to unbundle their basic and enhanced services, with only the basic services being subject to tariff conditions. This would assure that basic service providers which also offer enhanced services would not gain unfair advantage over enhanced service providers which did not own underlying transmission facilities. Under Computer II, AT&T and the BOCs were made subject to the structural separation requirement for the offering of enhanced services, while the requirement was removed from other underlying concerns.

This pro-competitive policy was designed to guide the nation's telecommunications structure as it evolves in the new technological era and directly affects the emergence of ISDNs in the U.S. The reliance on a multitude of service providers subject to varying degrees of regulation for the provision of basic and enhanced services requires that ISDNs for the U.S. be designed to accommodate multiple service providers. This must be reconciled with the objective of ISDN planners to create uniform standards that would permit national ISDNs to interconnect and form a worldwide system that would promote ubiquity of terminal equipment.

In August, 1983, the Commission issued a Notice of Inquiry in the matter of Integrated Services Digital Networks. Its purpose was to gather background on ISDN developments to date, to examine related telecommunications policy, and to discuss various issues raised by the potential implementation of ISDNs. Comment was sought to determine the appropriate role for the FCC to play in the ongoing ISDN standards development effort and to determine how the Commission could best continue to promote its pro-competitive policies during the ISDN development period. The Commission's attitude toward ISDN standards development was expressed, as follows: ". . .we are cognizant that international and domestic standardization efforts involving the U.S. government, telecommunications providers, large users, and foreign entities are presently underway. We do not intend that this proceeding hinder or conflict with these ongoing efforts, rather, this NOI seeks to complement existing efforts. In addition, the NOI will serve as another vehicle to apprise the Commission and the public of concerns which U.S. service providers, equipment manufacturers, users and the public may have about implementation of ISDN networks."

As a result of the inquiry, the Commission released a First Report in April, 1984. The Commission affirmed that its Computer II rules and policies were established and properly should be accommodated in ISDN planning efforts. It noted that the CCITT draft recommendation for categorizing ISDN services accommodated the basic/enhanced service dichotomy of the United States. In the Report the Commission concluded that, ". . .it is important that the overall ISDN concepts which evolve appropriately accommodate the Computer II policies, that a multiple-vendor environment in the United States not be foreclosed through technical recommendations or concepts directed toward a single-vendor environment, and more generally, that sufficient flexibility be built into the ISDN concepts to accommodate the broad range of different market structures which exist in various nations." The Commission went on to say that, "We also have a potential regulatory interest since, under the Communications Act of 1934 as amended, FCC certification may at some point be required that the public convenience and necessity will be served by the construction and acquisition of many facilities by facilities-based carriers to implement the ISDN concept." The FCC affirmed its role in the planning process as one not of making decisions on the technical characteristics of facilities and services, i.e., designing ISDNs, but of identifying policy issues as they arise and ensuring that they are adequately addressed in the technical and other judgments to be made. It concluded that the advisory committee processes of the Department of State and the processes of voluntary standards organizations such as the ECSA/T-1 Committee provide an opportunity for the policy implications of largely technical ISDN planning judgments to be addressed, and that through informal participation in these processes by the FCC, the policy issues can appropriately be resolved. The Commission rejected the option of subjecting ISDN planning to formal rulemaking; however, the ISDN docket continues to remain open to serve as an information forum.

Based on experience gained from the application of the Computer II decision, on regulatory modifications resulting therefrom, and in recognition of the vastly different telecommunications industry existing at the time from that during Computer II, the Commission, in July, 1985 adopted a Notice of Proposed Rulemaking for the Third Computer Inquiry (Computer III). The Commission noted that there was increased competitive presence in the telecommunications marketplace after the divestiture of AT&T and with the rise of other parties, notably IBM. It tentatively concluded that regulatory protections short of structural separations could control potential anti-competitive behavior. In addition, the Commission noted that the structural separation requirements appeared to be denying the public the benefits of enhanced services that could not be efficiently offered unless they were integrated into the basic network. Another motivation for the Notice was the frequency with which the FCC was being called upon to clarify, interpret and apply the "enhanced services" definition. The Commission proposed a new regulatory framework that would permit carriers to offer both basic and enhanced services on an unseparated basis. This was to be accomplished by requiring the carriers to provide interconnection opportunities to competitors on an equal access basis, a concept called "comparably efficient interconnection" (CEI). In the proceeding, the Commission asked for comment on how these proposals might impact ISDN designs currently being developed.

In the Computer III Order, released June, 1986, the FCC established a new regulatory framework to govern the participation of the AT&T and BOCs in the enhanced services marketplace. This new framework replaced the structural separation requirements that were established in the Computer II proceedings with a system of nonstructural safeguards. In the Order, the Commission established conditions under which AT&T and the BOCs could offer a specific enhanced service free of the structural separation regulations. These requirements contemplated the filing of a plan to demonstrate CEI, which would be reviewed and approved by the FCC prior to a carrier offering of the specific enhanced service. The Commission stressed that carriers using ISDN architectures must conform to the CEI principles, but noted that the record indicated that there would be no unmanageable CEI problems associated with an ISDN architecture.

Parties to the Computer III proceedings had explored the possibilities of applying the CEI principle to a more expansive approach that would open the basic network to efficient access by enhanced service providers. In the Order, the Commission built on this to introduce an Open Network Architecture (ONA) guideline. This guideline extended to enhanced services generally the CEI goals of preventing discrimination and promoting efficiency; it permitted AT&T and the BOCs to offer such services on an unseparated basis without the need for the prior approval of service-specific CEI plans. Basically, the Commission concluded that a technological implementation of its CEI requirements could replace service-by-service regulation. Whereas networks as currently designed may require a number of quite different CEI implementations for different services, the Commission envisioned basic networks designed to satisfy ONA requirements that would be self-enforcing in controlling discrimination. These open networks would permit enhanced service providers to obtain access to the basic network equivalent to the access provided by carriers to their own integrated enhanced services operations. In the Order, the Commission declined to adopt a specific network architecture, but instead set forth general requirements that ONAs must satisfy, including provision of unbundled basic service elements to others on a tariffed basis and a technical design satisfying the CEI requirements. It left to the private sector the development of specific network designs.

Under this flexible approach to network specification, the FCC is leaving to private standards organizations, such as the T1 Committee, a major role in resolving relevant standards issues. The Commission is watching the ONA development process very carefully and retains the responsibility for reviewing and approving specific ONA plans after public comment. Each BOC and AT&T must submit a comprehensive ONA plan to the Commission by February 1, 1988. While yet unproven, this approach should provide incentives and marketplace flexibility for competitive ISDN development. By permitting equal and open access to the many and varied ISDN related vendors, competition should be enhanced and development of new and innovative services encouraged. In an ONA atmosphere, it is reasonable to expect the emergence of a viable integrated network of diversified services.

During the last two decades the Commission has continued to explicitly and implicitly recognize that common carrier communications technology is moving in the direction of a multiple service, integrated system based largely on digital technology. Two Commission policies became increasingly clear when studying the various FCC proceedings and decisions in this area. First is the Commission's insistence on an open access, pro-competitive environment. Second is the Commission's preference for a limited role in the implementation of the new technology and the standards for it.

IV. STANDARDS POLICY

ISDNs, virtually by definition, require standards to ensure proper interfacing with each other and with a variety of interconnecting equipment. The Commission has signalled an intention not to hinder or conflict with, but rather to complement, ongoing industry efforts to arrive at acceptable standards. The Commission does not intend to usurp the role which the service providers should play in the setting of standards. It believes that everyone's objectives can best be served with the FCC in an 'observer' role, permitting the interaction of users, carriers and equipment manufacturers to develop the direction for the most feasible, efficient, and practical standards.

The Commission recognizes that, internationally, the orientation is somewhat different. The monopolistic structure of telecommunications industries in many countries forces the Commission to assume a more active role to assure that the pro-competitive policies in the U.S. are accommodated in international deliberations. While the FCC seeks to have representative U.S interests put forth in the CCITT activities, obvious attention must be paid to national interests as they are developed through open dialogue. While the Commission will continue to work towards the adoption of mutually acceptable standards, one of its primary objectives internationally is to ensure that regulations are not adopted which would inhibit in any way the implementation of multiple, competitive ISDN networks in the United States.

A major outstanding issue arising from the Computer III proceeding is the requirement for carriers to comply with the ONA requirements. While this is a means of overcoming inefficient structural separation requirements, it nevertheless imposes a need for vigilance by the Commission to monitor compliance with the ONA requirements to prevent cross-subsidization and discrimination. The Commission will monitor the entire ONA process and will, if necessary, engage in rulemaking to overcome any abuses or deficiencies.

In conclusion, the Commission strongly believes that standards are vital to establish viable telecommunications networks. Hopefullly, the standards development process should serve as an inducement to innovation and efficiency. To meet this objective, ISDN standards are best and most efficiently set in an open marketplace manner, by consensus and with minimal government interference. At present, there do not appear to be any ISDN standards issues which would benefit from direct FCC intervention.

Dr. Stanley has served as Chief Engineer of the FCC's Office of Engineering and Technology since February 1986. He has served in a number of other postions in the Office of Engineering and Technology (formerly the Office of Science and Technology) since joining the FCC in 1981. Those positions included Acting Chief Scientist, Deputy Chief Scientist for Operations, and Chief of the OST, Technical Planning Staff.

From October 1982 until June 1983, Dr. Stanley served the FCC as Assistant for National Security Telecommunications, to assist the designated Defense Commissioner and the Managing Director. Prior to his work with the FCC, he was also affiliated with the Institute for Defense Analyses, the U.S. Army Signal Corps, and Bell Telephone Laboratories.

Dr. Stanley received his Ph.D and MA in Electrical Engineering from Princeton University and his BES in Electrical Engineering from Johns Hopkins University.

ISDN Standards Policy: A Legislative Perspective

F. Matos

U.S. House of Representatives
Legislative Staff

Paper will be provided at the conference.

INSTITUTIONAL OBSTACLES TO ISDN DEVELOPMENT

Robert J. Keegan

Massachusetts Department of Public Utilities

This presentation raises concerns about the ability of the local telephone operating companies to be sufficiently competitive to develop the resources necessary to grow the public switched network into an ISDN. Among the obstacles they must overcome are substantial depreciation underaccruals, existing pricing anomalies for their services and the line of business restrictions contained in the MFJ.

1. INTRODUCTION

The integration of digital technology into the public swithced network holds the promise of bringing a wide array of new services to the general public. The opportunities for increasing the intelligence of the public switched network and fully developing the services which end-to-end wide-band digital communications systems offer will also provide benefits in terms of opening new markets for our economy and improving our competitive position in international markets.

The development of an integrated services digital network ("ISDN") will provide the capability for a single terminal to simultaneously transmit data, conduct a voice conversation and receive information on an incoming call. The potential for the full integration of voice, circuit-switched data, packet-switched data and telemetry at the same network interface and over a single access line offers the opportunity to develop services which have yet to be defined. But these potential benefits will not be fully realized and will not penetrate as deeply into our society as desired unless ISDN technology is ubiquitously integrated into the public switched network. I therefore see as important unresolved questions the role which the Bell operating companies will play in the evolving Information Age and whether that role will provide them with the resources necessary to make ISDN universally available.

The Bell operating companies are in a position to be a major force in the telecommunications market of the Information Age. Technological developments are creating a demand for a greater array of telecommunications services which can be provided through the innovative use and upgrading of existing local networks. The potential for an open network architecture, with private vendors accessing the central office with a comparably efficient interface, provides the local operating companies with the opportunity to play a central role in the growing markets of transporting and processing information. But these new and exciting opportunities which technology presents for the local operating companies do not come without significant risk: The risk that the local operating companies may not be in a position to meet the growing competitive challenges which they confront. Their inability to meet these challenges is an issue of significant concern to regulators and others, because it will affect the architecture of tomorrow's public switched network and determine the degree to which the general public will have the opportunity to participate in the benefits of the Information Age. I will therefore focus in this paper on what I believe are the most important factors limiting the ability of the local operating companies to participate in these developing markets.

I will first discuss certain characteristics of the predivestiture telephone industry which I believe continue to affect the local operating companies in a way that threatens their ability to succeed in the post-divestiture environment. I will indicate what I believe are the major challenges currently confronting them, and give a regulator's view of what must be done to ensure that the local operating companies have the opportunity to be sufficiently competitive to ensure that ISDN benefits become universally available to the general public.

2. HISTORICAL PERSPECTIVE

Since the Communications Act of 1934 was signed into law, the goal of universal service has been the bedrock for most public policy decisions in the area of telecommunications. Concerns about universal service led to the consolidation of most segments of the telephone industry into a highly integrated service, provided by a single monopoly entity, subject to pervasive cost of service regulation on both the intra- and inter-state levels. In that environment, where affordable rates were viewed as the essential ingredient for achieving universal service, capital recovery through depreciation rates was extended for periods as long as could be justified. The pace at which technological innovations were integrated into the network could be controlled, and capital recovery periods discouraged that pace from being as rapid as was technologically feasible. Market segmentation and product differentiation were used as tools to help the monopoly provider maintain affordable local rates while continuing to meet its revenue requirement. Prices were established on the unquantified concept of the value of a service as opposed to its cost.

These policies worked reasonably well in the predivestiture market and were largely responsible for the industry's success in achieving what was viewed as the paramount goal of universal service. At the same time, the industry succeeded in developing the most sophisticated and reliable telecommunications network in the world, with an emphasis on quality and service beyond the reach of most competitive industries.

3. TECHNOLOGY AND COMPETITION

If the success we experienced in the past was so pervasive, why then are we now in an environment where the network has become less integrated and pressure on local exchange rates threatens to make service less than universal? The answer, in a word, is "technology". The technology of information systems has led to significant customer demands on the network for services which were not required for voice-grade communications. Technology has also provided larger consumers with the means to develop their own communications systems, should they determine that their needs will not be met by the public switched network. And, finally, advances in technology have diminished both physical and economic barriers which had previously precluded the possibility of effective competition with the services provided by the existing network. In short, technology has made competition a reality, and in doing so has altered the standard used by the industry to measure success by focusing away from universal service concerns in favor of the bottom-line profit figures of a competitive industry.

In general, I view these technological advances and their potential for introducing competition into segments of the once-monopoly market as a positive development. Pervasive economic regulation is an inefficient surrogate for competitive market forces, and the possibility of substituting competition for such regulation holds great promise. Indeed, the Massachusetts Department of Public Utilities expressed such a view in its recent decision to allow competition in the intra-LATA market beginning in December 1986. In that case, which was docket number D.P.U. 1731, we stated the following:

> ...Competition...will provide incentives to competing carriers to minimize costs and optimize the efficient use of resources.... [It] will provide a market environment where the introduction and exploitation of technological innovations will be encouraged. In many market segments, competition

will promote a greater sensitivity to specific customer demands and result in an increase in the diversity of service offerings and options available to telecommunications users.

But the challenge of competition also represents a significant change, affecting the entire structure of the industry, and it requires substantial modifications from the way business and regulation have previously been conducted. The recent developments in technology have created an increased demand for an ever-greater array of telecommunications services, and the market has become further segmented, as customers require products and services customized more and more to their individualized needs. Product differentiation has become a tool for competitors to carve a niche for their services in one or more of the growing number of segments of the market.

The needs of the public are becoming more and more diverse and must be viewed in terms of the individualized way in which customers wish to use the network and the market segment that that use places them in. The vast majority of subscribers still want only high-quality voice-grade communications from the public switched network, and also, I might note, efficiency and understanding from local operating companies in dealing with problems relating to billing disputes and new service connections. But the needs of the public are also likely to grow as further technological innovations penetrate more deeply into the day-to-day lives of our average citizens.

Therefore, I think, the critical question becomes whether the public switched network will develop in a way which will allow the general public to participate in the benefits of the Information Age, or whether those benefits will be available only to individuals who can afford to interconnect with one or more highly specialized but limited-access private networks. If new technologies are not incorporated into the public switched network, it will take a much greater period of time for services to be available to the low end of the market where the general public resides, and it will also limit the ability of the markets that do develop to access the general public efficiently.

4. OBSTACLES TO ISDN

The competitive threat to the role which the Bell companies will play in the new telecommunications market and which also threatens to impede the development of an ISDN on the local operating company level comes from two sectors:

1. From private network providers and other common carriers, particularly AT&T, for originating and terminating access services (bypass) as well as toll service between densely populated traffic-intensive load centers (cream-skimming); and

2. From terminal equipment providers who will be marketing customer equipment as a substitute for services which could be more efficiently built into the network.

In meeting this competition, local operating companies face two kinds of constraints which threaten their ability to succeed in developing a network incorporating the technology needed to deliver the ever-growing number of services required in the Information Age.

4.1 Predivestiture Obstacles

Constraints in the first category are imposed by conditions which grew out of predivestiture regulatory and business practices and which continue to affect the way local operating companies do business. The second type of constraints arise from the restrictions imposed by the Modified Final Judgment on the activities of the local operating companies. I will briefly indicate what these constraints are and outline the regulatory actions which have been taken in Massachusetts in an attempt to ameliorate their impact.

4.1.1 Depreciation Underaccruals

First, with respect to predivestiture regulatory and business practices, past depreciation rates which failed to account adequately for technological obsolescence have resulted in a significant capital recovery underaccrual. The ability to collect that underaccrual in a competitive environment becomes increasingly limited as prices are driven closer to cost by competitive forces. The local operating companies cannot deal with this situation as they have in the past, because the continued expectation that prices can reflect this underaccrual will seriously limit the pricing flexibility needed to compete in important sectors of this market.

4.1.2 Pricing Anomalies

Second, the past practice of value-of-service pricing has produced a situation where the underlying cost structure of the network and the industry is inadequately understood by both regulators and industry personnel. Existing service classifications and prices simply do not reflect the factors which impose costs on the network. In order for the local operating companies to compete effectively in those sectors of the industry which are becoming competitive, their services and prices must be restructured in a way which more accurately reflects cost causation. Only then will they be in a position to identify, from a competitive point of view, what the strengths and weaknesses of the existing network are, so that effective marketing strategies can be developed which will enable them to retain a sufficient share of the high end of the market to support the continued mordernization of the public switched network.

Other industries provide examples of similar circumstances for other corporate entities in the past. The way those entities responded to increased competition is instructive.

During the infancy of the photocopying industry, Xerox established a pricing policy of small monthly fixed charges for the rental of the copying machine and fairly heavy per-copy usage charges. That pricing structure simply did not reflect the high fixed-cost and low variable-cost nature of its operations. In response to competition, Xerox was forced to begin reflecting its underlying costs more accurately in its price structure, to the point where usage charges were significantly reduced and even eliminated in favor of increased fixed charges.

In the auto rental business, pricing policies underwent a similar transformation as competition entered the market. Initially, prices in the industry were characterized by low daily rental fees and significant per-mile charges. The high fixed-cost nature of that industry has produced a price structure of higher daily rental fees and virtually no usage charges.

Competition drove prices toward costs in both the photocopying and automobile rental businesses, and it will do the same in telecommunications. It is critical, therefore, that local operating companies quickly arrive at a better understanding of the underlying costs of their systems so that they can implement pricing decisions which increase their chances for success.

4.1.3 The Massachusetts Response

On the issue of depreciation underaccruals, the Massachusetts Department of Public Utilities arrived at an agreement with New England Telephone Company in the 1984 FCC depreciation represcription meetings. That agreement provided for an increase in NET's annual Massachusetts depreciation expense of approximately $63 million. This was almost double the amount being recommended by the FCC staff and was an effort to begin to deal with a significant ongoing problem.

In addition, the Massachusetts Public Utilities Commission undertook a generic investigation into the issues of whether intra-LATA competition should be authorized in the Commonwealth and what the appropriate degree of regulation should be in both the inter- and intra-LATA markets. In that case we decided to authorize intra-LATA competition beginning on December 1, 1986, because we recognized that, regardless of our decision, technology was making competition a reality. The issue became whether to limit the development of competition to the implementation of bypass technologies or to allow competition to be developed by authorizing other common carriers to enter the market. We decided on the latter course because it created an environment for promoting greater efficiency in the industry.

In ariving at this decision, however, we recognized that there were a number of unresolved issues and several potential problems.

1. Not all segments of the telecommunications market would become competitive at the same time or pace, and, indeed, some segments of the market may not become competitive at all. It was therefore necessary to develop a regulatory framework which would allow for increased regulatory

forbearance in those markets which become increasingly competitive while at the same time limiting the potential for the local operating company to cross-subsidize its competitive activities through unwarranted price increases to its monopoly customers.

2. It was also necessary to allocate depreciation underaccruals and the cost of technological obsolescence between services so that competitive markets would bear their fair portion of those costs along with monopoly customers.

3. And, finally, it was recognized that the local operating company, if it was to have a chance of succeeding in competitive markets and of precluding uneconomic bypass, would have to restructure its service offerings and prices in a way which would better reflect its underlying costs.

To address these concerns the Department:

1. Established a dominant-nondominant framework for regulating companies, with virtually no cost of service or price regulation for nondominant carriers and the exercise of regulatory forbearance for dominant carriers for those services offered in a market which was demonstrated to be sufficiently competitive, e.g., facilities-based Centrex pricing; and

2. delayed the start of intra-LATA competition for approximately one year and required New England Telephone Company to file fully embedded cost of service studies and incremental cost studies to support any proposals to restructure its prices and service categories.

The fully embedded cost studies have been reviewed and the Department issued an Order on July 31, 1987, indicating its approval of certain cost allocation method and establishing the framework for determining appropriate rate classes and services. The Department will next examine the company's marginal cost studies and establish rates which send more accurate price signals for the company's services to the market. The desired result will be a more competitive price structure for the dominant carrier in the intra-LATA market and a cost allocation structure which will facilitate costing and pricing decisions necessary to provide nondiscriminatory access to bottleneck facilities in the open network architecture environment envisioned in the FCC's Computer III decision.

4.2 MFJ Restrictions

The second source of constraints on the ability of the local operating companies to meet the growing competitive challenge threatening their ability to develop a ubiquitous ISDN is the Modified Final Judgment, which contains conditions and restrictions constituting the ground rules for divestiture.
Under these ground rules, the local operating companies cannot offer customer equipment except through a separate subsidiary; they cannot manufacture equipment; they cannot provide important information services; and they cannot provide inter-LATA or interstate service.

AT&T and the OCC's are not similarly constrained by these restrictions. On the interstate level, although our experience with competition to date is limited, we are beginning to see evidence emerging which indicates that AT&T is a powerful competitive force in the market. Its existing worldwide communications system, and its ability to manufacture equipment and provide information services, give it a significant competitive advantage. Indeed, it is the only entity which currently has the ability and is authorized to provide worldwide, end-to-end telecommunications services. It appears that AT&T not only benefits from economies of scale associated with its existing network but also enjoys economies of scope across business lines which allow it to take advantage of the synergy of an integrated system.

The local operating companies, on the other hand, are denied similar synergistic benefits by the restrictions contained in the Modified Final Judgment. In addition, I believe it is becoming clearer that the promise of competition in many segments of this industry can be realized only if the local operating companies are able to enter the market free from the artificial barriers which now limit their activities.

The core prohibitions on the activities of the local operating companies, however, were premised on legitimate concerns over the existence of bottleneck facilities and the potential for both anticompetitive behavior and cross-subsidization. It is likely that the court will require that its initial concerns be sufficiently addressed before it seriously contemplates removing prohibitions against manufacturing equipment and providing interexchange services. It is necessary, therefore, that the industry and regulators move aggressively to implement equal access requirements and establish the framework required to minimize the potential for using bottleneck facilities in an anticompetitive manner and for cross-subsidizing competitive services.

The structure which appears to offer the potential for alleviating many of these concerns involves requirements that a comparably efficient interface be available through an open network architecture and the implementation of adequate cost allocation procedures. The sooner the local operating companies and regulators place the industry in a position of being able to provide these competitive safeguards, the sooner it is likely that the onerous lines of business restrictions will begin to be removed. Only then will we be in a position to determine the telecommunications business segments that are truly competitive versus those which continue to exhibit monopoly characteristics of scale economies and barriers to entry. Such a determination is essential if we as a nation are going to develop a telecommunications system which reaches its full potential. Whether that system can be developed in a fully competitive environment, which is the preferable course, or whether the underlying cost structure continues to require regulation of many industry segments are questions of critical importance which must be determined as expeditiously as possible.

5. CONCLUSION

The stakes involved in this challenge are significant. The ability of the local operating companies to compete will determine what the architecture of the network will be in the future. If private vendors are successful in continuing to segment the market in a manner detrimental to the development of the switched network, the general public may well lose the major benefits which technology and competition were expected to provide. If the local operating companies are unable to retain a significant share of that segment of the market demanding digital services, then the composition of their customer base will not justify upgrading the network, and when the opportunity comes to enter other lines of business, they will be severely handicapped.

The first step in maintaining that market share is to gain a thorough understanding of the underlying cost structure of the business, and to establish service categories, prices and marketing strategies accordingly. I believe that the overall goals of the local operating companies and state regulators responsible for protecting the public interest coincide to an unprecedented degree in the current environment. But regulators also have a responsibility to protect the rates of those customers who look to the public switched network solely for voice-grade communications. It is, therefore, our obligation to ensure that the risks associated with investments designed to serve more competitive segments of the market are not placed on that segment of the market which continues to exhibit monopoly characteristics. Consequently, the success of the local operating companies in retaining market share will determine how the network develops in the future: whether it becomes a critical element in delivering to the general public the full array of services which an ISDN promises, or whether that potential will fade into plain old telephone service and an antiquated public switched network.

COMPUTER IV: Regulating the National Public Information Fabric

A. M. Rutkowski

Telecommunications Magazine
Norwood, MA 02062
Massachusetts Institute of Technology
Cambridge, MA 02139

A few years from now, after attempting to administer the tenets of *Computer III* and dealing with the intrinsic deficiencies of its outdated conceptual model, the FCC could initiate a *Computer IV* proceeding in which the Commission would develop a regulatory framework for a new conceptual model, a "national public information fabric." Necessarily, the proceeding would confront technological reality - that public telecommunication networks are nothing less than distributed computer systems. Such a framework would likely rest on physical and logical implementations of open network architectures applied to four kinds of public entities: 1) facilities, 2) low level transport capabilities, 3) primitives, and 4) basic teleservices. Preemptive federal regulation of these entities is discussed, which includes a modified Comparatively Efficient Interconnection concept for concurrent access to physical and logical resources by all parties, at well-defined, stable interfaces, while allowing the provision of almost any other information capabilities or services by any party on an unregulated basis.

Introduction

Just as Star Trek IV followed III, so too is *Computer IV* destined to follow its predecessor. Radical changes are now occurring to telecommunication networks. Processing/switching and transmission technologies and costs have changed so dramatically and the functions merged so completely that most contemporary telecommunication networks, public and private, are now simply geographically distributed computer systems optimized for information transport.

Within professional network planning forums, domestically and internationally, these changes have manifested themselves in the form of an emerging aggregate environment of interoperating public and private ISDNs, dedicated networks, disparate terminals/CPE, signalling and OA&M systems, and applications. This aggregate is hereafter referred to as the "national public information fabric."

Traditional regulatory concepts and approaches underlying the FCC's *Computer* trilogy, and ancillary proceedings like the ISDN inquiry, are inadequate in this environment.

This paper explores the nature of the transformation taking place and seeks to infuse new concepts into the ongoing regulatory dialogue. It is divided into three parts. Some of the attributes of today's regulatory regime are discussed as a prelude to a *Computer IV* proceeding. This is followed by a hypothetical description of issues that might arise in such a proceeding. A potential *Computer IV* regime is presented that seems to comport with existing and emerging network environments as well as important strategic U.S. interests.

The Road to Computer IV

Perpetuating the Dumb Network Concept. The regulatory regime for telecommunications in the U.S. consists of a hodge podge of provisions, oriented around a world that existed five decades ago - plain old telephone and telegraph services plus a handful of radio services. As telecommunications became more complicated during the subsequent decades and new technologies became available, the tendency was simply to employ judicial and regulatory edicts that partition out anything but the old services. As Peter Huber recently quipped in his Geodesic Network Report,[1] the network became frozen in time at a primitive level of intelligence.

The FCC's most recent regulatory regime for the public network, *Computer III,* relies heavily on a largely unexplained concept of open network architectures (ONA) which is characterized in terms of the equal availability of an undefined set of transport-related functional capabilities that the Commission called *Basic Service Elements (BSEs)*.[2] In doing so, the FCC embraced a well-established network concept and innovatively adopted it as a national mandate.

The bad news is that *Computer III* ignores technology and perpetuates outdated, simplistic notions about contemporary telecommunication networks. One of the more obvious is a belief that all services can be divided into basic and enhanced, and that somehow the underlying supporting functionalities can be similarly partitioned or attributed to specific services. Even before the regime became effective, the Commission embarrassingly decreed protocol processing to be "enhanced," a throwback to the old dumb network concept, which sparked a vigorous dissent by one Commissioner.[3]

Technology Not Considered. Equally disturbing was the failure to address the implications of open network architectures, or what kind of network functionalities or primitives are encompassed by the requirement, the relationship to service provisioning, or at what interfaces the open architectures would be manifested.[4]

Abstract economic doctrines concerning competition and deregulation, embracing an outdated conceptual model of the network, serve as the basis of the *Computer III* regime, rather than objective analysis of technological trends and strategic national needs, an understanding of how information networks function, or the vision of a national public information fabric.

The Months Ahead. Thus, what will unfold in the months following February 1988, when initial ONA plans are filed, remains highly problematical. A reasonable, lowest common denominator set of Basic Service Elements with some regional embellishments will no doubt be submitted to the FCC by the Bell Operating Companies (BOCs). Major unknowns are: the extent to which the BOCs will demonstrate that each BSE meets each of the *Computer III* Comparatively Efficient Interconnection principles; whether the FCC will use these plans as the basis for allowing BOC entry into *all* enhanced/information services without a separate subsidiary requirement; or what the ongoing process of devising and approving BSEs will look like.

Of course, BOC provisioning of information services also will be affected by Judge Greene's court. Another significant unknown is the role of the state regulatory commissions will play in the process of defining and tariffing BSEs, or how power will be shared with the FCC.[5]

What is clear about the coming months, however, is that some forum or forums will be called upon to make many complex decisions, and hierarchal bodies will be required to adjudicate the disputes. The FCC will be near the top of this pyramid.

Significant Regulatory Change is Constrained. Although it might be theoretically possible for the Commission to engage in considerable innovation in implementing *Computer III,* the concepts and policies underlying the decision seem constraining. Already, since the *Computer III* Phase I Order was adopted in May 1986, planned evolutions of the public networks are more dramatically different than in the previous 130 years of their existence. These changes, layered on top of the underlying deficiencies of the decision when it was adopted, will inevitably bring about a *Computer IV* proceeding.

The Difficult Issues Can Be Anticipated. The costs of openly providing the necessary network interfaces and BSEs, especially on a nationwide scale and with older equipment, can be enormous. No guidelines presently exist as to how to separate the necessary from the frivolous, nor to decide what is funded out of the existing regulated rate base versus what is derived from other sources of revenue, nor how to separate those functionalities which are employed for interstate versus intrastate services.

A minimal set of seamless national information services and capabilities is very important to most users. Little cognizance of the importance of the matter has been manifested by the Commission, much less a discussion of the competing considerations and how they might be balanced.

Other significant questions will involve differentiation among transport, signalling, and OA&M functions. The application of *Computer III* requirements to the last two of these "planes" will be very complex and difficult, made even more so by the increasing blurring of boundaries among these functions. Network security and liability considerations loom large in this environment, and should be accorded substantial weight vis-a-vis competitive goals.

Lastly, and most significantly, is the matter of establishing BSEs. It is the cavalier treatment of BSEs in the *Computer III* proceeding that may do the most to facilitate the emergence of a *Computer IV.* Indeed, a Trojan horse has already appeared under the guise of true "virtual collocation," i.e., the concurrent residence and operation of non-BOC applications programs within BOC central office hardware with equal interrupt priorities.[6]

Contemporary information systems - in the form of today's public and private telecommunication networks - consist of physical and logical entities in various layers or levels.[7]

At the lowest level - physical (also called *real*) systems - the world is fairly simple. There are: input/output devices like display screens and keyboards that act as information translators between the human and electronic/photonic worlds; processors that manipulate and switch the information; memory that stores the information; and transmission paths through which information can be moved. What happens is determined by executing real time or pre-programmed instructions in the form of software, firmware, etc.

It is logical (also frequently referred to as "virtual") reality that matters. These are the functional capabilities or services that are actually provided to other machines, networks, and people, and that enable the provision of information services. Such functions can include an incredibly varied assortment of information access, transport, storage, administrative, and delivery capabilities.

Because such functions are generally layered - from the primitive to the sophisticated - disputes over access, control, and pricing are likely to be endemic to the public facilities environment, although the ability to effect many of these functions outside the public network should provide a useful check.

The dominant and constantly recurring public policy questions will focus on: what functions should be performed at various levels or aggregates and by whom. It is not clear, however, who will resolve these questions, or by what methods. The levels of complexity, the dynamics, the required knowledge, and the efficacy of traditional policymaking processes seem to exceed considerably the resources and abilities of public bodies in the U.S. It is not surprising, therefore, to find that telecommunications policymaking in the U.S. has been characterized by: 1) a shift from the public to the private sector, and 2) monumental growth in the amount of activity.

The Forums. The forums and activity for implementing ONA will continue to increase dramatically. This includes, most notably, the T1 committees and the CCITT (International Telegraph and Telephone Consultative Committee).[8]

Most of the disputes can be expected to be resolved within these forums where the procedures generally provide good due process, especially considering the enormous amount of complex matters that must be resolved in very short periods of time. The interests of smaller participants or users may not, however, always be effectively infused into the decision making. In addition, there are presently significant deficiencies in top-down policy making in the U.S. forums, as well as ineffective administration of the work.

Inevitably there will be continuing disputes that percolate up to the FCC. In addition, ONA/CEI plans or tariff authorizations will come before both federal and state commissions. It is not clear how these bodies will deal with the complexities before them when the regulatory model being employed is so at variance with reality. The recent NARUC resolution on ONA makes it clear that there is unlikely to be agreement on even the basic functions of the network. It is patent, however, that resolving these disputes in terms of a basic/enhanced dichotomy or "multiplexing exceptions" will not produce meaningful results.

A Computer IV Proceeding

Motivating Factors. After several years of struggling to administer the tenets of *Computer III* and dealing with its intrinsic deficiencies, it seems probable that the FCC may initiate a *Computer IV* proceeding.

Another factor leading to this result may include an enhanced sensitivity to the strategic nature and importance of a national public information fabric.

An entirely new generation of highly innovative and aggressive specialists in digital processing and switching techniques are taking existing and anticipated electronic and photonic technologies and collectively developing transport architectures that represent a radical departure from the past with orders of magnitude improvement in capabilities. They are fashioning a universal, optical, super high speed network in which all information - data, voice, or video - is packaged and transported to any location in the most efficient possible fashion.[9] However, effective implementation may require very integrated operation, administration, and maintenance of all constituent networks at the transport level.

All the major industrialized nations are extensively involved in these broadband developments - many with considerable direct government funding and strategic planning. At the recent Hamburg meeting of CCITT's Study Group XVIII the Asynchronous Transfer Mode using labelled multiplex techniques was adopted as the global imperative, toward which all nations will plan.

In the next decade, optical fiber will begin to be deployed extensively in the "last mile" to business and residences, bringing with it the prospect of broadband services - especially digital television of exceptional quality. At the same time, local central office facilities, local area networks, customer premises information systems, and work stations will offer ever more powerful capabilities at rapidly descending cost. This kind of complex interoperation between private and public domains may result in synergistic uses of and demands on the public networks in ways not even anticipated today.

These revolutionary and strategically important changes will constitute a significant additional dimension in considering national policy options for information technologies.

In addition, there is the lingering basic/enhanced dichotomy. If such a distinction made little sense in the past, it appears to make no sense today. Virtually every manufactured active electronic product is operated by a microprocessor with associated memory under the direction of stored programs. We live in a world of digital intelligence interconnected with other digital intelligence through defined protocols.

Some of this interconnected intelligence is brought together in networks optimized for one particular information service: transport. The point is that we are dealing with technology that is common to all information services. Today everyone is axiomatically in the information services business.

A More Appropriate Conceptual Model. Significant storage, processing and low-level formatting functions will exist in the public networks as a powerful national resources. A dominant issue in a *Computer IV* proceeding will be how such resources should be used.

In such a proceeding the Commission may seek to develop a regulatory framework for a new conceptual model, a "national public information fabric." Necessarily, the proceeding would confront technological reality - that public telecommunication networks are nothing less than distributed computer systems. Embarking from this perspective, some fundamental questions would then be posed regarding how such public networks should be employed.

Important Objectives. Building on the open network architecture concept and the work of the network planning community, the Commission might seek a "return to technology" and declare its intent to achieve uniform, stable, characteristics for a ubiquitous, minimal set of interfaces, functionalities, and seamless cost-effective offerings that are critical to an effective and efficient national - indeed global - network fabric.

A Computer IV Regime

It would be presumptuous at this time to detail a U.S. regulatory regime for the network of the 90's. However, many of the factors to be considered have been raised above. Most prominently, this would include the directions and extraordinary rate of change in technology, the large investments involved, and the strategic importance to the nation.

One would expect knowledgeable discussions of such matters as: reconciling the frequently divergent goals of fostering competition, deregulation, standardization and innovation; the country's need for a national information fabric for the world of the 90s and beyond, as well as low-cost basic universal services, and how these goals should be most equitably achieved; and how to foster advanced research and development, and remain a leader in these technologies.

ONA would be regarded is much more than a means of controlling the BOCs. Instead it would be viewed as a useful tool to promote flexibility and innovation in network architectures and in the provision of information services.

The Commission would find that public telecommunication networks are nothing less than distributed computer systems and proceed under Title I jurisdiction.

To be compatible with all the above factors, it would seem necessary for a *Computer IV* regulatory framework to rest on physical and logical implementations of open network architectures applied to four kinds of public entities: 1) facilities (i.e., real systems), 2) low level transport capabilities, 3) primitives, and 4) basic teleservices. A modified Comparatively Efficient Interconnection concept for concurrent access to physical and logical resources by all parties, at well-defined, stable interfaces, while allowing the provision of almost any other information capabilities or services by any party on an unregulated basis. Significant initial work along these lines has already occurred today.[10]

In a national public information fabric, all physical and logical resources are by definition interstate. The roles of State regulatory commissions in this area will significantly change. What their new role should be needs serious study.

The public policymakers of tomorrow cannot be micromanagers, but they must be substantially cognizant of the technology, its implementations, and strategic employment, and infuse that understanding into the regulatory models and decisions that are adopted. The nation needs organization and leadership to facilitate the emergence of a national public information fabric to allow the U.S. to have a future in an information age economy.

FOOTNOTES

1 Peter W. Huber, *The Geodesic Network: 1987 Report on Competition in the Telephone Industry*, U.S. Department of Justice, Antitrust Division (1987).

2 *Notice of Proposed Rulemaking* [in CC Doc. No. 85-229], FCC 85-397 (adopted 25 July 1985); *Report and Order,* FCC 86-252, 104 FCC 2d 958 (adopted 15 May 1985); *Supplemental Notice of Proposed Rulemaking,* FCC 86-253 (adopted 15 May 1985); *Memorandum Opinion and Order on Reconsideration,* FCC 87-102 (adopted 26 Mar 1987); *Report and Order* [Phase II Order], FCC 87-103 (adopted 26 Mar 1987), with Separate Statement of Commissioner Patricia Diaz Dennis (released 30 Jun 1987); *Erratum,* DA 87-841 (released 8 Jul 1987).

3 *Idem.*

4 In some measure, the Commission rectified this failure in the Reconsideration Order by deciding that BSEs need only be created for enhanced services providers, but once that occurs under federal tariff, anyone can obtain access to them for any use. See *Memorandum Opinion and Order on Reconsideration, supra* at para. 106.

5 See, e.g., Resolution Regarding Open Open Network Architecture, National Ass'n of Regulatory Utility Commissioners, (adopted 22 Jul 1987).

6 The term "true" is used to differentiate the customary usage of the phrase "virtual collocation" by the information systems community from the use of the phrase by some parties in *Computer III*-related events in describing an alternative physical collocation scheme.

7 *See* An Architectural Framework for Private Networks, European Computer Manufacturers Ass'n, Doc. ECMA/TC32/87/84 (Apr 1987) [recently introduced in T1D1 as Doc. T1D1/87-200.]

8 The work of international networking planning and standards bodies like the CCITT are strongly influenced by national and regional governmental policies and representatives, with the U.S. being the major exception.

9 *See, e.g.,* Minzer (Bellcore) Broadband User-Network Interfaces to ISDN, Proceedings of the IEEE Int'l Conf. on Com. '87, Seattle, June 1987 (ICC87) at 11.2; Gohara et. al. (Hitachi), A New Distributed Switching System Architecture for Media Integration, ICC87 at 11.4; Coudreause & Servel (CNET), Prelude: An Asynchronous Time Division Switched Network, ICC87 at 22.2; DePrycker & Bauwens (Bell Belgium), A Switched Exchange for an Asynchronous Time Division Based Network, ICC87 at 22. *See also,* Report of the CCITT, Study Group XVIII Broadband Task Group Meeting at Hamburg (July 1987), Doc. COM XVIII R__.

10 *See,* especially, An Architectural Framework for Private Networks, *supra.* The present version of this ECMA Report notes that "the material is equally applicable to scenario's involving public environments...[however, it] is restricted to data applications only. It is the intention to include non-data applications in a later version." *Id.* at 2. *See also,* Fleischman (Bellcore), Operations Systems Strategic Plan for Exchange and Exchange Access Operations, ICC87 at 4.2; Gopal & White (Bellcore), A Paradigm for the Modular Decomposition of Distributed Telecommunications Switching Software, ICC87 at 33.1.

Open Network Architectures and Broadband ISDN: The Joker in the Regulatory Deck

Richard Jay Solomon

Research Program on Communications Policy
and The Media Laboratory
Massachusetts Institute of Technology
Cambridge, MA

Proposed Broadband ISDN standards and the 'Asynchronous Transfer Mode' (ATM) for switching and transmission will have a revolutionary effect on how telecommunications networks are perceived and used, creating novel policy issues not heretofore faced by regulators. Open network regulatory concepts such as the FCC's ONA and the EC's ONP will have to confront the creation of powerful virtual user networks, the result of the functional merger of user CPE with carrier circuits. Other regulatory problems will emerge because B-ISDN standards as proposed leave little role for satellite carriers. But fiber carriers will find it easier to share OA&M network control with some customers, creating a new quasi-carrier class.

1.0 INTRODUCTION.

A funny thing has happened on the way towards implementing the FCC's call for Open Network Architecture — the technology has changed yet again. Technology change has been behind all of the FCC *Computer Orders* (I, II, III, and surely IV, V, and VI) [1], but in none of the decisions has it been explicitly recognized that the 'computer' in the title is not merely a new application, but a permanent, restructuring of how people (and machines) communicate.

Computer I and *II* — with their emphasis on such trivialities as attempts to distinguish between 'basic' or 'enhanced' services or the role of the carrier in data processing [2] — illustrate the general misunderstanding policy-makers have in dealing with computers, what they do, and how they network. [3] With the introduction of computers into a network, the conceptual models of how analog systems really work are no longer valid. This is due to two simple facts:

1) stored-program digital computers inherently communicate (they could not work any other way); and,

2) by definition, stored-program machines are *recursive* — they are capable of modifying their own programs — and so modified they *become* different machines (or in this case, different networks). [4]

There are profound implications of this when the computer is extended outward towards the rest of the world: *self-modification is not something found on conventional networks!*

1.1 Degree of computerization about to change.

Up to now, despite heavy computerization for switching, the true power of the digital computer has rarely been applied to networking, for reasons ranging from politics to ignorance. Computers that merely switch circuits may permit interesting new services, such as calling party identification, or user-to-user signaling, but these are not going to force a radical change in the way we deal with telecommunications. Yet new ways of thinking about networks are gaining ground — large, virtual, distributed data processors which transport and concurrently manipulate voice, data, and video following dynamically self-modifiable programs. That type of system could usher in services which may permanently alter both telecom and the industrial infrastructure it serves.

A lot may depend on how governments and regulators treat the new technologies. The political and economic problems are much more complex than the engineering issues. Whether the new technologies prove catastrophic or enlightening, depends on one's viewpoint.

1.2 Broadband demand not a prerequisite.

The first mental hurdle which must be vanquished about broadband digital transmission is the idea that broadband services have to be offered in order for the new technology to be successfully implemented. *Demand for long holding time, high bandwidth traffic has little to do with basic implementations of the new broadband technologies.*

Implementation of broadband is happening anyway as carriers add new optical fiber plant, and retrofit old distribution plant; the major policy questions will be how, or whether, to speed up the retrofit, and what to do about the anomalies which will confuse the regulatory atmosphere. In this paper we discuss why broadband — in the form of integrated virtual networks — differs from conventional, present-day telecommunications, and the workings of some of broadband's anomalies.

1.3. Market Aspects.

Though this is not a paper on the market potentials of broadband, the most significant aspect of a wide bandwidth, public switched telecom network is that for the first time telecommunications would be working at speeds equal to or greater than that of most computer processors. This creates new opportunities for services quite different from those found today on conventional telecom networks or even local area networks. The technology would affect every form of computer/communications applications, from database access to new forms of video utilizing intelligent receivers. (We discuss this briefly in section 4.2.3.)

Universal implementation of networks offering broadband would also encourage services such as switched video over the telephone network. And broadband to the home could solve spectrum allocation problems for high-definition video. [5]

2.0 BROADBAND AND VIRTUAL NETWORKS.

The new technologies for broadband transmission and switching will be a true shift in thinking for telecom engineers and communications planners — what historians of technology would call a paradigm change, a different *zeitgeist*, or a new way of modelling reality. [6] The operative word for the new networks is *virtual*, as in virtual computer memory, or virtual computer networking — two sides of the same coin. In a virtual digital network, as implied by proposed CCITT standards for the Integrated Services Digital Network (ISDN), [7] the physical network becomes transparent to the user, only the logical network can be perceived (unless access is given to system control). It is difficult to get meaningful legal definitions when dealing with things in terms of logical things — the mathematical definition of logic — especially things that are self-definable according to the whims of programmers, and are physically based on elusive electrons or photons.

For example, how do you enforce non-discriminatory clauses for computer network interrupts? [8] How would you explain this to a programmer setting up an interrupt table? Will we have FCC designated, non-tamperable chips on each and every network interface with fair-minded tables in ROM establishing interrupt priorities?

It doesn't help that words for technical concepts often do not mean what they say (hence we use a lot of quotes around technical terms in this paper). ISDN is a prime example: because telecom carriers initially approached the digitization process with a view of finding new things to sell to their customers, *services* was put in front of *digital*; the other way around would be a better indication of what ISDN has become — an end-to-end (customer premise equipment to CPE) network of digital (computer) processors. The key ISDN word is *integrated*..

2.1 Switching integrated with transmission

In integrated digital networks, switching (which is computer data processing, no matter how you want to cut it) and transmission work in conjunction. Though perhaps non-intuitive to a mind used to working with precedents and analogies to older technological forms, there is *no* way to separate the switch DP process from its transmission process. It is all a matter of electronic timing circuits internalized by the network. The link is merely an extension — a *virtual extension* — of the switch or node's electrical (or optical) bus, just like on a computer motherboard. While this may be obvious for a packet-switched system, it is also true of conventional circuit-switched, all-digital network.

2.2 Control.

Control of the physical components in an integrated network is what is important, and that gets very complicated. An end-to-end digital system, such as ISDN, is a totally different kind of animal than today's analog, or hybrid networks. In an analog network, digital signals (in the form of binary bits) are merely carried in analog forms for later reconversion back to bits; not much can be done with such signals on or by the network. But in an all-digital system, the bits are buffered by computers using timing signals no different that those found inside of a desktop PC, until they reach their 'destination,' which can be another computer, another digital network, or a digital device which translates these signals into a form a human (or another machine) can understand. [9]

2.3 Timing.

System timing is crucial to the understanding of control, for at any one slice of time (which can be a very small slice, indeed) control of the signal may extend halfway across the world, or just down to the edge of your coiled phone cord. Digital machines are designed to give over control to other machines (hence 'interrupts'), depending on complex 'states' of their programs. That is what happens all the time on a PC when it sends data to the printer, the screen, checks the keyboard, or the modem, or beeps. The same concept applies to end-to-end digital telecom networks, but it is much more difficult to pinpoint at any single point in time.

Averages do not mean much here, such as statements which attempt to indicate 'dominant' modes for 'most' of the time. And there is little point in trying to discern what the control programs are doing, since whatever they do, they do millions or billions of times each second. Furthermore, computer science teaches us that there is no way to know what has just taken place, nor what will come next just by looking at any one time slice; even in the context of a large program, a human programmer (if there is one) has difficulty in determining what is happening at any single point in time. [10]

Again, those looking for analogies to current networks for regulatory purposes will quickly find themselves going around in circles.

2.4 Broadband ISDN and regulation.

Now just as the regulators have begun to understood some of these implications for ISDN-like networks (enough to attempt an 'open network' plan, thinking that this would be like promulgating open hardware architecture for computers), the technicians have come up with another surprise — *Broadband* ISDN. [11]

B-ISDN turns out not to be just a wider bandwidth extension of the ISDN that we have discussed for over a decade, but an even more elusive creation taking the power of the recursive, stored-program digital computer to its ultimate level. conventional ISDN (if there ever was such a thing) was to be an upgrade and overlay on today's analog systems. Plans were to do this by applying new software and digital drivers to much of the equipment already out there — digital switches and the existing copper local distribution plant.

ISDN's two major problems are that its defined channels are of fixed bandwidth and its distribution circuitry is *narrowband* — less than 1.5 or 2 megabits per second. [12] While that sounds fast to users of 1200 bps modems, it is not fast enough to carry and switch certain services such as video, nor it is fast enough for certain types of user-shared network control nor for direct interfacing to local area networks or computer central processors themselves. ISDN is best characterized as an carrier computer-driven, intelligent network basically emulating today's voice and slow-speed data systems.

With proposed B-ISDN standards, we find this concept turned upside down: carriage becomes somewhat passive (though perhaps even more lucrative), with bandwidth and delay options (and some level of routing, to boot) being driven by the user's computers — that is a big difference. This permits true virtual networking, and will create definitional problems for regulators beyond mundane distinctions of enhanced vs. basic, or bearer vs. teleservices. *Comparatively Efficient Interconnection* in B-ISDN will mean that it will be *impossible to tell which is the carrier and which is the customer.* That goes beyond the issue of not being able to determine who is trunkside and which is lineside in narrowband ISDN.

3.0 B-ISDN VERSUS NARROWBAND ISDN.

How did we get to this state of affairs? B-ISDN has quietly sneaked up on us even as narrowband ISDN has just begun to be implemented. Indeed, the promise of B-ISDN technology is so powerful and overwhelming that narrowband ISDN increasingly may be seen as obsolete before its time. This is due to the curious fact that the physical elements which make B-ISDN practical — fiber optics and certain types of passive switching called 'cross-connects' [13] — are rapidly being installed in the telephone plant in any case. Furthermore, new forms of broadband switching, based more on data processing techniques than hardware, is closely coupled with the use of these optical links.[14]

(Hardware cannot be ignored, for superconductors and optical processors may revolutionize switching nodes yet again. But hardware moves slower than software, and the superfast node may well wait for second-generation B-ISDN. [15] The first generation is enough to create policy problems.)

Hence, the much-acclaimed gradual 'evolution' of the world's existing telephone and telegraph systems to ISDN instead may turn into a rout.

3.1 Source of B-ISDN technologies.

The biggest surprise is that the new technologies come right out of the international committees which had been establishing the narrowband ISDN standards since the mid-1970s. [16] In trying to cope with setting interface standards for the growing hierarchy of high bandwidth (150 megabit to gigabit) digital optical trunks, it quietly dawned on the standards makers (principally of the past two years) that an approach radically different from that of interconnecting narrowband digital coaxial cable and wire pairs was necessary. The technique which evolved is based on computer database methodology, discussed further below. It appears to be effective in overcoming numerous problems with ISDN, from user control of network administration and maintenance, to the ability to handle video, distributed data processing and requirements for flexible, and instantaneously variable bandwidth and connectivity. [17]

3.1 Fiber capacity the key.

When ISDN standards-making began at the beginning of the decade, 'single-mode' fiber was thought to be impractical to manufacture, but about three years ago fiber economics changed drastically. There are predictions, based on current trends, that single-mode line costs will drop below that of wire pairs before the decade is out.[18] But more important, for all practical purposes, single-mode offers the unique feature of "almost unlimited capacity on a single fiber," to quote a recent AT&T report. [19]

3.1.1 Capacity to play with.

Here we have a revolutionizing concept — for first time, there is *bandwidth to play with!* This huge excess bandwidth permits a large fraction of the digital transmission rate to be used for 'overhead' bits to route, and keep track of the 'payload' bits. For some types of connections, it may be possible to bypass the central office by using an optical form of the digital cross-connects now replacing the main wiring frames in modern COs [20] The first B-ISDN applications will likely be using cross-connects for virtual 'dedicated' circuits on customer-controlled and reconfigurable networks. (In fact, this is happening now with T1 circuits.)

Though the main emphasis by the broadband study groups is to simplify interfaces so that a new type of central office can control all interconnections, programmable cross-connects, may make it possible that switching will be on the fiber trunk itself. (Eventually the concepts of cross-connects and COs will merge.) Switching may be under the dynamic control of the user or shared with the carrier. Surely this is a unique regulatory proposition in itself, perhaps obviating any need to dictate ONA or CEI procedures. Access to the cross-connects or COs could be tricky; perhaps the stored-program control could be time-shared, a much more useful concept than even sharing the hardware.

Therefore, it matters not whether the user needs or wants broadband *per se*, for carriers and customers use broadband as a matter of course for trunking: the current, fast-expanding applications of fiber optics in the intercity and local distribution plant already permit the carriage of extremely wide bandwidths (as compared to the ISDN basic rate of 160 kbps), upwards of 150 to 600 megabits/sec.[21]

The primary constraints for single-mode capacity in the near future are dependent on the type of laser drivers used.[22] AT&T, for example, quadrupled the capacity of their Philadelphia-Chicago lightwave line this year by upgrading the repeater electronics and lasers, and not laying any new fiber.[23] (Fortunately AT&T chose single-mode for this line when it was built in 1983; earlier optical trunks used multimode fiber which cannot be easily upgraded.)

3.1.2 Multiplexing on fiber.

Initially, the reason to use fiber was to economically combine (multiplex) numerous narrowband voice and data channels on trunks and local distribution loops. Typical is the 1.5 megabit/second T1 circuit, the mainstay of most inter-office connections in North America, in Japan, or the 2 megabit primary rate used in Europe. T1 interleaves the digital signals in time (time-multiplexed) to carry 23 standard (56 kilobit) voice circuits and one overhead channel for operations, maintenance, and elementary circuit routing. Primary digital circuits originally were engineered to be transmitted on two ordinary copper pairs,[24] later were carried on coaxial cable, and today are mostly installed with fiber optics.[25] It is this subtle, quiet conversion to fiber which has changed things.

Single-mode's enormous capacity offers extraordinary opportunities for network design which were not obvious with the first digital time-division multiplexed trunks. The ratios are critical: while T1 uses about five percent of throughput for overhead data, at the 150 Mbps (H4) rate,[26] about 10 percent of the fiber throughput can be allocated for overhead and still yield a 'payload' of about 135 Mbps. H4's 15 Mbps overhead is 10 times larger that the entire capacity of T1, and yet its payload is about 90 times larger.

It has only dawned on the standards-makers within the past two years, that with 15 Mbps to play with on an integrated system, this huge overhead can be used to build different, and potentially superior, flexible network architectures.[27] Such order-of-magnitude changes in the ratios of overhead to payload permit complex routing and switching to be manifest on the circuit, instead of only at a node. Junctions can be simpler, networks can be extremely robust (since they can be designed in the form of redundant webs instead of trees or stars), and the user can take control of routing and level of service.[28] It doesn't matter if today there is no perceived customer demand for broadband services, this flexibility comes with the territory.

Network architectures are further modified when 'intelligence' gets distributed around the network. Such patterns do not fit preconceptions assuming physical separation of carriers and subscribers, nor ideas about in which jurisdictions bits may be buffered, routed, stored, or replicated.

Recognition of this strange new telecom world is given by its latest generic name from CCITT Study Group XVIII — the 'Asynchronous Transfer Mode' (ATM).[29]

3.2 Transfer Modes

It is a sign of rapidly changing times indeed when language cannot remain stable for more than the span between standards-making committee meetings — about 3 months! Earlier terminology calling ATM merely a 'fast-packet' network has been dropped as insufficient to describe its true capabilities: *ATM/B-ISDN is neither a new form of packet switch or a mere extension of switching as we have learned to understand it.* Just what it is will begin to emerge as applications and technology are put into place in the next few years.[30]

ATM was not planned to be a revolution, just a new interface. The original intention of the various broadband planning groups had been to harmonize the different primary (1.5 to 2 Mbps) rates used in various countries so as to permit a gradual upgrading from narrowband ISDN.

The interface for wideband fiber required a different method of handling the packetized bits than that used on current packet networks or for on conventional digital transmission lines, such as primary rate T1 or ISDN's basic rate 2B+D circuit.[31] Unlike earlier packets with framing bits just for synchronization and addressing, ATM packets are 'envelopes,' with the data framed in the form of a matrix. [32] The matrix is constructed, with pointers or labels, as if it were a relational database. And like a relational database, this pointer technique is very powerful, permitting routing and data massaging as part of the transmission procedure itself. Pointers and envelopes can be contained within other envelopes with more pointers, expediting the ready, dynamic construction of virtual networks invisible to the carrier. Therefore, ATM's overhead contains its own signalling as part and parcel of its interface framing.

Initially, it was assumed that broadband would be switched on some form of hierarchical network similar to today's telephone systems, or at least parallel to that for proposed 'conventional' ISDN. A combination of 'fast (virtual) circuit' and 'fast packet' switches had been suggested to route digital packets, depending on the type of service required.

The discussion over the past few years in the broadband committees on the circuit switched vs. packet net dichotomy caused a number of key technical issues to surface, especially the notion of whether broadband should be synchronous extensions of narrowband digital lines, or asynchronous. It now appears that the overall standards issue have been settled in favor of an asynchronous, and therefore all-packet network [33]

A synchronous transfer mode (STM) would have its virtual data path, bandwidth, and arrival times remain the same and predictable throughout a 'call' or for a virtual 'private' circuit. But asynchronous modes have the attraction of extremely flexible and robust networks, with powerful economies based on dynamically variable bandwidth utilization, and self-routing packets.

Synchronous proposals assumed future development of fast-packet digital switches optimized for virtual circuit routings — a superfast version of today's digital telephone switches. Though STM offers the possibility of building less complex physical switches and terminal equipment, it requires more network memory and call routing capability. ATM, on the other hand, offers flexibility and potential network control savings due to built-in self routing algorithms working with the envelope pointers Hybrid schemes — combining circuit and asynch packet switching — have been rejected as having the worst of both without any distinct benefits.[34]

At one point in the development of transfer mode concepts, it appeared that STM would be more appropriate for television transmission where frames show up at regular intervals and loss of data might cause annoying picture dropouts. Careful modelling of both ATM and STM, plus some innovative research by the Bell Telephone Mfg. Co. in Antwerp [35] has shown that television can be successfully transmitted with asynch packets.[36] In fact, the dynamic bandwidth allocation feature of ATM permits some very clever compression of video signals using the network itself.

Like any good stored-program technology, virtually all synchronous services can be emulated with ATM envelopes and pointers, as long as the asynch envelope runs faster — or has more overhead — than the derived synchronous service. This includes 'private' or dedicated lines (see section 4.2, below).

(We offer a simplified visualization of how ATM works in section 5.0.)

4.0 IMPLICATIONS OF ATM FOR REGULATORS.

4.1. B-ISDN under the New Transfer Mode has some totally unique features clearly demanding different network architectures, and possibly different regulatory regimes. It remains to be seen if either the FCC's Open Network Architecture [37] concepts, or the European Commission recent proposal for 'Open Network Provisions'[38] will meet these new requirements. But here are some ATM anomalies and parameters to contemplate, with few parallels or analogies to what we have today:

4.1.1. The virtual, connectionless network becomes the guiding premise of B-ISDN under ATM. Nothing like it exists today save for a poor analogy to some local area networks. With connectionless systems, it does not matter if the node drops a connection, if circuits can be re-established in less time than the application can detect the drop. This could be a negotiated sequence for level of service between the user and carrier, possible because the payload transfer rate is significantly less than the total circuit rate.

4.1.2. Self-routing envelopes means that the network need not track a call or connection eliminating the need for extensive — and carrier proprietary — call auditing, state tables, and huge tracking machinery required by common-channel signalling systems, such as CCITT No. 7.[39] Though narrowband ISDN required SS7 for out-of-band operations control and accounting, the new transfer modes *must* have an in-band system, under shared user control, to prevent delay times from becoming excessive. In fact, SS7 investments may become redundant, adding one more issue to be confronted by regulators as carriers are forced to depreciate technology outdated before installation can be completed!

4.1.3. Network architecture might resemble a logical, seamless web, instead of the hierarchy of current local and toll systems.[40] Non-hierarchical systems make it difficult to distinguish the boundaries between networks.

4.1.4 The ATM is intended to be a true open network, with the user gaining a high level of control. It may not be necessary to promulgate regulations, or ask for carriers to file ONA plans, for B-ISDN/ATM makes openness a fundamental *basic* service by virtue of the transfer interface. OA&M (operations, administration and maintenance) become complicated, because the customer may now have something to sell the carrier in terms of network state data. If OA&M, can be easily shared, do customers have to file a tariff in order to connect their internal network to the 'public' network?

4.1.5 Circuit delay times become much more important than on today's networks, because switching could become more dependent on the processing capabilities of customer premise equipment (CPE) instead of on network storage capacity. Shifting intelligence to the CPE could make a network less expensive to use and easier to maintain. The subscriber will be able to order the network to control delay, with the network dynamically predicting and feeding back state information in realtime — like on well-designed computer operating systems. It may be useful to begin to think of B-ISDN as a gigantic time-sharing network with computer software efficiently interfaced with ATM protocols, calling for close coordination of DP with telecom standards-making.

4.1.6 Choosing a circuit by bandwidth becomes a user, not a network, signalling function. Since bandwidth can dynamically vary in ATM, statistical algorithms can be applied by customer equipment to efficiently compress full-motion video, voice, and bursty database transfer. This will enhance distributed processing demand, placing more network intelligence in the CPE where it can be used for critical memory-dependent artificial intelligence or expert systems applications.

4.2. Policy Issues.

These characteristics lead to an initial set of public policy problems:

4.2.1 The concept of a 'private' line becomes meaningless, because all connections in ATM can dynamically request permanent or semi-permanent time slots or use the pointers to establish packet envelopes within envelopes.[41] A 'permanent' circuit can be established and torn down within microseconds. This is true 'connectionless service' and unlike the never resolved argument over datagrams for X.25 packet service, connectionless is inherent in the proposed B-ISDN interface standards. A private line or any type of circuit connection can be emulated or discarded at user option. Since 'real' private lines today are no more than reserved time slots on digital multiplexed links, such as T1, ATM connectionless reserved slots is just the next step. Should carriers be permitted to drop private line tariffs as a redundancy?

4.2.2 Satellite technology is a casualty of fiber and the new transfer modes. This will be dictated by the growth of applications using low-delay time connectionless systems, having dynamic, variable and very high bandwidths, and demanding very low bit-error rates (BER). Fundamental B-ISDN principles adopted a year ago by the CCITT [42] require a BER between 10^{-7} and 10^{-10}, and a packet length of 125 microseconds, both of which effectively eliminate satellite links for burst traffic due to inherent equatorial orbit delay. [43] There is nothing satellite carriers can do about this since the speed of light is a physical constant that cannot be exceeded. But there will be select niches for satellites in B-ISDN, and interfacing optimally to these links using innovative software will surely become a critical policy issue, particularly for international regulators.

4.2.3 With B-ISDN transmission and routing an order of magnitude or more faster than that of the fastest, non-optical local area networks being sold today, coordinated standards for optical LANs becomes desirable. [44] Identical standards — down to memory word addressing — would enhance the connection of computers at the direct memory access (DMA) level. If network speeds are the same order of magnitude as computer central processing units, true cross-network 'peer-to-peer' communications would be feasible.[45] Networked microcomputer applications could finally implement sophisticated artificial intelligence with its requirements for enormous memory. Narrowband ISDN primary rates, and packet switching as we know it today, are simply too slow to be effective for DMA, expert systems, and AI. But there is a dichotomy today between standards setting for public networks and computer manufacturing, though often the same people sit on both sets of committees. Do we want regulators to get involved in PC standards?

4.2.4 Carrier 'services' could be reduced to the most elementary: simply transporting packets and reporting their 'state.' Though ATM is more like a gigantic, open-architecture computer bus, than like a contemporary telecom system, even synchronization between intelligent customer equipment and the net may be irrelevant, as we detail below. Can we finally drop discussion of 'basic' versus 'enhanced'?

4.2.5 Simple traffic-sensitive charging in B-ISDN/ATM won't work — a major user fear for narrowband ISDN tariffing.[46] The anomalies of the system would encourage heavy users to grab as much network resources as possible during peak periods. Usage charges will have to be based on some combination of number of call connect instances (there can be many during one call), time of day, and network resources accessed. Charging for network resources beyond marginal levels will encourage the user to apply more intelligent customer premise equipment. A built-in disincentive for regulation. Similarly, re-sale and shared channels will be impossible to monitor with envelopes inside of other envelopes, and because in ATM there are no circuits (and no 'users' *per se*) just self-routing packets.

Entirely new ways of charging and cost allocation will have to be worked out.[47] New theories of network economics may be exhibit philosophies found in charging for computer time-sharing. How does one set marginal prices for components shared between the carrier and the user? Will carriers sometimes have to pay their customers? Will customers then have to file tariffs? Such concepts will not be easy to put before telecom regulatory commissions.

4.2.6 Since the network should be able to accurately predict demand in realtime, feedback will permit user terminals to schedule traffic in microsecond increments. It may then be easier to develop tariffs which can optimize network investment for both carriers and users. Could CEI/ONA be reduced to just the provision of network state data? What about customer state data which affects other network users as well?

4.2.7 ISDN standards committees have had great difficulty in adopting satisfactory OA&M procedures for users that carriers feel would not interfere with network security. With ATM, all the network sees are unintelligible bitstreams, routing, and administrative flags, while the customer has access to sufficient overhead bits to run any type of virtual network OA&M needed. Certain types of security is likely to be much higher for ATM than on other digital networks: AT&T Bell Labs has shown mathematically that routing algorithms could be designed to prevent fraud. Tampering or accidental misrouting of a frame would cause them to self-destruct.[48] And self-routing frames offers potentially greater robustness for the user, and would be difficult to track, even by internal network procedures. Security is a serious issue that cannot be left just to the marketplace, the user community, or the carrier. What role is there for regulators?

4.3 Summation of ATM features.

With B-ISDN using new transfer modes, virtual networks become more important than physical networks. Experience tells us that it is likely that the larger the network cross-section, the more robust and cheaper operations become. But, while customers only perceive virtual networks, regulators and accountants think about physical networks. The problem will be compounded if virtual networks spread over multiple jurisdictions in a seamless set of interconnected, asynchronous, broadband systems.

Multiple systems may raise the issue of natural monopoly once again, but this time the technology argument becomes pervasive: as we have noted, it will be difficult to determine who is the customer and who is the carrier!

5.0 A SIMPLIFIED VIEW OF ATM.

There are numerous technical descriptions of ATM, and of variations on the theme. Some are listed in the footnotes. For those less technically inclined, we offer a simplified visualization of how the new transfer mode would move digital bitstreams, using the analogy of a railroad system carrying relatively short trains of varying length:

5.1 Closely coupled.

Visualize a railroad track where each train runs quite close to the next one, with a very accurate signalling system rigidly controlling (synchronizing) spacing and preventing collisions. Most cars on a train carry a piece of information called 'packets' which can be used to recreate a voice, data, or a video signal; a few of the other cars carry information about packets, the train of packets, or the network itself. This is the overhead. (Actually, each car may have its own overhead section, as well.)

5.2 Unique indicators.

So far this sounds like an analogy to a normal packet switching network. But with ATM, a very important feature is the way all the trains and their cars of packets are numbered by the system as they enter the track. These numbers indicate precisely when the train of packets — the envelope — should arrive at an interchange. The numbers are created by combining data about the delay parameters that the customer can accept (found by the system at reserved locations in the overhead cars) with information about the current state of the network's tracks and switches — a relatively simple database exercise. Some cars in the train have *pointers* to other cars, indicating where they should be switched and when they have to arrive. Certain short packet trains carry envelopes with information pointing to the arrival times and destinations of other trains running behind them.

5.3 Matrix envelopes.

An inherent part of the numbering scheme is the way they are carried in the cars. If the numbers are written down as the cars go past, *starting a new line when each number shows up,* the resulting table forms a predictable, rectangular matrix which can be temporarily stored in the interchange or nodes' memory.

This matrix being regular and rectangular, permits array or database processing to be efficiently performed by computers embedded at the nodes, or by customer equipment at the entrance to the network (or both). For example, when the pointer part of the packet envelope goes past certain nodes, they drop off their advance listings of the other packet trains or envelopes. As the envelopes on that list begin to arrive at the nodes, the node is already prepared to pull them out of line and to switch them to a different path; some of the packets in that envelope may be given a new list of pointer numbers before they are sent on their way. In this fashion, storage at the node can be minimized. The packets can be re-routed in less than one CPU cycle, unlike packet assembler and dis-assemblers in current packet networks which require relatively large, and expensive memory.

5.4 Switching controls.

And so by this method nodes and customer premise equipment become switches by rebuilding arrays of numbers for the next node down the line. The envelope arrays are being designed with user options, so that they can contain information requisite to their baggage that other networks can extract; this information can be modified at the interchanges or at the terminals, or by the packet envelopes themselves using DP techniques.

5.5 The self-programming network.

The envelope/packet arrays are truly recursive, making maximum use of the stored-program computer's power, and processing information while being transported! Processing, then, is a *basic* ATM 'service.'

In New Transfer Mode the envelopes are the trains of bits or 'packets,' moving past the interchange at currently standardized speeds of hundreds of millions of bits each second. The next step upward will be an order of magnitude jump to billions per second. That is fast enough to carry the necessary information to recreate several high-definition video digitized signals, for example, or upwards of hundreds of thousands of simultaneous telephone calls. At those speeds, local memories on desktop microcomputers can be filled up in less time than it takes to transfer data from built-in hard disks. With such a network, virtual memory can be very far away, permitting novel applications online unlike anything proposed in current ISDN standards.

6.0 SUMMARY.

Many constraints found in current circuit switching, private lines, and low-speed packet networks disappear with B-ISDN working under the labelled transfer modes. Though, ATM manipulates connections and transmission in a totally different way, all current networks can be perfectly, and transparently emulated if need be.

That's the bright side. On the downside is the assumption that most everything out there, save the fiber trunks being installed today, may have to be replaced — user terminals, everything analog, probably most narrowband digital switches, and perhaps anything that connects to satellites. There have been some suggestions that existing digital switches, because they are fundamentally general-purpose stored-program machines, could be re-programmed to serves as B-ISDN network controllers.[49] In any case, this will not be an easy transition, and one that adversely affect the financial posture of more than one carrier and major manufacturer.

But like all revolutions which flip the world upside down, there is no turning back: existing and abortive ISDN techniques will become functionally obsolescent as quickly as fiber elements and faster switching devices are installed for the normal modernization of today's networks. At some point in the next decade, unless the telecom community recognizes this built-in transition and effectively plans for it, a catastrophic level may be reached which will tip the balance towards use of the broadband elements in the network, particularly by major users.

The major fallacy of the FCC's ONA proposal, as well as Open Systems Interconnect (OSI) models, is the belief that networks can be segregated piecemeal from technological change. Regulators must not forget the truism that when you change one thing, you affect everything on an interconnected network!

REFERENCES.

1. *Computer I:* 17 FCC 2d 587 (1969) and Docket No. 16979 (FCC 66-1004)*passim.*; Computer II: 77 FCC 2d 384 (1980) and Docket No. 20828 (FCC-76-745, July 29, 1976); and *Computer III* (Open Network Architecture): Docket No. 85-229.

2. More extensive treatments of this subject by the author are in: .I Pool and R. Solomon, "Intellectual Property and Transborder Data Flows," *Stanford Journal of International Law,* Summer 1980, pp. 113-139; R. J. Solomon, "Vanishing Intellectual Boundaries: Virtual Networking and the Loss of Sovereignty and Control," *Annals of the American Academy of Political and Social Science,* January 1988; (forthcoming) and "WATTC Continues Never-ending Story: Telecom Regulation and DP Bad Mix," *International Networks* , May 15, 1987.

3. For what computers do, a good lay introduction is in J. Weizenbaum, *Computer Power and Human Reason*, Freeman, 1976, esp. chaps. 2 and 3.

4. See P. McCorduck, *Machines Who Think*, Freeman, 1981; Weizenbaum, *op. cit.*

5 . The market demand for switched video on broadband is a chicken-and-egg problem in market economics beyond the scope of this paper. See C. Judice, et. al, (Bellcore), "Video Demand: A wideband Service of Myth," IEEE paper CH2314-3/86/000-1735, 1986.

6. The author has discussed these implications in R. J. Solomon, "Changing the Nature of Telecommunications Networks," *Intermedia*, May 1986.

7. See figure I.340 in proposed Rec. I.32x on "connection types for ISDN network architecture." Reproduced and annotated in CCITT XVIII/4 TD 12, p. 15, Hamburg 28 June-17 July 1987; "Draft Rec. I.1yy - Framework for Providing Additional Packet Mode Bearer Services," CCITT TD42(XVIII/1), Hamburg, 9 July 1987, esp. p. 17, fig. 8 "Mapping of Virtual Call Control and Data Transfer Elements Across an IWU." For a good background to the ISDN process, see A. M. Rutkowski, *Integrated Services Digital Networks*, Artech, 1985.

8. H. Hellerman, "Interrupt," in Ralston, ed, *Encyclopedia of Computer Science,* 1st Ed, 1976, pp. 733-736.

9. Such connections are explicitly recognized by the CCITT Red Book Recommendations for ISDN. See Rec. I.112 and "ISDN Network Functional Principles," Rec. I.310.

10. D. R. Hofstadter, *Godel, Escher, Bach*, Basic Books, 1979, p. 18 *ff.*

11. Unfortunately, we are stuck with the word 'broadband' which means something quite different when dealing with co-axial cables. *Wideband* would be a better term to describe a circuit which can carry a large amount of information.

12. See J. Turner in *IEEE Selected Areas in Telecommunications*, November 1986 for specific technical critiques of current narrowband ISDN standards.

13. J. Nagel, "Photonic Switching and Automatic Crossconnect Systems," *Telecommunications (International Edition)*, May 1987, pp. 43-46.

14. A. M. Rutkowski, "Further Advances in Switching Technology...," *Telecommunications*, May 1987, pp. 88-98; and "The View from Copper Mountain: T1D1's Summit on Advanced Digital Networks," *Telecommunications*, July 1987, pp. 75-82.

15. Some of the possible technologies are covered in K. Phillips, "Applications of ISDN: An International Perspective on User Research," in *ISDN in the year 2000*, conference held at the European Patent Office, Munich, 24 June 1987; also see for a broad technical overview.of work ongoing in the EC's RACE projects, *GSLB — Seminar on Broadband Switching*, (Groupe Special des Communications a Large Bande), CEPT, Albufeira, Portugal, Jan. 19-20, 1987

16. A good history is in S. Minzer (Bellcore),"Broadband User-Network Interfaces to ISDN," International Conference On Communications, Seattle, June 1987, pp. 364-368; technical status can be followed in current draft reports of the American National Standards Institute's T1D1.1 subworking group on Broadband ISDN, the ITU's CCITT Broadband Task Group, and the Conference of European Post and Telegraph Administrations (CEPT) GSLB (now NA2) broadband planning group.

17. "Framework for B-ISDN Standardization," (chairman's report) CCITT SG XVIII/7, TD23, Hamburg, July 1987; and T1D1.1 doc/86-59, "Broadband aspects of ISDN — Services and Bit Rate Requirements," Apr. 1986.

18. D. Snelling, "Fiber to the Living Unit," Southern Bell Telephone Co presentation to the Eastern Communications Forum, May 1987.

19. "Economics of Fiber as Feeder Cable," *AT&T Technical Journal,* Jan.-Feb. 1987, p. 109. Assuming that drivers for each 1.3 nanometer-wide 'color' can be built, a very large number of light beams, each with a gigabit/second or more of capacity, can drive one hair-thin fiber. This report emphasizes that for the backbone "we pay the lightguide and outside-plant costs only once because the installed lightguide base can be upgraded to higher bit rates at only the cost of higher speed electronics."

20. Nagel, *op cit.*.

21. Gigabit speeds are being planned for the near future on intercity links, see ref. 18.

22 *ibid*

23. *ibid*

24. For various reasons concerned with installation practices, 'four-wire' rarely means 2 copper pairs but is normally a coaxial cable with digital transmission.

25 . See ref 18.

26 . M. Beckner (Bellcore), "A Broadband Channel (H4) Bit Rate and Format Proposal," T1D1.1/86-046(R1); Minzer, *op cit*; Nippon Telegraph and Telephone, "Proposed Basic Frame Structure of Higher Rate Synchronous Network Node Interface," a summary of various national proposals, Brasilia, Feb. 1987, CCITT COM XVII D.950.

27 . Minzer, op cit; and R. J. Solomon, *Intermedia, op. cit.*

28. P.E.Green, Jr., and D. N. Godard, "Prospects and Design Choices for Integrated Private Networks," *IBM System Journal* (to be published), advance copy, July 6, 1986; H. I. Peck & M. B. McLaughlin, "Broadband Services - A Network Evolution View," paper presented at the Eastern Communications Forum, May 1987, Stamford CT.

29. Report of the Broadband Task Group, XVIII/TD 55, 56, 57, Hamburg, July 15-17, 1987.

30. The various proposal for how ATMs might work all have in common two factors: fiber and software-defined interfaces. The technical references herein give a broad overview of these techniques. See esp. ref 13.

31 . CCITT 1984 Red Book, Rec. I.412.

32. The name for this new concept has gone through several changes, but at the Brasilia CCITT meeting in February the BBTG adopted "the term asynchronous transfer mode [to be] used as a generic term [replacing] new transfer mode [NTM] and asynchronous time division [ATD]," SG XVIII/TD130, p. 29. ATM/B-ISDN fiber interfaces are also known in the literature as 'labelled transfer mode' (LTM) which refers to the matrix explicitly; 'frame-relaying network'; 'cell-based label multiplexing technique' (LMT); and 'SONET,' Bellcore's proposed 'Synchronous Optical Network Interface' which in its latest incarnation is no longer strictly synchronous. Other terms used are even less illuminating. See CCITT, "Report of the Broadband Vocabulary Correspondence Group," WD 18 (BB), Brasilia, Feb. 1987; "A Proposed SONET Based Frame Structure for the B-ISDN User-Network Interface," draft T1D1/87-138, 20 May 1987; "Draft Optical Rate and Format Standard," Rockwell International, T1D1.1/87-291, May 15 1987.

33. CCITT TD 55-57, Hamburg, *op cit.*

34. Green, *op cit.*

35. A subsidiary of Alcatel.

36. W. Verbiest, "Video Coding in an ATD Environment," in *GLSB, op cit.*.

37. Computer III Report and Order, May 1986., 104 FCC 2d 958

38. Commission of the European Communities, *Towards a Dynamic European Economy,* 'Green Paper on the Development of the Common Market for Telecommunications Services and Equipment,' COM(87) 290 final, Brussels, 30 June 1987, pp. 69-73.

39. CCITT 1984 Red Book, Rec. Q.701-795

40 . The partial implementation of similar technology in the out-of-band signalling system for narrowband ISDN and conventional toll systems has already downplayed hierarchical network design.

41 Rutkowski, *Telecommunications*, July 1987, *op cit.*

42. CCITT XVIII-COM-16, June 1986.

43. R. J. Solomon and L. Anania, "Open Networks: Is There a Role for Satellites in a Fiber World?," *Telecommunications*, June 1987.

44. J. McCool, "The Emerging FDDI Standard," *Telecommunications,* May 1987, pp. 54-62.

45. Such as IBM's Logical Unit 6.2 interface.

46. R. J. Solomon and L. Anania, "Open Networks: Pricing Access and Interconnection," *Telecommunications*, March 1987, pp. 30-34.

47. R. J. Solomon and L. Anania, "Open Networks: Paradoxes and Puzzles of Digital Networks, Part 2," *Telecommunications*, Feb. 1987, pp. 28-30.

48. N. Batis (AT&T), "Fraud,Vulnerability and Misrouting of Frames in a Frame-Relaying Network," T1D1/87-127 (T1D1.1/87-206R), May 18-22, 1987, Copper Mountain, CO.

49 Phillips, *op cit; GLSB op cit.*; J. Bauwens & M. DePryker, "Broadband Experiment Using Asynchronous Time Division Techniques," *Electrical Communication*, Alcatel, No. 1, 1987, pp. 123-130.

NETWORK VIEW OF ISDN SERVICES

MICHAEL L. BANDLER

PACIFIC BELL, SAN RAMON, CALIFORNIA

ABSTRACT

Telecommunications networks around the world are evolving toward an Integrated Services Digital Network (ISDN) as the infrastructure to provide both existing and new service capabilities. This paper discusses the motivation for ISDN from an evolutionary viewpoint. It describes several end-user advantages in the context of the proposed ISDN service features. Further, this paper discusses Pacific Bell's network evolution plans and identifies several critical issues which require resolution in order to realize the benefits that ISDN can provide.

1. INTRODUCTION

Telecommunications networks around the world are evolving toward an infrastructure that has the capacity to support existing service capabilities, extensions to existing service capabilities, and new service capabilities. This infrastructure is known as an Integrated Services Digital Network (ISDN). The evolution to ISDN, which is considered inevitable, is being driven by perceived market needs that are fueled by advances in technology.

2. MOTIVATION FOR ISDN

Initially, telecommunications and information processing technologies evolved independently according to the separate demands of analog communication and digital information processing. However, for the past twenty-five years, new telecommunications technology has evolved based largely upon the use of digital signals for both transmitting and switching customer information and for controlling and operating the network. The use of digital technology is seen in the current widespread application of T-carrier, fiber and digital radio transmission systems, stored program control and time division switching systems, common channel signaling, and numerous automated operations systems. This proliferation of digital telecommunications technology is complementary to information processing technology and together, they have stimulated the development of end-to-end digital networks that are tailored to meet specific customer needs. Examples include private line networks that support Digital Data Service (DDS) and High Capacity Digital Service (HCDS), and switched networks supporting Public Switched Digital Service (PSDS) and Public Packet Switched Service (PPSS).

ISDN is the next major step in the evolution of current telecommunications networks. It is the logical end result of the network's continued use of digital technology to economically provide transmission and switching functions. In this light, ISDN can be viewed as a plan for organizing digital technology to provide both voice and non-voice services economically and efficiently. One of the key elements of this plan is the support of a wide range of services in the same network. This affords the ISDN certain economies of scale that will lead to lower costs for customers. Another key element in the ISDN is its use of digital technology to provide operational advantages, extensions to existing service capabilities and the introduction of new service capabilities. Thus, ISDN is seen as the vehicle that will allow customers to optimize their access to and utilization of information.

3. SERVICES

ISDN has the capacity to support existing service capabilities, extensions to existing service capabilities, and new service capabilities. This translates into cost effective solutions for customers through the integration of services, access to a rich menu of network services and features, increased customer control and flexibility, and a graceful transition from existing to future network services. The following paragraphs discuss this capacity by describing the major service features supported by ISDN.

3.1 Single Socket Network Access:

ISDN access is independent of the services to be provided. This means that a customer can access a wide range of network services over a single access link. This leads to the concept of single socket access or universal ports. The realization of this concept will provide a significant increase in flexibility and control to the end-user. This increase in flexibility and control will allow customers to increase the utilization of their communications facilities to attain economies. Another customer benefit is the use of existing premises wiring and local loop facilities to access both voice and non-voice services. This benefit will avoid expensive recabling of customer premises and will reduce the coordination with the network provider when rearrangements are considered.

Single socket access will also influence the development of multifunctional terminals, such as integrated work stations. This will provide a significant improvement in a customer's ability to use various communications modes to access and utilize information.

3.2 Interworking With Existing Services:

The ISDN will not only support new services, but will also support the services currently available on existing networks. As such, ISDN users will not be precluded from communicating with users served by existing networks that support voice, circuit-mode data, and packet-mode data.

3.3 Customer Control:

ISDN will significantly increase the customer's ability to specify service mixes for particular communications needs. For example, customers could utilize their communications facilities for voice and interactive data applications during normal business hours, and they could then use the same communications facilities to transport bulk data during evening hours.

3.4 Availability of High Speed Digital Access:

The provision of high speed access, 64 kbps, will significantly enhance many applications. Some of these include secure voice, high speed data, faster facsimile, rapid text transfer, high speed graphics, and videotex.

3.5 Out-of-Band Access Signaling System:

The out-of band access signaling system provides additional capability which allows the customer greater freedom in organizing communications. For example, customers will be able to:

- setup, control and release additional simultaneous connections;
- change mode during a particular connection;
- communicate with the network to activate or deactivate facilities, such as call redirection, on a per call or longer term basis;
- communicate with, and control, internal network storage and processing facilities;
- use the signaling channel to communicate user-to-user information;
- send and receive non-traditional types of network signaling information.

While the out-of-band access signaling system provides many new capabilities to the customer, it may pose opportunities for security violations. Therefore, either end-user limitations or specific security measures will be necessary to insure the integrity of both the network and the customer.

3.6 Provision of Storage and Processing:

While message storage and information processing are not inherent attributes of an ISDN, they are considerably facilitated in an all digital environment. For example, message store-and-forward, which is essential for text services and potentially valuable in a telephone environment, could be provided on a content independent basis for voice, data and text. Information processing, coupled with out-of-band signaling, could provide directory functions which would be useful for format and protocol conversion, or other similar applications in which "in stream" processing may be applicable.

An important point is that these capabilities could be provided by either the network or by service providers using external systems. In the later case, access to end-users could be provided through a comparably efficient interconnection.

3.7 Low Speed Services:

The packet data capability of the signaling channel can support a variety of low speed services such as telemetry, surveillance, meter reading, and remote control. This would be independent of the use of other main access.

4. NETWORK EVOLUTION

The realization of ISDN is generally viewed as an evolutionary development resulting from the continued use of digital technology within existing telephony integrated digital networks. This evolution, which is expected to span a period extending over one or more decades, will by necessity make use of existing and future network capabilities. The rate at which this evolution takes place will be dependent on customer demand, availability of standards, technological developments, realization of operations efficiencies, capital availability, and legal and regulatory constraints.

Since the evolution to ISDN is dependent on such a diverse set of factors, many of which are not fully understood, Pacific Bell has developed a strategy for evolving the network that is based on the most likely scenario of events. This strategy, which is discussed in the next pages, envisions the evolution to ISDN as occurring in three overlapping phases: Transition, Intermediate, and Mature.

4.1. Transition Phase

The transition phase positions the network for a graceful evolution to ISDN. It consists of the continued deployment of digital technology to provide both switching and transmission functions and the introduction of transition services.

Pacific Bell has been deploying digital transmission technology in the network for over twenty years as a cost effective solution to meet customer needs. To date, approximately 85% of our interoffice facilities are based on digital technology. In recent years, this deployment has expanded and includes the use of digital technology to provide switching functions (both circuit and packet) and to upgrade the local loop. The deployment of digital technology has advanced to the point where the core network has many of the elements necessary for ISDN implementation.

The introduction of transition services is a key step in the evolution to ISDN. Transition services are the products of new capabilities deployed in the network to satisfy identified end-user needs for both existing and new applications. Examples of transition services include packet switched data services, circuit switched digital data services (56 kbps), customer control network management services (reconfigurable private lines), and call management services (CLASS, 800, and Automatic Bill Services). The key role of the transition services in the evolution to ISDN is twofold; they serve as a vehicle for market development, and they also seed the network with technology that can be upgraded to provide the ISDN capabilities.

4.2. Intermediate Phase

The intermediate phase involves the initial offerings of ISDN-based services in selected areas. It is characterized by "ISDN islands" and the general availability of ISDN compatible equipment for both end-users and network providers. A critical activity necessary to evolve to this phase is conducting initial ISDN technology tests.

Pacific Bell is currently involved in a phased ISDN technology test that is planned to extend over a period of approximately three years. The initial phase consists of a digital adjunct front-ended to an existing analog stored program switching machine to provide ISDN Basic Access. The initial services include voice, circuit mode data (56/64 kbps), packet mode data, and Centrex features. In addition to providing these services over an ISDN access, we will also evaluate the capability of interworking with existing voice and data networks. The second phase will see the addition of two digital switching systems that have been upgraded to provide the ISDN capabilities. The services and interworking capabilities will be similar to those of the first phase. An addition capability for this phase will be the support of ISDN Primary Rate Access for all three switching nodes. The third phase involves the application of Common Channel Signaling #7 to the ISDN switching nodes.

4.3. Mature Phase

The mature phase will see the interconnection of the ISDN islands deployed in the intermediate phase. This interconnection will result in national and international ISDNs to bring about all of the benefits touted for ISDN. The key element that will trigger this phase is the availability of network-network signaling and the development of intelligent nodes to provide advanced service capabilities. This will allow both network providers and service providers to quickly introduce new services that can cross network boundaries.

While this phase characterizes the maturity of narrow-band ISDN, it also marks the beginning of a new era known as Broadband ISDN. Broadband ISDN will bring to the customer a host of new service capabilities that will be based on transmission rates that are greater than the current H1 rates (1.536 and 2.048 Mbps). It is envisioned that transport channels of approximately 140 Mbps will be available during this era. It is also envisioned that the user-network interface will have an access structure that will accommodate up to four broadband channels for a total transport capability of approximately 560 Mbps. These transport capabilities will support applications such as high definition video, high speed data, and LAN interconnection.

5. ISSUES

The realization of ISDN is dependent on the resolution of several issues. These issues cannot be resolved by a single industry participant; participation of all segments of the communications industry is required to bring about the development of ISDN. Following are several key issues of ISDN development.

5.1. Technical Standards.

Resolution of numerous technical standards issues at both the user-network and network-network interfaces is critical to the realization of the benefits of ISDN. Some of the more significant standards issues involve layer one of the Basic Access interface, supplementary service definition, call control procedures, and network-network signaling.

The specification of a layer one standard at the network side of the NT1 (sometimes termed the "U" reference point) is rapidly nearing completion. The capabilities that will be provided by this interface include full-duplex transmission of the 2B + D channels on a single twisted pair, an operating range of approximately 100% of the non-loaded loop plant population, and enhanced maintenance capabilities such as performance monitoring. By simplifying provisioning and maintenance processes, these network capabilities will result in cost savings and improved performance for customers.

The importance of specifying the services that an ISDN will support is related to terminal portability. That is the ability to use a terminal on multiple ISDNs and also the availability of a set of services that will be provided by all ISDN providers. The services that need to be specified include both basic services, (e.g. voice, circuit-switched data and packet-switched data) and a rich set of supplementary services, (e.g. call forwarding and call waiting) for both voice and non-voice services. This work is receiving high priority in both the national and international standards arenas and has progressed significantly since the 1984 CCITT Recommendations were issued.

The completion of the layer three call control procedures is necessary to eliminate vendor dependent interfaces and ensure terminal portability. The standardization of these procedures is a priority and has been focused through the identification and definition of an initial set of essential services. The national standards committee responsible for development of these procedures has focused their work on basic call control (circuit-switched), packet-mode using X.31 procedures, and supplementary services.

The issues regarding the network to network interfaces involve the interworking of the D-channel layer three protocol, Q.931, with the Common Channel Signaling

#7 protocols and security issues. The completion of these standards will provide a major step toward the interworking of individual ISDNs. This will result initially in national ISDNs and eventually international ISDNs.

5.2. Transmission and Performance

An ISDN must be responsive to a wide range of transmission rates and telecommunications techniques to support the routing of the various information elements required by the user.

When using telecommunication facilities, transmission errors are a major source of degradation in terms of lost information, which for data increase the response time by reducing the overall throughput.

A characteristic of some information processing systems is the exchange of many short messages between one location and other widely dispersed locations. In such a system the ratio of connection establishment time versus the time required to transmit a message becomes a critical factor in overall system efficiency. Thus, the ISDN should minimize the connection establishment time.

In addition, the ability to transmit information to a desired location without being inhibited by the underlying telecommunications network is essential. The user must be able to reroute messages, prioritize their delivery, and, in effect, manage the information flow within an ISDN for his own system requirements. Overall network performance must be pre-specified and deviation limits defined and published.

The planned ISDN should meet the requirements of many information processing systems, particularly when compared with the higher data transfer rates that will be offered for voice and non-voice services.

5.3. Tariffs, Pricing and Costs

A common objective in the establishment of tariffs should be to avoid discriminating against any class of users. Cross subsidization between various services should be minimized, and pricing for services should be related to the costs of providing that service wherever possible. Above all, the principles of pricing should be known and any cost interaction clearly indicated.

In an ISDN there should be no technological distinction between the various types of information. Thus, it will be easier to develop cost related tariffs for telecommunication services within the ISDN.

5.4. Operations And Maintenance

The capability for management of the networks by the network provider is an inherent attribute of the service. In the ISDN environment this function needs to serve the needs of information systems. The ISDN network is only one segment of such an information system. The primary responsibility in a data/voice communications environment for overall user system function and performance is the joint responsibility of the service provider and the user.

Inadequate diagnostic functions may adversely impact the user's ability to maximize the system's effectiveness, therefore the opportunity to utilize any new techniques available is the key.

Network initiated testing and maintenance functions should be designed to provide maximum system availability. Network maintenance procedures should be standardized to permit the provision of sophisticated diagnostic facilities by customers.

Today, users have a choice of telecommunications services to satisfy a range of customer requirements such as performance characteristics, transaction respone times, transmission rates, and end-to-end protocols. This must be carried forward to an ISDN and the first step is for network providers to understand users' needs.

5.5 User Migration

Users will only migrate to ISDN when the network services are widely available, are proven to meet suitable performance objectives, have acceptable pricing structures, and have perceived additional benefit over existing services.

Since the build-up is expected to be slow during the initial implementation phase, users are most likely to move over to ISDN in a controlled migration in accordance with availability. Several elements will be important in the user's decision:

- The disruption of the changeover, which will involve new interfaces and signaling protocols to be implemented in the network attachment. Currently installed network attachments either will have to be changed by field replaceable units or new units will have to be installed. In either case, a period of important change will occur which is subject to interface definition and standard.

- The date at which these changes should occur will demand careful planning. It may have significant business consequences if the dates of service availability are not firmly committed in advance.
- The attractiveness of changeover will depend on existing or improved services with new innovative pricing structures. Pricing is essentially made of two elements: the tariffs of the transmission facilities, which should be known in advance, and the cost of the new hardware/firmware/software required in the network attachments.
- The existence of interim services can provide a bridge between existing services and the future ISDN. However, the interim services could create a new generation of attachments. These may differ from both existing and final versions and their life, whether short, long, or imprecise may create difficult business decisions when considering investment in new network attachments.

6. CONCLUSION

The opportunities for the future ISDN are exciting. The demand for more advanced telecommunications is increasing and will continue to increase as we move into an era which is becoming more and more dominated by the information society and by information technology. The means by which future demands can be met are now available, and our task is to solve the various issues with which we are confronted so that telecommunications can continue to be an important element in our national and collective growth.

The price advantages of today's digital technology are driving the network providers to migrate their networks to ISDN as rapidly as possible. However, in order to maximize the benefit that can be realized from this important national resource, the following actions should take place;

- The customer should be consulted as to their requirements and the price/performance they might expect.
- Standardization that allows network interconnection and simple attachments on a worldwide basis is essential. This would facilitate international trade in all types of telecommunications services and support facilities.
- Discussions, open to all interested parties, should continue to take place to consider ISDN requirements and establish national and international positions.

Much has already been accomplished in the establishment of Recommendations for an ISDN by the CCITT. However, much remains to be done.

A Business Oriented Assessment of Technology

Ralph Krichbaum

Westinghouse Electric Co.
Pittsburgh, PA

This presentation describes the wide area (inter-site) networks that are currently implemented and planned. Westinghouse Electric currently has a backbone voice/data transmission network which utilizes digital T1 facilities where possible and cost effective. Three services utilize these backbone T1 circuits; the corporate voice network (WIN), the circuit switched data network (WESLINK), and the packet data network (WESPAC). The sites on this private dedicated network enjoy many of the functions of the planned ISDN network, plus the availability of channels at speeds lower than the 64 Kbps rate. The voice network uses ADPCM at 32 Kbps. For site data networking IEEE 802.X LANs are recommended. Providing network function, quality and cost effectiveness is people intensive in our current environment. Products and services designed to standards such as ISDN should make it easier for us to meet our objectives in the future.

6a. Future Network Direction

Strategies

- Combined Voice and Data Network -- based upon evolving Integrated Services Digital Network standards (ISDN)
- Optimum use of wide band service
  Voice and Data T1 = 1.5 megabits
  T2 = 6.3 megabits
  T3 = 45.0 megabits
- Justification based on economic benefits at each stage

1986
Voice/Data Network

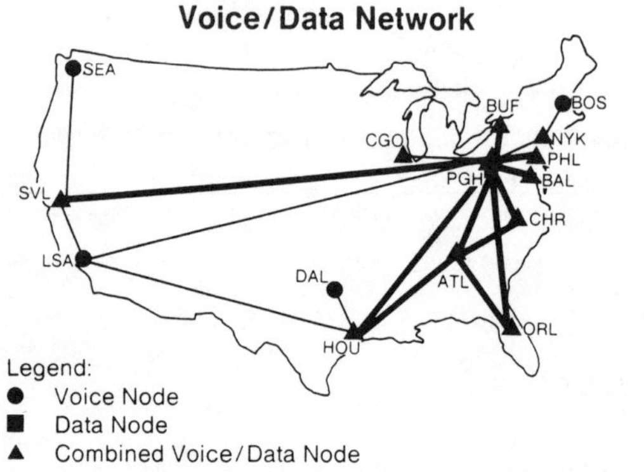

1989
Voice/Data Network

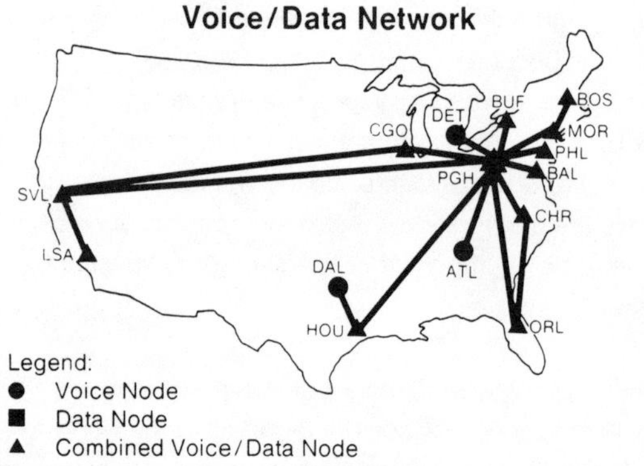

Integrated Voice/Data Network

Plan

- Improved cost performance
- Emphasis on quality, reliability, availability, and security
- Integration of voice/data switching
- High speed digital facilities
- Predominant use of fiber technology

6b. Ⓦ Integrated Services

This program is intended to provide a transport mechanism for the evolving information requirements of the company by implementing a Digital Network to support multiple services including but not limited to voice and data.

This will be accomplished by:

- Closely monitoring the evolving CCITT ISDN standards
- Favoring the utilization of digital transmission facilities for both the network backbone and access lines to Ⓦ
- Favoring local Ⓦ site telecommunication systems
  - Interface T1
  - Provide 64 kilobit switched digital connections
  - Utilize out-of-band signalling (SS7)
  - Permit multiple 64 kilobit connections between ports

Anticipated Corporate Benefits are:

- Economies and flexibility which the integrated nature of the network should permit
- Enhanced ability to provide support for new and changing requirements
- Economizes by avoiding redundancy and minimizing obsolesence

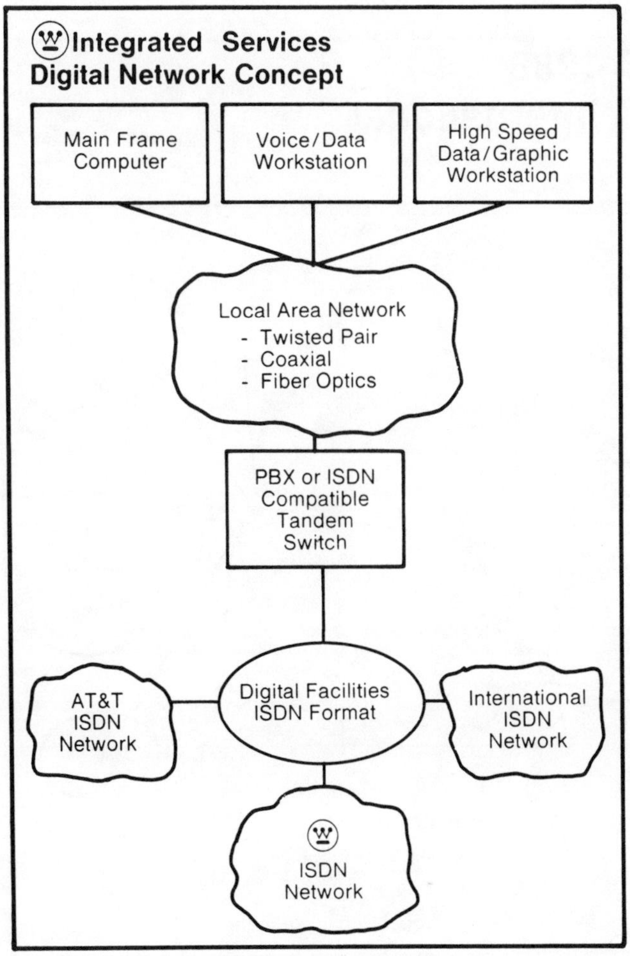

WESLINK

In 1984, Ⓦ installed its first regional node at the Ⓦ facility in Orlando to concentrate voice and data circuitry. Significant economic advantage was derived by combining these requirements in a common digital network. During '85 & '86, this integration continued with the installation of additional nodes in Sunnyvale, Chicago, Charlotte, Buffalo, Livingston, & Philadelphia.

Concurrent with these efforts to integrate voice and data, efforts were also initiated to integrate previously separate Ⓦ data networks by adding modern multiplexing equipment, not only at shared voice/data nodes but also at locations where voice/data integration was not economic. Deriving dedicated data circuits in this manner has grown so dramatically in the past two years that a name for the service seemed appropriate and thus **"WESLINK"** was established.

WESLINK provides point-to-point dedicated data service to Ⓦ Business Units at significantly reduced costs. Savings range from 20 to 50% below commercial equivalents. By year-end 1986, the network has expanded to nearly 100 multiplexors, from which Ⓦ derives more than 450 dedicated data circuits. Total network bandwidth now exceds 6.75 million bits per second.

WESLINK expansion will continue at a rapid rate during 1987. As the corporate voice/data digital backbone is extended to ever greater numbers of Ⓦ locations, **WESLINK** will also be expanded.

Field Support

This function has the responsibility of providing trained full-time technicians, equipped with the proper test equipment, at major network node locations.

Their duties are to coordinate installation of new services or equipment, perform first-level maintenance, assist in the isolation and elimination of network problems, and assist at remote locations as required for major system cutovers or changes.

In addition, the services of other technicians will be contracted on demand at smaller network locations. Due to the criticality of the networks, on-call service will be made available at any time.

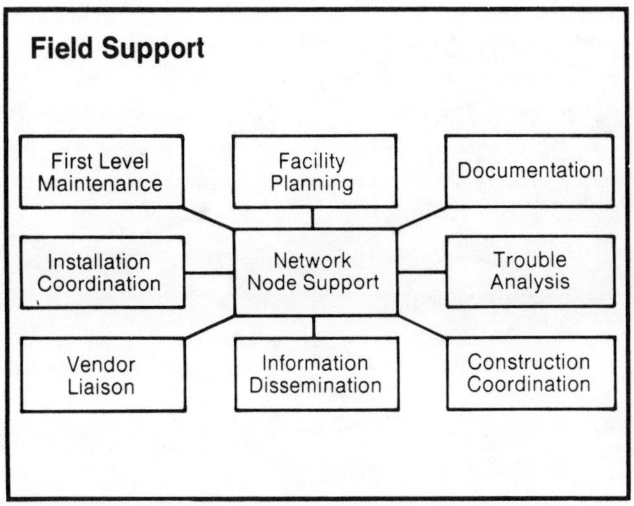

WESPAC

Competitive Advantage through Data Communications

1986 marked the first full year of operations for WESPAC, the Westinghouse private packet switched data network. WESPAC is a corporate data network providing simple, cost-effective communications for Ⓦ Operations world-wide. Based upon widely accepted standards, it is now used for information exchange with key Ⓦ customers and suppliers.

By year-end '86, the network provided reliable access to more than 250 Ⓦ host computer services. WESPAC is curriently accessed by Ⓦ personnel and customers in 37 different countries around the world. The WESPAC network backbone boasts of reliability in excess of 99.9% measured across 24 hours/day, 7 days/week.

WESPAC's value is greatly enhanced by connection to the public data networks. A continuing objective in 1987 is the diversion of public network traffic to WESPAC to reduce Ⓦ cost. WESPAC traffic currently exceeds public network traffic by seven times at approximately one-third the cost. 1987 expansion programs have also been scheduled to encompass a larger percentage of the Ⓦ interactive traffic with particular emplhasis on enhanced IBM protocol support.

A 24 hour per day operations staff supports WESPAC users via a modern Network Control Center. Skilled professionals attend to user problems in real time, working closely with Data Center Operations groups around the corporation.

Electronic Mail

Plan

Encourage the various Business Units to increase the use of the Corporate E-Mail system for Competitive Advantage to Westinghouse, through the increased use of the system by Customers and Suppliers.

Major Issues

- Growth of the system will continue, however at a slower rate than in the past
- Interfacing to other mail systems will continue to be a key issue
- Need for additional data base information will continue
- Move to a new computer operating system will become a requirement
- Upgrading of existing computer systems will be a heavy consideration

Major Strategies

- Implementation of a data base capability in conjunction with the E-Mail system
- Focus expansion of user base on Ⓦ customers and suppliers
- Continue to develop integration strategy with other E-Mail systems
- Develop a replacement for the existing TopCat access system

WESPAC 1986
USA Backbone & International

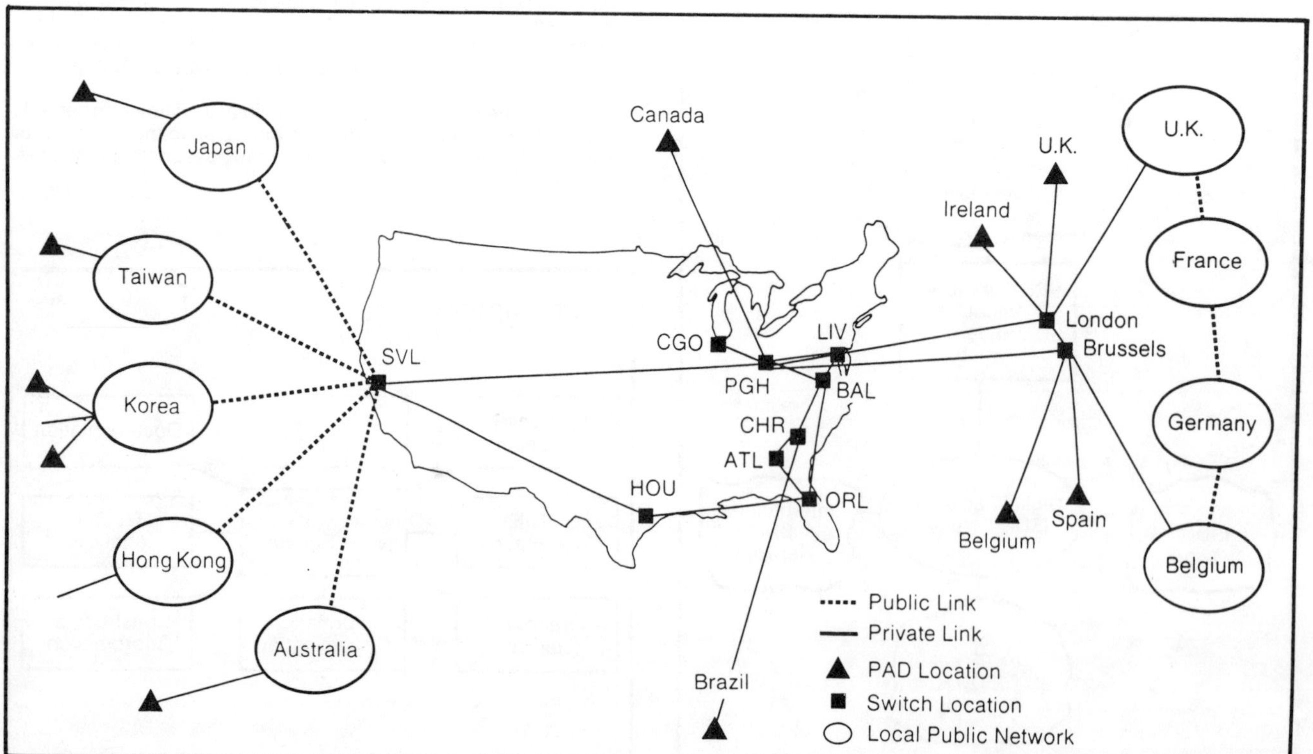

ISDN Regulatory and Business Planning Issues

Philip M. Walker

Telenet Communications Corporation
Reston, Virginia

ISDN technology offers many potential benefits to telecommunications suppliers and users, but it also raises several important economic and regulatory issues which must be resolved before its full potential can be realized. This paper focuses on two classes of economic/regulatory issues: (1) The cost of ISDN versus its benefits, and the consequent financial impact upon the telephone user and the carrier; and (2) the impact of ISDN upon competition in the intercity telephone and enhanced services markets. Two significant competitive issues with respect to IDSN implementation involve unbundled provision of D-channel packet data access to competing enhanced service providers, and ISDN numbering plan allocations to competitors.

Over the past ten years, ISDN has received much attention in technical publications, in the trade press, and at conferences such as this. Most of the discussion has focused on technology and standards issues, rather than economic, regulatory, and business planning topics. This paper examines two of the latter topics, likely to be the subject of much debate over the next several years. First is the cost/benefit aspects of ISDN, and its implications for the telephone ratepayer and for the telephone company's earnings picture. Second is the potential impact of ISDN upon competition in basic telecommunications services and enhanced computer/communications services. Consideration of competitive impact factors requires early resolution of issues regarding unbundled provision of D-channel packet data access to competing enhanced service providers, and allocation of space in the ISDN numbering plan to competitors.

1. ISDN COST/BENEFITS

Pundits sometimes irreverently referr to ISDN as a "solution in search of a problem," or "integration subscribers don't need." One of the reasons for such comments, I suspect, is that it's not entirely clear that the benefits of ISDN will justify the cost. To the extent that ISDN is viewed as technology for technology's sake -- carrying multiple channels on a single subscriber loop, but offering little in the way of new services -- it is fair game for skeptics to wonder about its cost/benefit tradeoffs and its implications for future telephone rates. ISDN is not cheap. Billions of dollars have already been spent by vendors developing ISDN technology and products, especially central office switches. Full deployment of such technology into the U.S. telephone network would involve investments of many tens of billions of dollars. The magnitude of such investment requires especially careful thought and consideration.

Where telephone plant expansion or modernization is clearly needed -- for example, replacing an obsolete electromechanical switch to meet increasing traffic volumes or to obtain needed new capabilities such as "equal access" (1+ dialing for inter-city calling via competitive carriers) -- the local exchange carrier (LEC) probably ought to purchase state-of-the-art switching technology. In a few years, this will probably be an ISDN switch. On the other hand, the decision to replace or upgrade a perfectly adequate non-ISDN electronic switch solely to obtain ISDN capabilities is much harder to justify. Thus, complete implementation of ISDN cannot occur overnight and, in fact, will probably take many years.

Regulatory questions regarding ISDN cost/benefits will most likely arise in borderline cases where the LEC seeks to install ISDN equipment to replace or modify existing switches that are not clearly obsolete or inadequate. To meet its public interest mandate, the PUC must satisfy itself that the LEC is not merely attempting to "gold-plate" its rate base in order to increase its earnings from the provision of existing services.* The LEC will need to show the PUC that its ISDN investment (a) will pay for itself by reducing the costs of providing existing services, and/or (b) will enable the LEC to offer new services for which a market clearly exists, at prices sufficient to recover the investment. There is little evidence to date, however, to support such a showing.

From a service standpoint, the benefits of ISDN lie primarily in the integration of multiple channels of communication -- voice, data and video -- on a single subscriber loop. This integration offers certain potential cost savings to large users, for example, by avoiding the need for additional intrabuilding wiring to support an increasing number of voice/data workstations. The ISDN digital loop will also support a wider range of data transmission speeds than is generally available today, and the Signaling System No. 7 (SS#7) capabilities associated with ISDN will enable telcos to offer additional call management and signaling features that will be valuable to some users.

Yet it is not clear how much the user is willing to pay for such new capabilities, and the telco runs a real risk that ISDN may not offer sufficient incremental revenue potential to recover its costs. Current and planned ISDN trials are focused on the demonstration of ISDN technology rather than evaluation of the market for ISDN-based services, and thus are unlikely to clarify this uncertainty.

The deregulatory environment surrounding the telecommunications industry today will also affect carriers' ISDN investment decisions. Currently regulators are considering new methods of price regulation that could replace the existing system of "rate base/rate of return" regulation for the LECs and AT&T. Such alternative proposals, including "social contract" and other price-cap approaches, would give carriers more pricing flexibility to deal with an increasingly competitive telecommunications marketplace. Rather than tying a carrier's earnings to the size of its plant investment, as is done today, under such approaches the carrier's earnings would depend upon its ability both to sell services and to control its costs. In such an environment, ISDN investments require increasingly careful business planning and market assessments, because a revenue shortfall would directly affect the carrier's bottom line.

In addition to the regulatory changes afoot, another factor affecting ISDN investments is the availability of less costly alternative technologies. One of the prime objectives of ISDN is voice-plus-data capability on the local loop, yet a variety of "pre-ISDN" derived-channel technologies have been developed recently. These technologies may be capable of offering all the voice-plus-data capacity needed by many customers, at a fraction of the cost of ISDN. Lack of standardization of such data-over-voice (DOV) systems has hampered their attractiveness to date, but their deployment is growing -- particularly for Centrex enhancement. Some LECs may find that such technology relieves much of the pressure on them to introduce ISDN prematurely.

2. IMPACT OF ISDN UPON COMPETITION

Achieving maximum efficiency in use of the local loop, combined with end-to-end digital connectivity and improved signaling, has been the principal impetus for the development of ISDN technology. Technological improvements in and of themselves are often judged to be in the public interest. In a monopolistic marketplace (as exists in most countries abroad) this is probably the case. The U.S., however, has determined that telecommunications services in this country will be offered competitively wherever feasible. Competition complicates the ISDN goal of maximizing efficiency in the local loop.

In the U.S., the LECs enjoy a monopoly in local exchange and (in most states) intraLATA toll service, while interLATA telecommunications and enhanced service markets -- both of which depend upon the LECs' local facilities for access to the customer -- are open to full competition. Further complicating this picture is the FCC's recent Computer III decision to eliminate its separate-subsidiary rules for Bell Operating Company (BOC) provision of enhanced services, requiring instead a system of accounting controls and the development of Open Network Architecture (ONA) to prevent anticompetitive conduct by the BOCs. At the same time, the BOCs are seeking permission to enter the information services and interLATA telecommunications markets under the Modified Final Judgment (MFJ) -- arguing that the FCC's new "nonstructural" safeguards (accounting and ONA) provide adequate protection for competitors.

As the BOCs begin to provide competitive services (such as enhanced data services) which depend on the local exchange "bottleneck" facilities, it becomes critical to ensure that other competitors are given equal access to such facilities and that the scope of the BOC monopoly is not extended beyond the bottleneck into the competitive domain. Consider the case of enhanced, packet-switched data network services. Such services are provided today by a number of "value-added networks" (VANs), on a nationwide as well as regional basis. These VANs, like other enhanced service providers (ESPs), are dependent upon the BOCs for the "last mile" of their networks, because only the local exchange carrier can provide the needed connections to the customer's premises.**

Heretofore, VANs and other ESPs have been able to obtain transparent circuit-switched and private-line connections from the BOCs for access to their customers, enabling them to support a wide range of data communication protocols as dictated by market demands. As the BOCs have begun to enter the packet-switched service market in competition with the VANs, they have utilized the same dial and private-line facilities for access to their customers. They have been required by the FCC to pay "comparable" tariff rates for such facilities -- adjusted for the fact that the BOC packet equipment is collocated in the central office (thus avoiding the need for outside plant loops to connect it to the CO switch) whereas the VANs are not so collocated.***

2.1 D-Channel Data Access

ISDN, as originally conceived, has the potential of upsetting this competitive balance in one important respect. The ISDN basic rate interface (2B+D) provides a facility for low-speed data transmission on the D-channel -- which has 16 kbps bandwidth and is used only sporadically for signaling messages. Since virtually all packet network connections to individual user stations utilize voice-grade dial or dedicated circuits today, and are thus constrained to 4.8 kbps (dial) or 9.6 kbps (dedicated), the 16 kbps D-channel offers greater capacity at a potentially attractive price. Current ISDN standards, however, do not provide for an unbundled interface to the D-channel handler in the C.O. switch, in order to allow subscriber data packets to be passed to a competitive VAN. Instead, all data packets on the D-channel would be passed to the BOC's own packet switching facility within the ISDN switch.

Such an arrangement would limit competitors to interconnecting with the BOC's packet network, passing their traffic through that network, in order to gain access to the D-channels running from the C.O. to their customers. This would represent an unnecessary and unwarranted extension of the BOC's bottleneck monopoly (here, the D-channel) into an activity that is currently competitive and should remain so (packet network facilities to which the D-channel is connected in the C.O.). Such an arrangement would also limit the functionality available to competitors and constrain their offerings to their customers, because data transmission protocol transparency would be lost. All customer data on the D-channel would have to utilize the ISDN standard protocol supported by the BOC packet network, rather than the multiplicity of protocols currently supported by competitive packet networks.

Finally, this restricted D-channel access arrangement would be at odds with the FCC's ONA policy, articulated in the Computer III decision, which requires BOCs to unbundle "basic service elements" of the network that are needed by competitive ESPs. Packet data transmission on the D-channel is just such a basic network capability and should be offered on a stand-alone basis, unbundled from the BOC's packet network service. Moreover, such an unbundled D-channel offering should support data transmission on as transparent a basis as possible, to enable VANs and other ESPs to offer a range of protocols to their users.

One proposal on the table to accomplish this unbundling calls for D-channel access to be provided that would be transparent above level 2 of the OSI hierarchy (standardized level 2 framing would be required for all data on the D-channel). Such access could be offered at the D-channel handler in each end office or, alternatively, data from multiple customers' D-channels could be multiplexed and trunked to a tandem office for hand-off to a VAN or other ESP. Recognizing the need for extensions to the ISDN technical standards to accommodate such an access arrangement, this proposal calls for the RBOCs to support expeditious action in the standards bodies, in order that switch vendors can begin implementation of this capability in their ISDN products at an early date.

It remains to be seen how this issue will be resolved -- through cooperative industry efforts in the standards arena, or through an FCC mandate as the agency evaluates the BOC ONA plans to be filed in February 1988. In either case, time is of the essence, because a substantial period may be required for (1) resolution of standards questions, (2) development and production of equipment capable of providing unbundled D-channel access, and (3) deployment of such equipment in the network. In the meantime, BOC packet services would enjoy an unfair competitive advantage in their superior access to the ISDN subscriber loop for packet data transmission -- a situation that cannot be tolerated for long.

2.2 ISDN Numbering Issues

A second area of discussion that may become an issue for regulators if it cannot be resolved by the industry involves the rules for assignment of numbering space within the international ISDN numbering plan to competing networks in the U.S. ISDN numbering issues have been discussed in North America for two years, first in the T1D1.1 committee and then in the ISDN Numbering Forum (INF). Yet several critical issues remain unresolved regarding the assignment of numbering capacity to VANs and other non-LECs. Five meetings of the INF have been held -- most recently, in September 1987 -- but these have not led to agreement among Bellcore, the LECs, and non-LEC competitors.

Issues in contention regarding assignment of ISDN numbering space include (a) whether a network must interconnect ubiquitously to all other ISDN networks in order to obtain such numbering space and (b) whether it must support an as-yet-undefined "minimal set of bearer services for basic rate/primary rate interfaces."**** Such pre-conditions for obtaining ISDN numbering assignments would deny unregulated, competitive network providers the right to make their business decisions regarding service offerings and interconnection matters based on economic and market considerations. Additional issues involve (c) the proper role, if any, for state regulators in assignment of numbering capacity and (d) the relationship between the ISDN numbering plan and the North American Numbering Plan (NANP) for the telephone network. If these issues are not promptly resolved, non-LEC competitors will be severely hampered in their ability to design and introduce new services which depend on numbering plan assignments.

Once the threshhold numbering issues are resolved, participants must determine which entity should be responsible for administration of the ISDN numbering plan: Bellcore, which currently administers the NANP, or a new independent not-for-profit organization created for that purpose. Bellcore and the BOCs have assumed that Bellcore will play this role but interexchange carriers and other parties do not necessarily agree.

3. CONCLUSION

ISDN technology can improve the efficiency and quality of telecommunications services and can make certain new capabilities available to users. It also provides equipment vendors with the basis for an entirely new product line. Thus, it is easy to understand the enthusiasm of carriers, vendors, and users with respect to the potential of ISDN. But ISDN, like any significant new telecommunications technology, raises economic and regulatory questions that must be resolved before its potential can be fully realized.

At the outset, the industry and its regulators must

understand the cost and benefits associated with investments in ISDN equipment. The industry must question whether technology for technology's sake makes good economic sense. Particularly in light of the trend towards partial deregulation of the telephone industry, carriers will have to take a much harder look at the cost and market implications of ISDN investments, and consider all available alternatives, in order to justify making them.

Moreover, the impact of ISDN on competition in telecommunications and enhanced services markets must be carefully assessed. The ISDN concept of a single digital "funnel" through which all subscriber traffic will pass contradicts the FCC's ONA concept of an "open" network -- unless all competitors are given equal access to that funnel on an unbundled basis. This implies, for the basic-rate D-channel, that ISDN implementations by the LECs must include provision for transparent transmission of subscriber data to the ESP of the subscriber's choice.

Finally, ISDN numbering plan issues must be resolved in such a way that competitors needing to utilize ISDN addressing are not artificially constrained and the LEC networks are not unfairly favored.

FOOTNOTES

* Historically, telephone operating companies in the U.S. have been regulated using "rate base/rate-of-return" methodology, under which a carrier's allowable earnings increase in direct porportion to the size of the carrier's authorized rate base -- i.e., the investment in plant and equipment needed to provide service to the public.

** Cable television systems exist in many localities, with coaxial cable into many homes and offices. However, such facilities are generally incapable of supporting switched, two-way data transmission and, thus, are of very limited utility to VANs at this time.

*** Competitors and others, including NTIA and the U.S. Department of Justice, have challenged this differential loop pricing policy and suggested that if the BOCs choose to deny C.O. collocation to enhanced service competitors, they should offer alternative equal-pricing arrangements for access to the exchange network. This concept, often referred to as "virtual collocation," is likely to be a topic of discussion for some time to come.

**** A T1D1 committee proposal, not yet ratified, would include a mandatory requirement that a network support D-channel packet data transmission by its users -- which non-LEC networks are incapable of supporting until they are given unbundled, transparent access to the D-channel, as discussed in Section II.A above. This requirement, if adopted, would have the unacceptable effect of denying ISDN numbering assignments to non-LEC networks until such time as the LECs provide unbundled D-channel access.

ISDN SERVICES: THE CHALLENGE OF IMPLEMENTATION

Gary J. Handler

Bell Communications Research, Inc.
Livingston, New Jersey

With the advent of the integrated service digital network (ISDN), the age of voice telephony, which has lasted over 100 years, will evolve to the age of information transport. This is the right response of our communications network to the needs expressed by the customers of the information age.

This paper will examine some of the services necessary for information transport and review alternatives within the architecture available to provide these services.

OBJECTIVE

With the advent of the integrated service digital network (ISDN), the age of voice telephony, which has lasted over 100 years, will evolve to the age of information transport. This is the right response of our communications network to the needs expressed by the customers of the information age.

Information, whether voice, data, or image, will flow over this network with the ultimate goal being efficient telecommunication. Here efficient telecommunication is defined as the successful interchange of information at a distance -- the required information delivered to the requested terminating entity at the right time. The technology is available to accomplish this now. The rate at which progress is made toward this goal will be determined by the degree of cooperative participation of the many players involved in providing telecommunications services.

This paper will examine some of the services necessary for information transport and review alternatives within the architecture available to provide these services.

WHAT WILL ISDN OFFER TO THE END USERS

What is ISDN?

CCITT coined the term ISDN to describe a general communication network concept that can efficiently carry all types of information in digital form. In addition, ISDN would allow flexible interconnection and interoperation with all other telecommunication and information networks and equipment. There are four key elements of an Integrated Services Digital Network architecture: end-to-end digital connectivity, integrated access, customer control, and international standards. In addition, ISDN is differentiated from today's network architecture by the use of a separate signaling channel which extends all the way to the customers' terminating equipment. One use of this channel is to allow the customers to interchange service control information with the network.

End-to-end digital connectivity provides better performance, and cheaper transport of all information versus analog. Integrated access offers opportunities to define and provide new services. Customer control enhances a customer's ability to manage telecommunications costs and features in a timely and reliable way. International standards ensure that all networks -- private, exchange carrier, and interexchange carrier -- can work in harmony to provide end-to-end-services. The use of a separate signaling channel enables customers to continuously send messages before, during and after call establishment, thus providing a very personalized control over the network.

What Are The Services?

For purposes of this paper, we will define a service within the ISDN architecture as a telecommunications service which involves network facilities that meet ISDN standards (i.e. have a 2B+d or 23B+D standard interface). Note that right away one faces the difficulty that some customers may not be interested in the channel construction as much as in the applications that ride over it. Thus, a service within the ISDN architecture is likely to have many components: the underlying channel, call management features, etc.

Who Will Offer The Services?

A typical service within the ISDN architecture will likely have five or six participants involved, in addition to the originating and the terminating customers. There may be two providers of customer premises equipment (CPE) at the two ends. These CPEs may be connected to the networks of two Local Exchange Companies (LECs). Finally, there may be an Interexchange Company (IC) and/or a Service Provider (SP) completing the picture.

Thus, to the question of who will offer services within the ISDN architecture one must answer: everybody. As complex as this picture is, one must add one more complicating factor: regulation. That is, even under the best of circumstances, the five or six players cannot simply cooperate and agree on standard technical approaches in a vacuum; their actions must be guided by rules designed by administrators guarding the competitive structure of the telecommunictions industry, and should be consistent with evolving technical industry standards. So the user, wishing to obtain telecommunications services quickly, must hope for the best: that all the participants in their endeavor still have the customer's best interest in mind.

SERVICE DESIGN ISSUES FOR ISDN

Figure 1 is a simplified view of the facilities involved in a service within the ISDN architecture.

Let us take a simple voice message that customer 1 would like to have customer 2 receive. As defined above, the success of the telecommunications service is dependent on achieving the customer requirements in a timely and cost effective manner. Let us examine what is needed in order to do that.

a) CPE Compatibility

Even assuming that both CPEs meet a network interface standard, there is no guarantee that they can communicate. Whether only one or both are ISDN CPEs, they may have compatibility requirements beyond the network interface standard. The question is, whose responsibility is it to provide compatibility?

b) Compatible Service Functionality

There are three networks involved in the call. One example of a service feature within the ISDN architecture is to let customers know who is calling. Will the customer identification be carried across the three networks? Will the CPE respond to the identity of the caller? Can the customers rely on the availability of compatible service functionality no matter with whom they try to communicate?

c) Network Management

Another important feature available within an ISDN architecture is network management. Suppose customer 2 is busy or not available. Whose responsibility is it to manage the call? Where will the call be sent?

THE ISDN CONNECTION

Figure 1

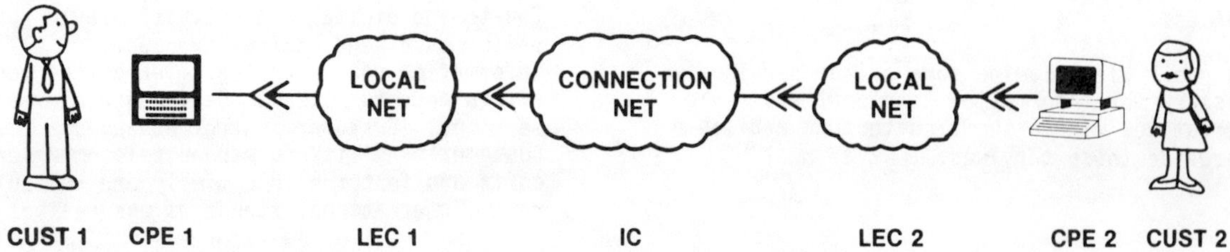

d) Call Completion

Remembering that efficient telecommunications services require not only the establishment of an end-to-end connection, but also the delivery of the message, who is responsible for accepting the message and delivering it later to customers?

Now consider a similar service need involving an information delivery which happens to be the result of a data base inquiry. Suppose customer 2 wants to retrieve information which may be stored in one of three data bases. Further, suppose customer 2 does this multiple data base access as a regular part of his/her business. The service needs of such an interaction are similar to the first example, except that in each instance they are more complex.

a) CPE Compatibility

The data bases involved should have no requirements forced on the accessing CPE. The data bases want to always be available to provide the requested data. Who is responsible to assure that different customers, with different CPE, can in fact access the data bases?

b) Compatible Service Functionality

The customer expects services to be available everywhere -- regardless of whether the data base is local or distant. Can in fact these services be made available everywhere?

c) Network Management

In the arena of data communication, network management is more complex. The integrity of the data is more sensitive to errors, the requirements for speed in call set up are more stringent, the need to complete the transaction tends to be more critical. With that in mind, one needs to ask again: who will supply the network management capabilities?

d) Call Completion

In the arena of data communications, call completion can also take on complex scenarios. The data may need to be stored for later retrieval, may need to be processed to a certain degree to ensure usefulness at either end of the communications path. Who will provide such services and can they be provided efficiently?

WILL THE SERVICES BE IMPLEMENTED?

As can be seen from the above analysis, the implementation of services within the ISDN architecture will pose complex challenges to all involved. The issue addressed in this section is whether in fact the environment is conducive for this implementation.

In this section the implementation challenges of the current environment are discussed. Success of the solutions to those challenges will determine the timely availability and the success of services within the ISDN architecture.

Challenge 1: Standards

With the number of participants involved in providing a service within the ISDN architecture, standards are critical; standardization beyond what has been experienced. For example, standards must encompass at least a minimum number of services that are provided and how these are provided. Otherwise, customers won't know what to expect on any particular call. All parties involved should work together to establish international standards. Standards when established should be voluntarily adhered to by all players -- major/minor, network/CPE, regulated/unregulated -- to maximize service efficiency.

Challenge 2: Integration of Information Media

ISDN's key word is "integrated". This implies that services and capabilities must be available for all types of information transport, regardless of whether the media is voice, data, or image. The regulatory trend in this arena has been contributing to the difficulty of this challenge. That is, regulation has tended to treat data communication differently from voice communication. For example, any necessary component of interworking was always accepted as an integral part of voice communication. In the case of data, protocol conversion has been singled out as an enhanced service. In ISDN, all information travels as digital signals, and voice call protocol conversion does not look any different than data protocol conversion -- the differences between voice and data have become blurred. Regulation needs to progress with the market and technology.

Challenge 3: Listening to the Market

One must always keep in mind the real reason for services: customer needs. If one is willing to listen, customers are fairly clear about what they want, and how much they are willing to pay for it. It is important that technologists and the regulators keep that in mind while creating an environment to fit the customer needs best, and not just to satisfy their own theoretical concepts, whether technical or regulatory. Thus, the challenges are to create the technical designs and to provide the regulatory support (emphasizing support) which bring about the solutions to customer problems. If all parties keep the customer in mind, there are many areas where they can work together to provide richer services, instead of limiting customer options to what one service provider has to offer. The work currently in progress related to the definition of basic service elements for an open network architecture is a concerted step in this direction.

Challenge 4: Complementary Functionality

To meet the needs of the customers of the information age all providers' services, even if competitive in nature, should be structured to be complementary with one another wherever possible. The accessibility of all providers to the basic service elements (BSEs) of the Open Network Architecture (ONA) will provide the complementary functionality necessary to meet the end-to-end telecommunications needs of the customer. For example, a customer who subscribes to ISDN might also be a candidate for voice mail service. If complementary functionality exists, a Message Waiting Indicator BSE could be accessed by the various providers to give a message waiting tone, thus providing a more convenient and complete voice mail service to customers. By using the concept of complementary functionality to create better services, total market potential can be increased, benefitting all providers.

CONCLUSIONS

The key to implementing services in an ISDN environnment is to recognize that the goal is to provide solutions to customer needs. The customer need may be as "simple" as communicating messages which originate in the voice medium, or as complex as providing access to multiple service providers who store and process data in large data bases. The issue is whether we are ready to accept the fact that the service these customers require is information transport, and that this country's voice networks must be transformed to information transport networks. If we are, then these networks must be designed, and be allowed to be designed, to manage the transport of information with the efficiency that the customer expects. This means that networks must provide open, standard interfaces which allow economic access by CPE, whether that be a simple digital telephone, or the sophisticated system of a Service Provider. This means that networks must provide the features which allow efficient, ubiquitous interconnection and communications, and that includes services that ensure compatibility, network management, and network control.

Without the removal of arbitrary distinctions among varous types of transport requirements, without the recognition that information transport is in fact the goal of the telecommunications services for the eighties, this country's networks will continue to slip in comparison to the rest of the world. Telecommunications has become a key component of the business environment. To ensure the competitiveness of American business, we must design and offer the best transport services.

ISDN SERVICES - MEETING CUSTOMERS' NEEDS

J. W. Seazholtz and R. F. Albers

Bell Atlantic, Arlington, Virginia

Technical feasibility, economics and customer choice should be the factors that determine the placement of functionality to provide communications services. Today's telephone related services provide such choices, although customer options have been constrained by regulation. Public policy initiatives to relax these constraints are underway, made possible not only by changes in the structure of the industry but in large measure by advances in technology. The introduction of ISDN capabilities represents a major advance in network technology that can - and should - result in more, not less, customer choice.

1. INTRODUCTION AND SUMMARY

The question posed for this session (ISDN Services - Embedded or Attached ?) unfortunately implies an either-or situation that would not be in anyone's best interest. Indeed, we feel strongly that the only answer is, ISDN services - embedded and attached. The world has changed far too much over the last couple of decades to allow any reasonable person to think that the future holds an environment of smart networks with dumb attachments, or dumb networks with smart attachments.

The forces of change are not only the explosive acceleration in technology development, which is what brings us to conferences like this one, but also in regulatory policy and, most important of all, in customers' awareness of technology and their expectations of its benefits. Customers' expectations, by the way, present some interesting dilemmas: on the one hand, our past successes have led to their taking technology very much for granted. As a result, those of us who have to deliver new services to them are sometimes amazed to see what they expect us to be able to do and how quickly they seem to think we can deliver it. On the other hand, those of us who understand the technology - for example, the digital connectivity and higher bit rates that we can deliver with ISDN, and the powerful concept of separate channel signaling - are sometimes a little frustrated by the fact that these same customers aren't falling all over themselves developing applications that will demand these new capabilities!

But, one thing about customer expectations does emerge very clearly: customers will not be patient with artificial limitations on their access to technology, nor to the services that the technology can provide.

2. SERVICES IN ISDN

One might begin by asking, what is an ISDN service anyway? The Red Book defines a service as, "That which is offered... to customers in order to satisfy a specific telecommunication requirement." Clearly, the two key words, if ISDN is to be successful, are customers and satisfy. Satisfy means to provide the complete service, if that's what the customer wants: it doesn't mean giving the customer only a commodity piece of the service, making him or her shop around elsewhere for the rest. Unless, of course, the customer wants to assemble the service from piece parts, in which case satisfy means that everyone who would cater to that customer ought to offer open interfaces with the option of piecing together the service - there's that word "service" again - with parts from multiple vendors.

The Red Book goes on to identify two major classes of service: Bearer service, which is,

> "the capability for the transmission of signals between user-network interfaces," (which sounds like what we called "dumb network" in the introduction)

and teleservices. These provide,

> "...the complete capability, including terminal equipment, for communication between users according to protocols established by agreement...".

Now, that sounds more like something that would satisfy a customer.

The protocols for use with ISDN are indeed being established by agreement - and the professionals represented at this conference are among those who are largely responsible for such agreements. The Standards give customers a great deal of choice about how they will satisfy their service requirements - choice which must not be limited by having regulators (or engineers, for that matter!) making arbitrary "either-or" decisions like the one implied in the session title.

3. SERVICES IN THE EVOLVING NETWORK

Let's look at embedded and attached services by drawing an analogy with today's telephone services. For years - decades, actually - we've been offering "custom calling" services, one of which is speed calling. This service allows a user to program a list of telephone numbers, and then to complete a call to one of those telephones by dialing just one or two digits. As an alternative to this network "embedded" service, a user may "attach" an automatic dialer, or a telephone that has built-in reparatory dialing. Customers have a choice of using either method, or even both, on any line. For example, many of us have a reparatory dialer at the office and subscribe to speed calling at home. At the office, we generally always use the same telephone instrument and might like the convenience of having the push buttons right next to the names on the list. At home, on the other hand, we may have lots of extension telephones, and prefer to use the network-based speed calling feature so that every phone can have the abbreviated dialing capability without having to buy an expensive automatic dialing telephone for every room. With ISDN, one could imagine having these same two choices, plus a third one: The "D" channel signaling message from the telephone set to the network could signify, "button number 14 has been pressed." And the network would know, because the customer programmed it much as is done today with the speed calling feature, that button 14 means, "place a call to 703-974-8006, using voice bearer capability, and include my name in the user data field of SETUP message."

ISDN is evolved from the telephony network. Wouldn't it be silly not to offer features like speed calling on ISDN lines (which are served from the same switch as the analog lines)? Imagine the success of the sales pitch for the ISDN access line: "Dear customer: We have a new ISDN capability to satisfy your every telecommunications need. P.S. - we just took away your speed calling."

4. NEW SERVICES WITH ISDN

So far, we've talked mostly about "backward compatibility" - being sure we don't take away any services when the customer changes from conventional access to ISDN access. But what of new services? We've already said that it would make no sense to withdraw previously offered services, and it would be equally senseless - perhaps even more so - to freeze the introduction of new services just when the underlying network "bearer" capabilities take a giant leap forward. The end-to-end digital connectivity, higher speeds and separate channel signaling offered by ISDN will make many new services possible.

One example of such services is encryption, for secure voice (and data) communications. A paper presented at ISS-87 [1] describes such a service, which entails combinations of B channel and D channel connections for challenges, authentication, and communications with servers that provide digital signature certification and session key generation. Since ISDN will allow users to connect to such servers inside or outside the network with equal fąCility, why should not the network provider offer such key management services? Who better to fill the "trusted agent" role for key management than the regulated local exchange carriers, for example? Large corporations, on the other hand, might find it desirable to maintain their own encryption servers for secure internal corporate communications. Again, customer choice.

Another important service in ISDN will be compatibility matching. Unlike telephone networks and services where open public standards are the norm, the data communications world has largely been characterized by proprietary, incompatible protocols and procedures. When the present networks are used for voice, the only compatibility problem is that the speakers must share a common language.

When the public telephone networks are used for data communication by means of modems, compatibility becomes a much bigger problem. In the US, value added networks and the private networks of large corporations solve these problems after a fashion: One common method is to allow the sender to store information in a host computer using one protocol/procedure set, so that the receiver(s) can retrieve it later, in a separate transaction, using a different speed, protocol or call establishment procedure. Procedures for communicating compatibility information with the service provider are often clumsy. For example, separate dial-in telephone numbers may be provided, one for each type of modem supported: the host may store a profile for each user, or might provide a menu to set options such as printer line widths and control codes, etc. Public data networks employing X.28/X.3 PAD services provide similar capabilities.

ISDN will go a long way toward resolving today's communications incompatibilities. Speeds and signaling procedures will be standardized. Information in the signaling packets will communicate bearer capability and service request information, using procedures established in public Standards. But the vastly increased range of services that can be supported on an ISDN access line will make the number of possible combinations larger. A customer might want to use a voice terminal with bit compression to provide a 7 kHz high quality audio teleconferencing capability, for example, but still would want to be able to communicate with analog voice subscribers, or ISDN subscribers who use ordinary PCM voice codecs. The terminal could be designed with dual codecs, but this adds cost and would not help in the situation where the customer desired a conference connection involving users with several different terminal types. Large corporate users might set up their own conference bridges with compatibility matching capabilities, but the smaller user would probably want a the matching service available in the network for occasional use.

All of which illustrates an interesting and important point: if services were only allowed to be provided by attaching servers operated outside the network, many services would only be available to the high end of the market - large corporations who could afford to set up and maintain the attached equipment, for example. Giving customers a choice of taking services from the network provider, or a third party, or providing it for themselves, will allow the broadest range of services to be available to the broadest base of customers.

5. WHAT ABOUT ONA?

In its Computer III proceeding, the FCC has required the BOCs to file plans detailing how they will provide an "Open Network Architecture" so that enhanced services can be provided efficiently by other parties as well as from within the Network. FCC approval of our ONA plans will allow us also to provide "enhanced" services from within the Network.

We believe that our network is already open, and getting more so all the time. For example, interface specifications for connection of customer-owned terminal equipment were published starting a decade ago. It's important to note that, while those specifications amounted to documenting the details of interfaces that already existed, we've assumed an ongoing obligation to publish any new interfaces before we start marketing any terminal equipment that uses them.

The steady increase in both the number and percentage of access lines served from stored program controlled switching systems has given customers and service providers more opportunities to increase efficiency and lower costs. Call forwarding, for example, provides a way for clients' calls to be delivered to an answering service without having to hard-wire an expensive loop to each client's line. We're starting to implement CLASS services, one of which will provide delivery of the calling party's number on an incoming call. Anyone can buy this feature when it becomes available in their service area, and we think it will be particularly useful to enhanced service providers. It will give them an opportunity to develop many new enhanced services, using the calling party identification to provide customized treatment to callers (not to mention billing them!)

The digital switches we're installing add to the growing list of open network capabilities. For example, call forwarding - busy line/don't answer on "POTS" lines as well as Centrex will be available. We're certainly motivated to keep adding useful new capabilities like these to the network. If our network service enables enhanced service providers to increase their range of services and their customer base, it's that much more network usage for us - a win for everyone. ONA, as we see it, is an ongoing, evolutionary, process - not a massive conversion to some special new network.

ISDN represents an important step in that continuing evolution. It allows network capabilities to be delivered to users and enhanced service providers more efficiently than can be done today, and provides useful new capabilities as well. An example of the former is delivery of the calling party's identity. Although this can be done today via expensive trunk-side connections or via the "CLASS" feature mentioned above, ISDN will replace today's several uses inband signalling methods with deliver of caller identity and other information about the call in a standard way, using faster digital signaling out of band. Examples of new capabilities include use of the D channel to enable new services such as meter reading, lighting of a "message waiting" lamp or inquiring about the availability of a particular terminal or terminal capability on an access line. All of these examples have appeared on our customers' "wish lists" when we've asked them about network capabilities needed for enhanced services, and all are things we can provide faster, better, or more economically with ISDN.

6. CONCLUSIONS

Economics and customer choice, not arbitrary policy decisions, should determine how a particular service is implemented. This means that customers should be free to use services delivered by their own terminal equipment, provided from within an ISDN, furnished by an enhanced service provider, or any combination. This includes using services provided by more than one of these on a single call, or different combinations for different calls or various stages of the same call.

We should all work together to use the rich potential of ISDN to give our customers more services and more options. And, in so doing, create more business for all of us.

REFERENCES

[1] O'Higgins, B., Diffie, W., Strawczynski, L., and de Hoog, R., "Encryption and ISDN - a Natural Fit," proc. International Switching Symposium 1987, paper A11.4

ISDN Services

Denis W. O'Shea & Thomas J. Pierce

IBM Corporaton

1. OBJECTIVE

This paper describes IBM's general views on ISDN and addresses the question raised for discussion at this conference session as to how ISDN services should be offered. User requirements, potential ISDN benefits and the roles of the network and other providers in satisfying those user requirements are discussed. IBM support of ISDN through standards, architecture and product planning activities is also described.

2. ISDN REQUIREMENTS AND BENEFITS

2.1 Potential Benefits of ISDN

ISDN promises to provide several potential benefits for customers. The implementation of a standard set of interfaces will give customers broad flexibility in the selection of vendors for ISDN customer premises equipment. ISDN might also provide significant economic benefits by allowing the sharing of a user's access lines to the network or by enabling reduced network charges for transmission and switching capabilities. In addition, ISDN might enable new integrated voice and data applications and enhanced services that could increase customer productivity.

The question then is not "IF" ISDN will happen, but rather, what will be the speed of customer acceptance of ISDN. The speed of acceptance will depend on how soon customers will be able to actually achieve the benefits of ISDN. Indeed, "WHEN?" is one of the most frequently asked questions about ISDN and one that cannot be answered easily.

2.2 User Requirements and Acceptance

In the area of standards, agreement has been reached on the standard protocols for transmitting user information over the ISDN Bearer (B) channels. However, additional agreement is needed before a range of products can be built to common standards. For example, the standards committees hope to reach agreement by the CCITT plenary session in late 1988 on the signaling and control standards for establishment of an ISDN call and a few supplementary functions like call hold and call forwarding. Plans will also be made at that plenary session for reaching agreement on other functions. Another factor related to standards is how and when uniform standards will be implemented by different networks to allow compatibility with customer premises equipment.

Customers will be able to begin to assess the economic benefits of ISDN after ISDN tariffs are filed by the local and inter-exchange carriers. The tariffs will describe both the actual ISDN services to be offered and the fees for those services -- both functions and fees -- with the services that are available today for their applications. Obviously, the greater the savings for the ISDN services, the sooner customers will want to migrate to them.

The speed of customer acceptance of ISDN will also depend on how soon new applications and enhanced services can be defined and developed. So far, enhancements to some tele-marketing and customer service applications have be described that make use of signaling information, such as calling party identification. Some customers may have such applications. However, other new ISDN applications which might benefit a broader set of customers such as those integrating voice, data and image, have not yet been described in sufficient detail to clearly identify ISDN as an enabler of many new applications.

3. NETWORK REQUIREMENTS IN PROVIDING ISDN SERVICES

3.1 Physical Connectivity ("Bit Pipe")

IBM's views on the use of ISDN in providing services are consistent with its previous positions on the use of carrier services. IBM has sought to promote the availability of innovative, cost effective services and the freedom to attach to and use those services efficiently. We continue to advocate that users must be able to use the underlying network in a flexible manner. It is, therefore, appropriate to evaluate the role of the network in satisfying user requirements. One such question is before this session today: will services over and above "bit pipe" or "transportation-specific" functions be provided to the ISDN user from within the network, that is "embedded", or will they be provided by service providers (including, but not limited to, the network operator) who use the network as a building block?

Such a question first requires definition of "bit pipe". In general terms, a "bit pipe" can be understood to be the transmission medium that provides physical layer (layer 1 of OSI model) connectivity. Related "transportation-specific" functions include the necessary signaling capability required to establish switched connections. Thus, in its simplest form, "bit pipe" or "transportation-specific" functions would embrace those core network capabilities that all users and providers are dependent on to enable a physical connection between end points. In ISDN terms this would include the signaling and transmission capabilities required to use a B channel in a clear or transparent manner. For these purposes, "transparent" access to a channel means that a) customers can use any end user protocol after an end to end connection is established; and b) that bit sequence independence is maintained by the network -- in other words, the very characteristics planned in the current standards for an ISDN B channel.

3.2 Services Over/Above "Bit Pipe"

After establishing that there are essential "bit pipe" capabilities that ISDN must provide, the questions of how and who should provide services above those core capabilities becomes easier to answer. Users are best served when a range of options are offered and selections made on the basis of an individual situation. Many providers, or users themselves, can and do build on these core network capabilities. Therefore, it becomes clear that the network capabilities should be available to all on equal terms to use as building blocks. All providers, including the ISDN operator, would then be free to provide a range of services over ISDN. Services that provide more than a "bit pipe" capability can include a broad range of optional services, from end user protocol support to storing and processing information.

In particular, services that provide interface support and conversion capabilities for end user protocols above the physical layer (Async, Bisync, X.25, etc.) can be offered by a range of providers, or a user can implement a private network solution with end to end protocol support in customer premises equipment.

4. REGULATORY REQUIREMENTS FOR ISDN SERVICES

4.1 Open Network Architecture (ONA)

The Federal Communications Commission in Computer Inquiry III has recognized that certain safeguards are required to allow enhanced service providers that do not own underlying transmission facilities to compete with enhanced service offerings of dominant carriers, since all are dependent on the same network facilities. In decisions relating to Comparably Efficient Interconnection for protocol conversion offerings (interim arrangements until ONA plans are filed), the FCC has determined that such "equal access" arrangements for enhanced service providers include access to underlying transmission and circuit switched services provided under the same terms and conditions to all. Such decisions promote competition and encourage growth in the important and rapidly expanding enhanced services sector.

In the ISDN environment, it can be expected that similar safeguards would continue to be required to preserve the competitive nature of enhanced services and the right of a user to develop private solutions. Thus, it can be assumed that FCC regulatory requirements will continue to be required to enable alternative providers of services based on an ISDN network operator's "bit pipe" to compete with similar services offered by an ISDN network operator. Such safeguards would promote competitive equity relating to access to ISDN and the terms and conditions under which core network services are obtained.

4.2 Additional ISDN Service Requirements

The FCC Notice and the Open Network Architecture discussions relating to ISDN have raised a number of questions regarding access to ISDN signaling information and access to customer data packets that may also be transmitted over the ISDN signaling channel -- the D channel. While the specifics of obtaining access to signaling information continue to be studied, some consensus is emerging relating to enhanced service provider requirements for a cost-effective means to receive data packets directly from their customers over that channel. Some enhanced service providers have also expressed their desire for an ISDN access which is transparent above the required D channel framing approach (layer 2 protocol). This would facilitate support for both X.25 as well as other networks and customer premises equipment. Certain extensions (for addressing, routing, etc.) to technical standards for the current layer 2 D-channel access procedure (LAP-D) protocol would be required to implement this approach.

5. IBM SUPPORT OF ISDN

5.1 IBM Standards and Customer Activities

One of IBM's telecommunications goals is to support international standards and we believe that

ISDN will be an important and widely accepted set of standards. IBM and ROLM products will, therefore, support the standard set of ISDN interfaces. IBM has actively participated in U.S. and international ISDN standards committees since 1980, submitting many papers and contributing to the progress of ISDN. We have also been meeting with our customers, discussing the potential benefits of ISDN and what is needed to realize that potential. Finally, we have established an ISDN system management organization within our Communication Products Division to coordinate our activities throughout the corporation.

5.2 ISDN Product Support

IBM is aware that the speed of customer acceptance of ISDN will also depend on the availability of customer premises equipment that supports the new ISDN interfaces. IBM is carefully weighing user requirements and the potential benefits of ISDN as it puts together plans to support standard ISDN interfaces on its products.

IBM's product support for the transmission of data using ISDN will be based on Systems Network Architecture (SNA). ISDN and SNA are complementary - ISDN forms a transport option for the SNA architecture. There are relatively few architectural changes required for SNA to support ISDN. However, product implementations will require significant effort because of the many IBM products that interface to telecommunication networks and because uniform standards have not yet been defined or implemented.

An important factor in IBM's development plans for ISDN is the strong belief that customers should be able to continue to manage and control their own networks. This will require that information about the network status be passed to users. IBM has made several submissions to standards organizations in support of this position, and its development plans continue to reflect the importance of providing its customers with user-controlled network management.

IBM continues to contribute to ISDN's evolution and to assess both the customer benefits and the work needed to achieve those benefits. We expect to have products available when our customers are in a position to achieve the true benefits of ISDN.

6. SUMMARY

ISDN offers technological advantages and several potential benefits to customers. It is important that ISDN standards be defined and uniformly implemented on the international level. It is equally important that those standards enable providers to develop solutions that both satisfy end user needs and enable users to satisfy their own needs. Effective participation in the provision of services and equipment by all interested parties would ensure innovative and cost effective solutions can be developed based on this new digital infrastructure. This can be best accomplished by providing "equal access" to the core transmission and switching facilities required by all.

ISDN -- AN APPLICATIONS MARKETING PERSPECTIVE

Charles E. Yates

Basking Ridge, New Jersey
AT&T

Abstract

Implementation of Integrated Services Digital Network technology means many things for both vendors and users in the telecommunications industry. But the ultimate promise of ISDN is the benefits it will bring to end-user customers. AT&T is working diligently to bring the benefits of ISDN to end-users through six key integrated applications of premise products and network services. Already, those benefits are being recognized and valued by customers throughout the country and around the world.

Introduction

When people today hear the words Integrated Services Digital Network (ISDN), they instantly imagine the new and innovative technology -- and the standards behind that technology -- designed to make ISDN a reality. But there is much more to ISDN than switching equipment, transmission lines and globally accepted communications protocols.

What will make ISDN as successful in the real world as it has been in the laboratory is customer benefits. In essence, ISDN's true worth depends on what the practical benefits will be to the business future and financial health of the end-user customers.

That's not to say that technology and standards are not important. Indeed, they are vital to the eventual success of ISDN for global telecommunications services. And the standards bodies throughout the world will ultimately define ISDN technology. That's their job and they are doing it well.

But it's important to remember that the people and organizations that make up those standards bodies are also vendors and users of communications services. The standards they develop will be based on their perception of problems within the industry, and on the ways that the solution of those problems can help assure their own future success.

The problems that businesses face today are many. In addition to the basic day-to-day business, companies must manage their customers, their own internal information and communications processes, and, in many cases, a host of suppliers and vendors. In addition, mergers, acquisitions, competition, new technology and the increasing globalization of most businesses bring even more challenges and stress to managers everywhere.

One of the key promises of ISDN is that it will make it possible to get the right information to the right person at the right time. That goal -- so simple yet so complex -- will ensure that corporate decisions can be made quickly, accurately and intelligently to insure corporate viability, and to permit managers to better control the highly competitive environment in which they work.

ISDN Benefits In A Competitive Environment

ISDN is designed to deliver some substantial benefits to this whirlwind environment.

For instance, ISDN's globally accepted standards will permit global connectivity and systems integration. No longer will end-users be held hostage by a single supplier of communications equipment. With ISDN, systems from all vendors will be compatible, and that will give end-users the freedom to choose vendors as they see fit.

ISDN also will help protect users' extensive communications investments by allowing a smooth migration from the systems in service today to those envisioned in the future. It will help companies plan for both the short and the long-term, with the knowledge that the information infrastructure so vital to business health will be there to accommodate their needs.

ISDN also will open the doors to more and more of the international marketplaces. With well-defined standards and the improved global connectivity that ISDN ensures, more and more companies will be able to take that often daunting leap and enter the multi-national competitive market.

Finally, ISDN will give end-users the promise of the future today. With the technology and the standards behind ISDN, customers will be able to define, develop and implement new and innovative services and applications almost at will, and without the constant, nagging fear that their communications system won't support the new application.

Benefits Of The ISDN Network

Integrated Services Digital Network is, in essence, a delivery path for a set of capabilities. There is no doubt that ISDN will provide a more powerful and flexible global information movement and management network for years to come. In fact, the definition of the ISDN network is broader than most network definitions in use today: AT&T envisions that the ISDN network will expand to include the desktop computing so indispensable today through wide area networks.

Standards for ISDN today include both digital access allowing for multiple, simultaneous channels, and the unique feature/function that a new protocol will provide. The effective use of these standards and protocols will depend upon the adoption of an architectural blueprint driven by communications-based, network intelligence.

Again, the benefits from the new ISDN network standards will be derived by the end-user customer. Some of the key benefits include:

-- Network design flexibility made possible through a dynamic circuit allocation environment that sells circuits and services on a call-by-call basis, or a dedicated basis, which ever is more cost-effective;

-- Improved reliability and performance thanks to virtual circuits and the added security made possible by the service node environment;

-- Growth and expansion through the ability to easily add both premises and service nodes to the ISDN network;

-- A cost effective customer network because ISDN will let end-users improve the utilization of existing services as well as protect the investment they have made in obtaining current capabilities;

-- Ease of use guaranteed through a standard international interface; and

-- Enhanced management and control through the availability of extensive statistical reports on how the network is being used.

-- Integrated applications through an added signalling channel that brings additional functionality to the applications environment.

The beauty of ISDN, perhaps its ultimate value for vendors and end-users, is that through communications-based intelligence, ISDN can make possible the easy handling of all transactions over a single system. Whether it be the simplest alarm circuit, or voice communications, or data traffic or full video transmissions operating at 386 kilobits per second, it can all be done via ISDN connections. Today, each different type of service requires a separate line.

ISDN And AT&T Integrated Applications

AT&T has identified six major integrated applications that drive the major thrust of business communication and networking needs today. They are:

- -- Telemarketing
- -- Voice Networking or network management
- -- Data Networking
- -- Disaster recovery, or crisis management
- -- Teleconferencing
- -- Electronic Order Exchange.

In tomorrow's ISDN environment, these applications undoubtedly will have increased value for the end-users.

In telemarketing with an ISDN flavor, a customer could call a business' order center and have their calling number automatically forwarded to the telemarketing service representative via the AT&T network. The customer data base would provide the telemarketing representative with information about the account even as the call is answered.

This type of ISDN enhancement should do two things for telemarketers. First, by giving quick access to account information, it will improve the productivity of the telemarketing representatives. They will be able to answer more calls more knowledgeably. Second, the added speed and knowledge will improve the telemarketing company's image to the caller and should help generate even more business. All thanks to ISDN.

With ISDN-enhanced voice networking, the end-user would be able to pool access to services over 1.5 megabit private lines rather than through separate trunk groups. ISDN would permit the services to be accessed on a call-by-call basis, and this would greatly reduce the ultimate number of channels needed, significantly cutting customer costs. As an added value, the customer's network configuration is improved dynamically.

ISDN also will provide better statistics on network utilization, permitting companies to better manage their communications networks through more extensive control over configuration and design. Those statistical reports also will make it easier for customers to add -- or delete -- facilities to their network on a cost effective basis.

Turning to data networking, ISDN will result in demonstrated improvements for end-users who depend on accurate transfer of data. For example, a terminal user could dial into a remote computing resource and the user-user information field of the Q.931 message could be used to send a password and login. That information, along with the station identification code, would serve very nicely as an initial security check. Besides improving security, this would reduce connect time and speed up access to vital information. The ISDN application would also reduce the initial input required by the end-user, thus making the data networking system easier to use.

In another example, ISDN would allow users to request bearer capability (B channels) of additional bandwidth to handle bulk data transfers. The ability to do this would increase the design flexibility of the network through call-by-call utilization of data networking facilities. It would also boost the productivity of both the system and the end-user by reducing the delay required to allocate facilities manually.

In both of these cases, ISDN capability leads to a decrease in the overall network usage time, benefiting the host, the network and the end-user.

AT&T believes ISDN also will play a key role in disaster recovery, or crisis management. ISDN's on-demand digital switching capabilities can provide instantaneous reconfigurations of network services and facilities, making it possible to reroute and restore communications services disrupted by natural or man-made disasters.

The digital circuit switching capabilities used for disaster recovery can also provide costs savings and efficiencies for customers using yet another application -- this one being teleconferencing. For example, video teleconferencing customers could use the ISDN circuit switching capability to set up such domestic or international links immediately, thus eliminating the need to reserve network time or to maintain separate, dedicated facilities that would be costly and that would often remain idle.

In the sixth AT&T-identified application, electronic order exchange, ISDN again will make a difference for businesses in the future. The user-to-user information and the station identification/automatic number identification could be used as a login, password and security check, similar to that used in the data networking application. In addition, the higher-capacity, on-demand switched digital services available through ISDN could be used to down-load the order at the proper location.

The end result: again, more efficiency, better service, better network control and management, and cost savings for the customer. All thanks to ISDN.

The Piece Parts

At AT&T, integrated applications are viewed as the ultimate communications solutions to customer problems. But, like any solution, applications are composed of a number of piece-parts. AT&T is working to bring ISDN capabilities and benefits to those piece parts as well as to the over-all application.

Today, AT&T is evolving all aspects of end-to-end ISDN intelligence from desk top to desk top. In the workstation environment, we are working on a number of products and services designed to benefit the end-user customer.

For example, by using terminal adapters, AT&T will protect customer investment in existing systems while still providing the functionality of ISDN. Workstations will be easier to use, thanks to push-button access to features on telephones, such as a leave-word calling button instead of a special code that must be punched in.

ISDN will enhance customers' existing investment in those systems, by providing switched access to multiple host on 3270 terminals. And productivity will be increased on the work stations, mainly through push button access to applications and features, as well as increased speed -- up to 64 kilobits -- on data applications.

Turning to distribution, ISDN will provide a standard interface connection: the 8-pin jack defined as a physical layer on the OSI model. There will be standard communications links -- either twisted copper wire or fiber optic cable -- for all devices, whether they be telephones, computers or terminals.

In early 1987, AT&T announced that its System 85 is the first premise-based private branch exchange with the ISDN primary rate interface. The PBXs are the intelligent premise node and support a precursor to the basic rate interface -- the 2B+d. That will provide 64 kilobits per second of clear signalling channels, with the final standard due in October of 1988.

Looking at computers, right now there are 120 licenses out to provide the digitally multiplexed interface to computer systems. Many of them use AT&T's ACCUNET (R) T1.5 Service for interconnection. As a result, customer will have 23 data channels taking up only a single port on their computer, giving them the ability to use more links on a stand alone machine. And, in the long run, this will give customers more independence from vendors since all suppliers will support ISDN through the universal standards.

In the service node area, the intelligent controller in the node will provide the ISDN primary rate interface to the premise locations.

Finally, we are bringing ISDN functionality to the local level through our No. 5 Electronic Switching System-- 5ESS (R) -- already equipped with ISDN technology.

In sum, then, AT&T's efforts -- along with international standards bodies -- are providing a smooth path for the ultimate evolution for customers large and small to the ISDN environment of tomorrow.

The Functionality Today And Tomorrow

AT&T is determined that ISDN will fulfill its promise of the future as quickly as possible. We have demonstrated a commitment to international standards in our implementation of the CCITT's ISDN PRI in both our network and in our line of premise communications products. We have proven to our customers that AT&T will protect their investment in data processing terminal equipment and that we will perform the needed conversions to ISDN BRI compatibility.

The promise of ISDN is really the working marriage of network and premise intelligence in a single coherent system. The premises bride brings to the marriage on-site distribution systems, integrated voice and data communications, gateway technology, multi-vendor support and a wide range of end-user applications.

The network bridegroom contributes as well, with integrated voice and data services, nodal services, routing control, network management techniques and network applications. The result of the marriage is ISDN: enriched signalling, access flexibility, services integration and applications development.

And, to take the marriage analogy one step further, the off-spring of the premises and network union is ISDN-based integrated applications.

AT&T is clearly the only vendor with both extensive network and premises intelligence and a demonstrated commitment to ISDN evolution that can perform and nurture the marriage of these vital components.

Today, AT&T's product line sets the standard for network management and soon we will unify such features as Advanced 800 Service Routing Control, Software Defined Network Service, DATAPHONE (R) II and ISN into a single network control system. AT&T also provides the most extensive network design support in the industry, with a staff of design specialists and the most widely used network design tool -- EINOS.

In addition, through our six key applications, we are bringing integrated solutions to our customers to solve their business problems. We are mixing and matching our network services and our premise products so that our customers will have the easy to use, cost-effective and efficient communications systems that they need to stay in business.

It is that ability to integrate and combine that sets AT&T apart. And that differentiation will be further enhanced as we continue to integrate premise intelligence and network intelligence into end-to-end ISDN applications and solutions.

That is the promise of ISDN. But there is a roadblock we must all attempt to remove. ISDN will be successful only if all vendors of communications equipment and services line up behind and commit to universally accepted standards and protocols. Then, and only then, will the promise of ISDN become reality.

ISDN: Impact On Carrier Marketing Planning

John H. Atterbury III

Southwestern Bell Telephone
St. Louis, Missouri

This presentation describes how Southwestern Bell is responding to the needs of the marketplace while conducting traditional market research activities. These activities are designed to allow the company to earn new sales revenues from the early deployment of ISDN while positioning the company as a market leader in the provisioning of new technology.

In the United States, the RBOCs, along with major vendors, are leading the way in the evolution to ISDN. New central office and customer premises equipment with ISDN capabilities are the focus of industry publications and discussions with investor groups, regulatory agencies and large business customers. And BOC trials are being mentioned daily in the same forums.

But ISDN is a reality not because planners are preparing for it, nor because vendors are manufacturing it and not even because telephone companies are deploying it... but because customers are buying it. At Southwestern Bell market potential is already being realized with the signing of three companies, Tenneco, Shell Oil and AT&T Network Systems, to long-term contracts that provide ISDN functionality. While planning still continues at Southwestern Bell, signed customer contracts prove that ISDN is not a technology looking for a market but a market, demanding services that the technology can offer. In any business, sales determine success, and Southwestern Bell has ISDN sales. We have *no* ISDN market trials, only ISDN customers and specific plans to ready ourselves to provide their service.

Tenneco is based in Houston and in February signed a 10-year contract for its 5,000 stations. Tenneco's profile includes multiple diverse subsidiaries all having heavy data requirements and multiple internal networks in various locations within Houston. The six primary locations are served by Centrex provided by Southwestern Bell Telephone Company.

Tenneco had their own perceptions of how ISDN could fit into their communications and business planning. It's probably best put by Louann B. Watkins, a manager of communications for Tenneco Incorporated, when she said, "The ISDN capability was selected because it provided the flexibility necessary to accommodate Tenneco's decentralization and diversification. In addition, Tenneco believes that the ISDN technology is on the edge of a new technology revolution in the field of telecommunications and it best positions Tenneco for all future technological advancements."

The second customer, Houston-based Shell Oil Company, signed a contract in March again for 10 years and approximately 5,000 lines. Shell's profile is different than Tenneco's in that their activities are concentrated in the petroleum industry. Although they are nationwide, they have two primary locations within the Houston area including a significant amount of leased space. But like Tenneco they have multiple and diverse internal networks and are an existing Centrex customer. Robert Kleinach, manager of telecommunications for Shell Oil, stated his company's reasons for choosing Southwestern Bell and ISDN when he said, "I want to use ISDN to integrate everything on the desktop that I can do based solely on today's economics. In addition, as I look farther down the road, I believe the ISDN alternative gets me into what I believe is the technology of the 1990s. Other currently available alternatives may not be able to accomplish that."

The third customer, AT&T Network Systems, signed a contract in January for their Southwestern Regional headquarters located in St. Louis. The contract for 2,300 stations will run for four years. Their profile is different from either Tenneco and Shell in that it is a single large location in excess of 400,000 square feet currently served by a Dimension 2000. Business functions including payroll and billing, procurement, marketing, and engineering for the Southwestern region are located here. Because of these functions it has extremely heavy data use and multiple networks. Douglas K. Jacobs, regional vice president for AT&T, stated his company's reasons for selecting ISDN when he said, "We expect this technology to significantly increase our productivity. This alone will result in an anticipated annual savings of one million dollars just in time and expense involved in moving data terminals." These three customers account for over 13,000 ISDN lines that will be in service in Southwestern Bell Telephone in 1988. Negotiations are also continuing with other customers. While the profiles of the customers are different, their reasons for buying are similar. They're looking beyond voice needs, for cost savings, flexibility, and an opportunity to avoid heavy investment of their own capital dollars.

How does Southwestern Bell provide this service without a tariff? Through our new Centrex custom digital offering. During 1986, Southwestern Bell Telephone Company reversed a trend that had affected its and other RBOC's older Centrex offerings. In 1986, SWBT led the industry by increasing our installed Centrex base by 10%. The Customer Specific Proposal process is based on generic tariffs that have been filed and are effective in each of our five states. These tariffs provide guidelines and parameters which permit the flexibility to design tailored communications systems for those large customers who have unique and specific communications requirements.

How is Southwestern Bell Telephone preparing to meet the service requirements? At Southwestern Bell the implementation of ISDN technology has two strategy focal points. The ISDN Network Implementation Team, chaired by the Network Department, is preparing to serve the first customer applications. Their goal is to complete the planning, network testing, loop qualification and other technical work that must be done to ensure that customers will have the technologically superior service they require. The Marketing-chaired ISDN Focus Team objective is responsible for all the activities necessary in delivering a general service offering to the marketplace. These activities include pricing and packaging decisions, promotions, training and market research.

The ISDN Network Implementation Team's primary activities include the establishment of the Advanced Technologies Laboratory and subsequent testing and evaluation of ISDN equipment. This process has been broken down into three phases. Phase One, which is just beginning, includes central offices in the lab to be tested as discrete islands to ensure compliance both to the standards that have been established as well as the specifications that Southwestern Bell requires in its equipment. Phase Two, which will begin later in 1987, includes interworking of the three switches between each other, between packet switches, Signaling System 7, local area networks and private lines. Phase Three, which will begin in 1988, includes interworking with interexchange network carriers. This phase is important because it will allow us to test interLATA centrex and equal access via ISDN.

Bellcore is supporting phases of Southwestern Bell Telephone Company's ISDN technical trials as it is the other RBOCs. Specifically Bellcore supports portions of the SWBT trial, include interworking of packet networks, interworking to interexchange carriers, interworking to local area networks, and interworking to nailed up private lines. The goal of these tests is to test technology functionality, and to develop methods that will be in place and ready for general deployment.

The ISDN Marketing Team efforts are directed at preparing for a general service offering. One of the first steps to this end was to establish demonstration centers and a demonstration strategy that would provide the ability to expose any stakeholder in the operating territory to a hands-on experience with ISDN. Practically speaking, demonstration centers provide the following four elements: first, exposure to ISDN to prove that ISDN is not just a concept; second, education based on the premise that you cannot sell what is not understood; third, as a sales tool to demonstrate applications; and fourth, as an applications lab giving customers an opportunity to experiment with the technology and to discover their own applications. Demonstration sites are established in a manner that will provide coverage to all customers in our territory. This includes permanent sites in Dallas, Houston, St. Louis and portable demonstration units that can be deployed throughout the territory.

The Dallas and Houston sites have been in operation since December of 1986. Through a joint marketing project with Ericsson Incorporated, the two sites are served by remote switching modules from an Ericsson AXE 10 host central office. Both sites are equipped with four positions that include 2B+D lines, 1B+D lines, Diavox 1000 sets with keyboards and displays, PCs, and terminals. Each location is also equipped with two group IV digital fax machines as well as one group III analog fax. These centers are unique in that they allow customers to transmit from Houston over 300 miles to Dallas over the network using a working ISDN switch.

The St. Louis location, which will be operational by early fourth quarter, will be provided technical functionality from the Advanced Technical Laboratory located in the same building. This lab will also give us the capability of using various CPE equipment. The Dallas, Houston and St. Louis sites are unique in that they are demonstration not simulation centers. All service provided at these locations will be off of working ISDN central offices.

In addition, four portable AT&T simulators have been deployed throughout our operating territory. This strategy provides maximum flexibility by allowing the delivery of an on-site demonstration of ISDN to any customer in our operating territory.

This is an aggressive demonstration strategy that will impact our customers in three ways. First, it dispels the myth that ISDN is a concept only. Secondly, the hands-on ability simplifies and reinforces understanding of what the technology is and what applications are available. Thirdly, and most important, it emphasizes that applications, and not technology, are what ISDN offers to the end user. ISDN, 2B+D, 23B+D are the buzz words that have been used to promote the technology. They are not what the customers will buy. Even integrated voice and data has no benefit as such unless it can be used to provide an application that has a benefit to customers.

Initial research indicates our early customers are the large and medium business customers segments and within that grouping, target segments such as education, health care, insurance, state and local government, and manufacturing.

This penetration is based on obvious benefits that these market segments may realize and on the fact that these customers are looking beyond voice services. First, cost stabilization, a system who's value does not increase from year to year but who's cost does is difficult to justify in a customer's mind. Second, one that has been widely touted is coax elimination, which should more appropriately be labeled cost economy. It is based on the elimination of multiple wiring plans with twisted pair which is the clear winner since it is the less expensive alternative and is already in place. Third, increased functionality and flexibility, which is at the heart of the ISDN technology, providing increased access to whatever lies ahead in the digital world. Fourth is system management, providing the potential for customers to control what their system does and when it does it. Fifth is to position for future services. ISDN provides the path that will allow our customers to access services not yet technically defined or legally permitted. Actually, many of these benefits that have been stated have a technological or vendor focused ring to them. Customers are looking basically for a service that can do it cheaper, that can do it better and/or that can do something new. Southwestern Bell is in time identifying the risks and rewards associated with delivering these customer-focused benefits.

Small business and residence customers may possess the greatest potential, but they are not leading edge. The deployment strategy that we intend to pursue and the technical maturity these customers possess dictates that they initially must be considered a secondary market. However, research is under way focusing on what the applications and demands will be in this lower end of the market. The keys are lifestyle improvement which includes entertainment, telemetering, home business monitoring and information services with a major focus on education.

Overall market development activities will include several applications development testbeds that will under our Discovery Program provide service to specific customers solely to determine applications that can be utilized by the broader customer base. Small business and residence also will participate in an information services testbed designed to provide customers in these segments with the ability to use these services, to evaluate them as far as benefits and cost, and to provide us feedback so that we can design the services that will be provided in the future.

Market research is our primary focus undertakings at this point in time as the unknown factors are identified. Through the use of large user focus groups and consultant studies, the risks and rewards of different pricing and deployment strategies can be quantified. Large user focus groups are being done cooperatively both independently and with vendor involvement. Customers who are purchasing ISDN as well as customers who have not yet selected ISDN as their service of the future will be participating. Consultant studies are based on the fact that outside direction and opinion are valuable in helping us design the services of the future. But in the mean time there are still many unknowns. What will our service offerings look like? Will it be 23B+D? 2B+D? B+D? or D? And what will the price of these service offerings or packages of services be? That has yet to be determined and over the next 12 months it is a primary task for all the RBOCs.

ISDN is not the total answer, but it is a major step. Since deployment will take some time we may not have it deployed soon enough for some of our heaviest data users. These users require broadband today while we are still struggling with basic and primary rate interface development. In addition, interim technologies such as C.O. LAN must be assimilated into our overall deployment strategy.

Our primary responsibility is to be responsible to rate payers but also ready to serve. Proper market planning allows us to meet these objectives. The bottom line is the customer must benefit through the deployment of improved applications functionality rather than the deployment of new technology itself. To sum up what Southwestern Bell Telephone Company's position is regarding ISDN, it would be easy to get excited about ISDN and its potential: in fact we are excited but we're also realistic. In the meantime, Southwestern Bell Telephone's approach is a "steady one" aimed at responsively committing resources to meet customers' demands and protect existing revenue streams, while at the same time being responsible to the ratepayers in our territory.

Blazing New Trails in ISDN: The U S WEST Customer Trial Experience

Robert C. Hawk

Mountain Bell

The purpose of all information technologies is to help customers communicate. That's the reason for the newest communications technology in our industry... the Integrated Services Digital Network, or ISDN.

For more than a decade, this network architecture has been a communications engineer's dream. A public telecommunications network that would not only integrate voice and data on the same line, but provide faster data transmission and greater flexibility for customers and international standards to increase the efficiency of information distribution worldwide.

That dream is now a reality. ISDN works. It works in the research lab, and we are learning how it works for our customers in their offices and work locations.

Mountain Bell and U S WEST launched the first customer trials in North America last year in Phoenix, when we placed the first ISDN commercial call on the public telecommunications network. But we did not rest on that milestone.

Since that first call we have made even more dust--nationally and internationally. For instance, we are nearing completion of the first phase of our six ISDN customer trials. Using four different switch makers and a dozen different customer equipment suppliers, we have been working closely with our customers--the Arizona State Government, Honeywell Bull, Intel, and U.S. Bank--to evaluate the viability of ISDN products and services.

In concert with these customers, we have been evaluating nearly 40 different customer applications including ISDN Centrex voice applications; private line replacement; packet switching on B and D channels; residential computer database access; home security and energy environment monitoring; SNA emulation; integrated voice, data and image; multiple terminals per access line; flexible adds, moves, and changes; personal computer networking; calling number identification; high-speed dial-up data interface; and access to multiple data bases, to name just a few.

These customer trials are giving us valuable experience in all phases of ISDN technology using digital switch protocols as well as ISDN analog adjunct switches.

Last March, we hosted a very successful technical exhibition at the International Switching Symposium in Phoenix. Nearly 4,000 delegates worldwide came to our exhibit and marvelled that ISDN really works. Many said they were amazed at how fast ISDN was coming to the marketplace. A telecommunications executive from New Zealand said, "ISDN is real. I'm going home to reassess my timing of this new technology. It's coming faster than I expected."

We have also completed extensive market research, focus groups, customer education efforts and product market evaluations to learn about the market, potential customers and the products and services that will be based on this new technology. Much has happened in a relatively short time frame.

One of the reasons we embarked upon the most comprehensive and complex ISDN customer trial effort in the world was that we wanted to do more than listen to ISDN rhetoric. We knew how much this new technology was being ballyhooed in the industry.

"Its significance rivals the development of the transistor, optical fiber and computer chip," we were told. But we wanted to do more. We wanted to put ISDN on customer trial. To learn all about the potential ISDN has to streamline communications systems and help customers manage their businesses more efficiently.

Now these efforts are paying off in the marketplace. 3M has signed an agreement with U S WEST and Northwestern Bell to provide custom phone service that is ISDN-equipped. The new system will replace five separate phone systems and provide 3M with communications efficiencies only dreamed about years ago.

Other customers are very interested in our ISDN efforts and are asking us for customer information and "test drive" demonstrations.

The key question is: will customers value ISDN enough to purchase products and services? We think they will. But for more important reasons then the Flash Gordon-type features it will provide.

ISDN allows customers to improve their current telecommunication systems. Key executives making purchasing decisions are interested in how this information technology can help reduce their overall costs of doing business or help increase their return on investment. This is why we feel it is important to partner with our customers to jointly develop ISDN's viability in the marketplace.

In each of our trials, our local telephone company is also participating as a customer. In Phoenix and Denver, Mountain Bell joins the Arizona State Government's Department of Transportation and Department of Administration, Honeywell Bull, Intel and GTE Communication Systems as customers. In Portland, Pacific Northwest Bell is participating with U.S. Bank of Oregon. In Minneapolis, Northwestern Bell is participating with a variety of customers.

Our customers have needs we think ISDN can meet. They want price performance. Will ISDN help them manage their businesses more efficiently? Will it be profitable? We believe the price performance improvement with ISDN should be in the area of 200-300 percent. While the price of an ISDN line will be roughly 1.5 times an existing Centrex line, customers will be provided with increased features.

Those features that now require separate lines for voice, facsimile and data transmission, can be provided over one ISDN line simultaneously.

Customers also want greater control over their communications systems. They want to move or change work locations simply and quickly. In our trials we have seen customers move their offices by easily unplugging, moving, and plugging-in at a new location--reconfiguring their networks based on need, time of day and other factors.

Customers want reliability. Not only faster, but more accurate transmission. ISDN digital accuracy minimizes re-transmissions caused by errors. Digital transmission is 99.9 percent error-free. It is clean, eliminating modems and other bottlenecks that can cause problems.

Customers also want ISDN to be available when and where they need and want. Not just ISDN islands, but national and international connections. During our ISDN customer trials we are transmitting packet data calls between Denver, Phoenix, Portland, Minneapolis and Seattle using ISDN technology from end-to-end. This will eventually allow customers to deploy ISDN products and services anywhere.

We are very active in pressing for internationally approved standards. This will help overcome the problems of incompatible equipment found in many industries. And with manufacturing economies of scales and resulting competition, the cost and price of ISDN products and services will decrease. Ultimately these standards will help bring ISDN to the marketplace quickly and provide a compatible global telecommunications network.

ISDN's viability will effect marketing plans. Telecommunications customers will not buy ISDN alone, just as a car buyer will not buy a V-6 engine alone. But in both cases, customers buy products and services based on a technology, whether that be ISDN or a V-6 engine, they perceive will bring them greater benefits and increased efficiencies.

For this reason, our market planning efforts are focused on customer needs. By providing customers and businesses with communications features they could get without buying their own expensive private systems, we will retain revenue and meet customer needs for efficient and effective information message transmission.

This philosophy, based on meeting market expectations and solving customer needs, will serve us well in a highly competitive industry.

For Mountain Bell, ISDN will help us meet the needs of our largest customers. If we can meet their demand for faster data speeds and integrated voice, data, facsimile and image transmission at reasonable prices, then they will look to us for transmission technology over the public telecommunication network rather than to expensive private systems.

ISDN may also allow us to provide increased features to the small business and residential market. In our customer trials, employees at Intel and Honeywell Bull are pleased with the results of using ISDN to work at home. This may prove to be a very attractive and lucrative market for ISDN services.

Additionally, we will be able to incorporate much of this technology internally to streamline our own communications systems. The bottom-line is cost efficiency for us and our customers and a more modern state-of-the-art public telecommunications network than ever before.

It is no secret that ISDN has the potential to bring new revenues to our companies. Our research and customer trial experience indicates opportunities exist to serve customers. And perhaps under an Open Network Architecture concept, ISDN may provide new revenue sources outside our traditional boundaries.

Other industries have experienced rapid technological change in the past few years, but the telecommunications industry has been relatively slow to adapt. ISDN will revolutionize our industry as never before creating new products and services and providing faster and more flexible communications solutions to meet the changing information distribution needs of our customers.

It is important that standards are adopted quickly, market opportunities thoroughly researched, marketing plans developed and alliances formed to develop technically efficient and cost-effective products in order for ISDN to meet the burgeoning needs of customers in a highly competitive Information Age.

ISDN is our key to the future. We are evaluating this technology in a fishbowl, per se, because of its importance to the world's telecommunications industry. ISDN -- the Integrated Services Digital Network, is tomorrow's communications today and Mountain Bell and U S WEST are ready to explore its potential for our customers.

ISDN Trials

Integrated Services Digital Network

USWEST

ISDN Integrated Services Digital Network

"High-Tech One-Stop Shopping"

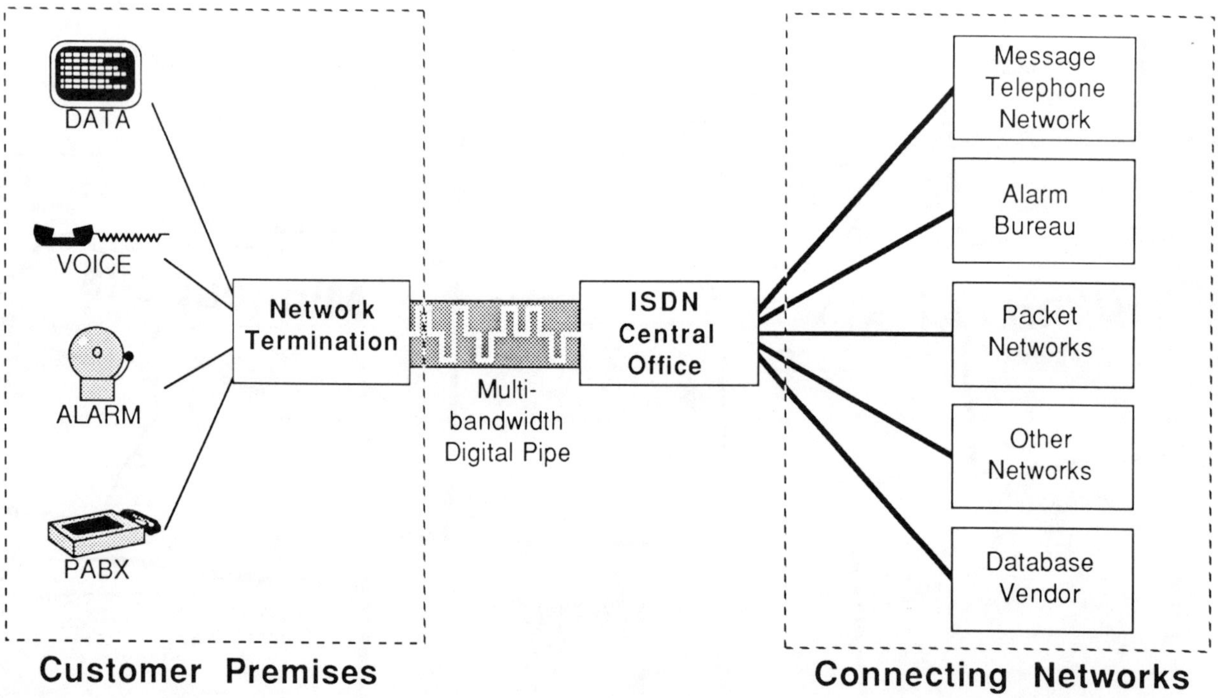

US WEST

ISDN *Integrated Services Digital Network*

ISDN Access and Application Examples

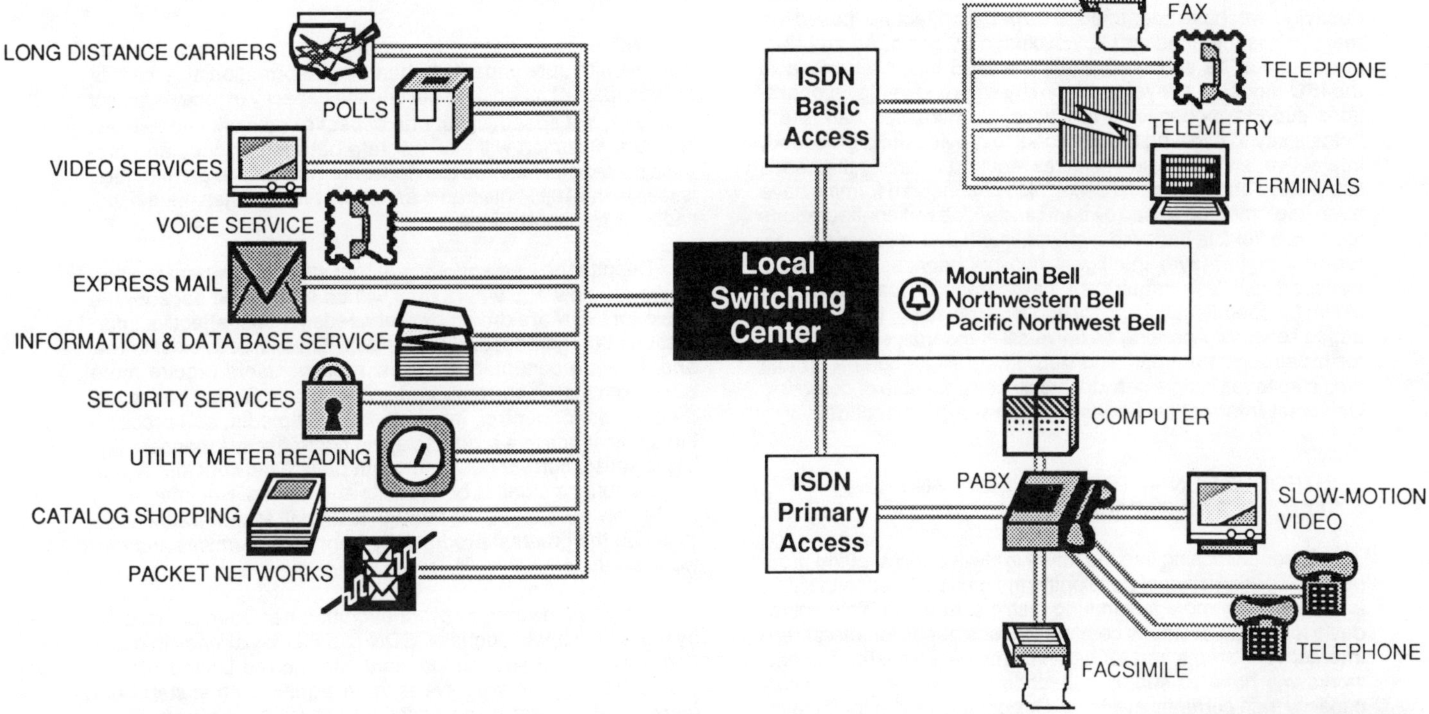

USWEST

THE FUTURE OF ISDN
An Architecture for the 1995 Time Frame

Dr. Peter K. Bohacek, Universal Information Services

AT&T Network Systems

ABSTRACT

ISDN provides an array of capabilities that will satisfy the information movement and management needs of today and tomorrow. Now that first-office applications have come into being, the pace of installing and expanding ISDN capabilities will increase rapidly over the next few years. In order to promote an orderly evolution that will result in greater information productivity, we have identified a target architecture based on realistic assumptions and key evolutionary goals. An architecture that can be achieved within the 1995 time frame follows the ISO model of a layered hierarchy where lower level operations support higher level functionality. Increased bandwidth holds a key to evolving the networks to provide the high-speed, interactive, and intelligent services required to satisfy the needs of business and residential users. Non-channelized formats have been implemented in narrowband and wideband configurations to provide flexible bandwidth management and to provide a platform for higher-layer intelligent network services. Bandwidth transport will also require an Asynchronous Transfer Mode (ATM) for operational simplicity and for extensibility to new value-added services. Agreeing on universal standards, such as ATM, for broadband transport and network intelligence is a critical next step to realizing a network architecture capable of delivering Universal Information Services within the next decade.

1. INTRODUCTION — TODAY'S NEEDS AND TOMORROW'S CAPABILITIES

Although looking into the future is always an intriguing prospect, the perspective of developing the next generation of ISDN is considerably more fascinating than just futurism. To meet today's identified business communications needs for integrated, interactive systems and for high-volume data transfer, the networks will have to supply subscribers with more bandwidth capacity than currently available. Customer pressure for increasing bandwidth will only grow stronger in the years immediately ahead. Therefore, some architectural issues we are currently resolving have to do with handling, processing, and moving information through the networks at rates ranging up to 150 or even 600 Mbps. At the same time, we are also looking at ways of introducing more intelligence into the networks and offering business users value-added features beyond mere transport.

What makes these network evolution plans so revolutionary is that today a network customer using a modem to transfer data is limited to a rate of about 9.6 Kbps. Even with modern data networking capabilities, we are still talking about relatively slow data rates. Today's data user is also confronted with numerous limitations due to incompatible systems. Therefore, data must often be converted to more complex formats carrying less information in order to be transferred over telecommunications networks.

Today's ISDN immediately increased network capabilities by providing 64 Kbps of bandwidth on a single channel. ISDN has also shown its capability of connecting otherwise incompatible data networking systems. But today's ISDN is only the beginning.

Within the next decade, we will see the evolution of a network architecture capable of handling information at rates four to five orders of magnitude beyond the capacity of today's fastest modems. Yet speed is just one aspect of network capabilities. Network evolution will also migrate both intelligence and processing capabilities into the networks. Therefore, architectural issues in the 1995 time frame are much broader than mere bandwidth or transport alone.

Despite the tremendous gulf that separates where the networks are now and where they will be in the next decade, the goals for ISDN are driven by real needs for cost-effective information management and movement. Those needs exist today, and they are continuing to grow rapidly. Users require more bandwidth than is currently and economically available to satisfy their needs for control, interactivity, multimedia, and processing applications in a productive and cost-effective manner. Trying to satisfy current needs through private network and equipment solutions alone is both costly and limited. But integrating public network capabilities through ISDN will allow network users to derive the greatest productivity from both premises and network systems.

Progress towards network evolution has been spurred on by first-office applications of ISDN capabilities already in place and operating in several locations around the United States. In less than a year, we have taken a significant first step from merely talking about the possibility of ISDN to where ISDN capabilities have been successfully implemented on production switches that are now serving business subscribers in Illinois and Arizona. ISDN is a present reality that represents a giant leap forward toward the architecture described in this paper.

2. THE PACE TOWARDS FULL ISDN QUICKENS

The pace of implementing ISDN is now quickening and will pick up even more so in the immediate future. By the early 1990's, we expect to see applications of Universal Information Services capabilities introduced to the networks. Universal Information Services is the name AT&T has given for a common set of goals in support of information productivity through the public network. Rather than a specific product or service, we see this as a feature-rich environment in which the network can provide any combination of voice, data, and image services with maximum convenience and economy.

Achieving the full potential of Universal Information Services benefits requires:

1. Broadband Services that can integrate voice with high-speed data and image.
2. Operations Migrating into the Networks to give end users a high degree of control.
3. Virtual Private Networks that behave exactly like hard-wired networks, except they will be software provisioned by the user through the switched network.
4. Intelligent Networks to support customer control and the virtual private network services, as well as the processing capabilities that can be provided more economically through the networks than by customer premises equipment.

The interests of everyone — end users, network providers, enriched services providers, equipment manufacturers, even standards bodies and regulators — are best served by network solutions integrated with customer premises equipment and private network alternatives. Everybody wins with network solutions. This network-oriented viewpoint also supports the rapid and cost-effective advancement of information productivity.

Target Architecture — Assumptions

In order to aim towards this potential requires that we make several assumptions about the target environment. Some of those assumptions are that:

1. We should see an overall decrease in regulatory constraints, and Open Network Architecture (ONA) will provide for a greater variety of services and service providers.
2. Integrated transport will give new impetus to end-to-end network services. Even subscribers who own private networks will start to integrate their own facilities with the service-rich capabilities of the public networks.
3. The use of fiber in the loop will be extended all the way to many business and residence end customers.
4. Photonics will support fiber cross-connect and selected switching applications. The effect of this will change bandwidth emphasis from compression to management and flexibility.
5. Although voice, data and computer applications will remain primary, video applications will become much more important.
6. Network control requirements will shape the architecture.
7. Significant advances in software productivity and techniques will have been achieved, supporting accelerated growth of both systems and applications.

Key Evolution Goals

Although current ISDN capabilities represent a major step forward, they are still quite limited. So far, ISDN deals only with narrowband services. It addresses access issues rather than end-to-end connectivity. And it provides transport which must be complemented with higher layer, intelligent, value-added services.

We must recognize that today's ISDN represents a first step in an ongoing evolutionary process. To move quickly beyond today's capabilities requires a set of goals that are driven by long-term objectives, urgent customer need, and available technology. Our goals start with current ISDN capabilities and address how to build on them to meet user needs in the future. We believe that these nine goals will guide the evolution of ISDN for the decade ahead:

1. To Facilitate Rapid Introduction of New Services by extending services network-wide from selected feature nodes and by allowing and encouraging network providers to create their own new services.
2. To Support Open Architectures for Network Services, in compliance with Open Network Architecture (ONA) requirements, encouraging external service vendors to offer their capabilities through the networks.
3. To Expand ISDN Data Capability, especially for data, by providing non-channelized interfaces at the ISDN primary rate (1.5 or 2.0 Mb/s) and higher interface to support bandwidth on demand and by migrating special services to ISDN for integrated access.
4. To Develop Virtual Private Network Features, including virtual private lines for voice and data, and software-defined networks that combine the control of private networks with the flexibility and efficiency of shared public networks.
5. To Introduce Capabilities for Broadband Services up to approximately 150 Mbps (optionally 600 Mbps) in accordance with emerging Broadband ISDN standards, including access, switching, transport, and operations suitable for video distribution and high-speed data applications.
6. To Automate Services and Facilities Provisioning so services can be established or re-arranged and underlying bandwidth allocated or restored across all interconnected network elements with minimal craft intervention.
7. To Offer Open Interfaces between network elements and Operations Systems (OS) and among OSs, while retaining the cost and performance benefits of closely coupled designs.
8. To Migrate Network Operations Traffic onto the Public Networks in order to use the OSs as sophisticated customers of the networks that will stretch and drive network capabilities.
9. To Continue to Improve the Transport Efficiency of network products, for example by incorporating synchronous optical transmission and integrated interfaces among network elements.

3. BANDWIDTH HOLDS ONE KEY

The narrowband BRI (144 Kbps) and PRI (1.5 or 2.0 Mbps) ISDN rates are becoming fairly well established. But future ISDN capabilities will require much more bandwidth. Some immediate applications for expanded bandwidth include:

- Video,
- LAN interconnect,
- Broadband CO-LAN service,
- CAD/CAM,
- And support of high performance work stations with remote file servers.

Up to this point, the focus of ISDN implementation has been aimed primarily at business users. Of course, business applications of ISDN capabilities will also be available to residential customers, expanding the concept of a home office to a level where support facilities and capabilities will be equally available at home as in the office. In addition, tests currently underway in a number of residential applications are beginning to indicate the efficacy of using fiber to distribute television to residential subscribers. Broadband ISDN could support television distribution at NTSC or better quality, with four active channels selected dynamically from an open-ended menu of programming available on demand.

One of the issues to be resolved in the 1988-1992 round of CCITT negotiations regarding ISDN will be the broadband interface rate. Current thinking sees a broadband rate of approximately 150 Mbps, with possible multiples of the 150 Mbps rate up to 600 Mbps and beyond. This interface could carry a broadband H_4 channel of about 135.168 Mb/s plus other narrowband channels, signaling channels, synchronization and operations overhead and network management information for controlling the use of the bandwidth. It seems clear that in the Broadband ISDN era, we can have an abundance of bandwidth, relative to our historical experience, and should therefore choose multiplexing formats on the basis of performance and robustness for unforeseen service demands, rather than for simple bit efficiency.

4. FUTURE MULTIPLEXING FORMATS

Further unresolved issues regarding future ISDN standards have to do with formats for multiplexing the broadband signal. SONET operates in a synchronous time division multiplexing format. However, the entire SONET frame can be organized into a non-channelized format called Asynchronous Transfer Mode (ATM). ATM is a packet-like system that can provide very fast and simple operations potentially extendible for higher-layer functions. ATM has the potential to support a wide variety of broadband and narrowband services on an integrated basis.

Many non-channelized broadband switching fabrics have one serious drawback, however. Unlike narrow and wideband non-channelized systems, which generally accommodate packets of varying length up to a few Mbps, many broadband fabrics achieve high speed at the expense of requiring fixed-length information blocks. Difficulty arises in selecting a length appropriately flexible and efficient for all types of traffic — ranging from extremely narrowband monitoring and call control functions to real-time voice traffic, to full-motion video or interactive CAD/CAM.

A recent AT&T innovation provides an answer to this particular difficulty. This approach would allow an ATM to consist of an arbitrary number of fixed-length cells chained together (Figure 1). ATM parcels with different numbers of cells could be used to carry:

- Signaling and control information following current ISDN Q.931 protocol formats;
- Operations information regarding network status, customer identification, billing, and so forth;
- Voice, data, video or other messages to be transported to appropriate user nodes along the network.

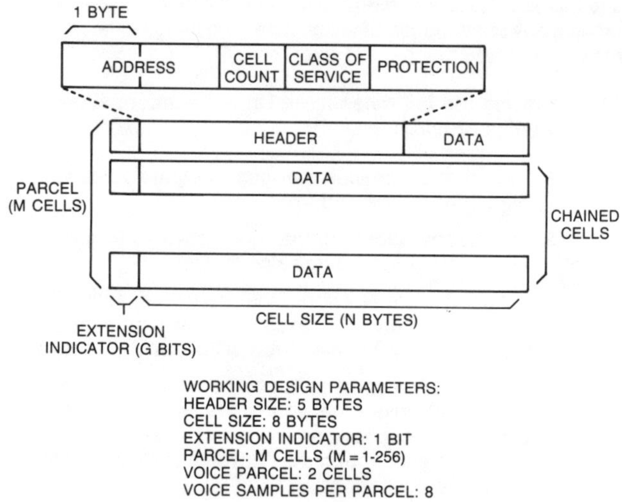

Figure 1. Proposed ATM Fixed-Length Cell.

This AT&T cell approach to ATM blocks is just one of several under consideration. As the evolutionary process moves forward and standards are adopted, one or more specific solutions will move to the forefront.

5. AN ARCHITECTURAL OVERVIEW — 1995

As we continue to progress in evolving the networks to intelligent, highly responsive, integrated communications resources, the target architecture (Figure 2) looks forward to a day when the network fabric becomes a single tissue. Functionality will be highly distributed, with some aspects of functionality located in the network, in customer premises equipment, or in both, depending upon economics, regulatory requirements, and other considerations.

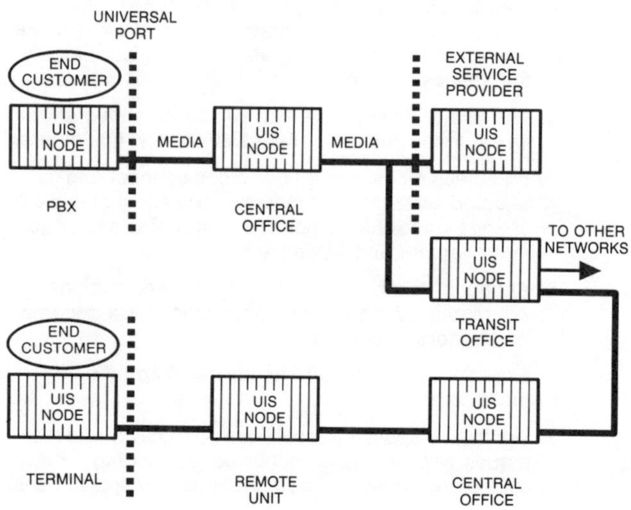

Figure 2. ISDN Target Architecture, ca. 1995.

Evolving beyond today's ISDN eventually leads to very simple architectural arrangements that consist of only three elements:

- Universal Information Services Nodes, connected by
- Physical Transmission Media, and accessed by
- Universal Ports.

The Universal Information Services Node would be an assembly of hardware, software, and possible human intelligence which can function with other Universal Information Services Nodes. A typical node might be a central office. However, any node which functions according to the architecture can be a Universal Information Service Node regardless of who manufactures it, who owns it, or where it is physically located. Thus compatible customer terminal equipment might be a Universal Information Services node. Enriched service nodes also constitute Universal Information Services Nodes regardless of whether they are located within a central office, at a remote switching site, or on the premises of third-party vendors.

The Physical Transmission Medium of choice in the future ISDN environment will be single-mode fiber. The use of a single-mode fiber for feeder in the loop and for long-haul applications is practically a given. In addition, current tests also show the efficency of using single mode fiber for distribution, to business and residence end customers. And current trends indicate that single mode fiber is being installed at an ever-increasing rate while the use of copper for new installations is decreasing rapidly. Two factors support these trends. First, costs for installing single mode fiber are decreasing quickly. Second, because of its tremendous bandwidth capacity, today's fiber installations will support any conceivable broadband capability well into the next century. In other words, fiber is an excellent and cost-effective transmission medium for the ISDN capabilities available now; it will also be an excellent medium for future capabilities. Thus, installing fiber cuts a path into the future of cost-effective, integrated communications.

Users would access and operate the network from one or more nodes through Universal Ports. The Universal Port will be a non-channelized interface providing direct connection to the network's native protocol. In addition to providing customer access, the Universal Port will also provide gateways to other nodes and networks.

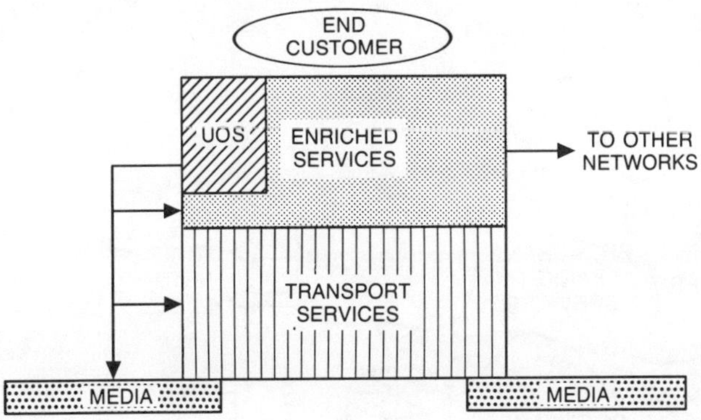

Figure 3. General View of the Universal Information Services Node.

The Universal Information Services Node

The functional architecture of a Universal Information Services node follows the ISO model of a layered hierarchy. Lower levels of functionality support the upper tiers. In the model of the Universal Information Services Node (Figure 3), we see two major tiers: Transport Services and Enriched Services, both of which sit on the Transport Media. Universal Operation Systems is included as an application in the Enriched Services tier, although one with a very specific role in the network.

The Transport Services Tier provides switched and non-switched communications among nodes. It converts the raw transmission media into a powerful and flexible transport network.

The Enriched Services Tier provides an open-ended set of software-driven, intelligent, value-added functions. It relies on and takes advantage of the Transport Services Tier. The Enriched Services Tier also converts the transport network into an enriched network which can be customized for each application.

Universal Operations Systems provides network operations support and end-customer network management for both Transport Services and Enriched Services Tiers. It is also a user of the enriched network and the transport services tiers to carry out its distributed processing functions.

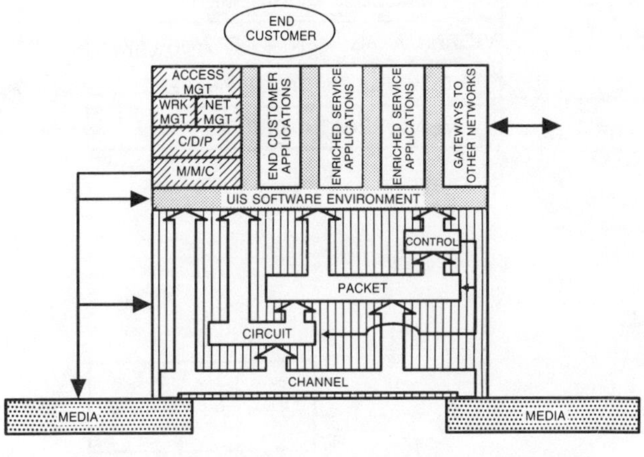

Figure 4. Universal Information Services Node Details.

A more detailed view of the logical structure of a Universal Information Services node (Figure 4) shows even further hierarchical divisions within the Transport Services and Enriched Services tiers and Universal Operations Systems.

- Transport Services are sub-divided into Channel, Circuit, Packet, and Control Tiers.
- Enriched Services consists of a Universal Software Environment (USE) and any number of applications, including various end customer and enriched services applications, along with gateways to other networks.
- Universal Operations Systems consists of Access Management, Work Management, End Customer and Network Management, Control/Diagnose/Prevent, and Measure/Monitor/Configure layers.

As a concept, we envision a future architecture (Figure 5) which is less rigid and far more open than any in place today. Single-mode fiber loops will support a full range of applications and interfaces — POTS, ISDN, broadband ISDN capabilities, video, and high-speed data. In addition to central office facilities, applications processors and feature nodes may be owned by and located with end customers and enriched services vendors; they will perform exactly like those in the central office.

A typical Universal Service Node might contain a core digital electronic switch, a photonic cross connect, and a multiplexer. Adjunct broadband switches will switch circuits of up to 150 Mbps and adjunct packet multiplexers and packet cross-connect systems will provide switching for voice and data up to 150 Mbps.

Applications processors will provide a wide variety of enriched services accessible from anywhere on the network. They will also support a software development environment to allow network providers to program new service capabilities they create specifically for their markets.

These Universal Services Nodes are interconnected by logical networks accessible from other nodes as well as locally. All of the logical networks will ride on the same underlying digital channels. Circuit switched and rearranged private lines will be supported at rates in multiples of 64 Kbps. Non-channelized ISDN Packet Mode (NIPM) networks will operate at 1.5 Mbps and non-channelized broadband services, supported by ATM facilities comparable to those described above, will operate at approximately 150 Mbps CCS#7 networks will provide signaling and control functions, supporting true network intelligence capabilities.

The total network will be supported by the Universal Operations Systems model. The implementation-specific "Measure, Monitor, and Configure" layer of the Universal Operations Systems may often reside in the network elements themselves, using the network as its distributed processing environment.

This target network architecture (Figure 5) is an exciting concept and quite doable within the projected time frame. More important, the evolutionary process has already begun with the networks as they are today. As the evolution progresses, it will essentially integrate all three basic tiers — Transport, Enriched Services, and Operations Systems — into a single end-to-end coherent network.

6. ARCHITECTURAL GOALS — SUMMARY

In targeting an architecture to be implemented over the next decade, the over-arching assumption is that local exchange

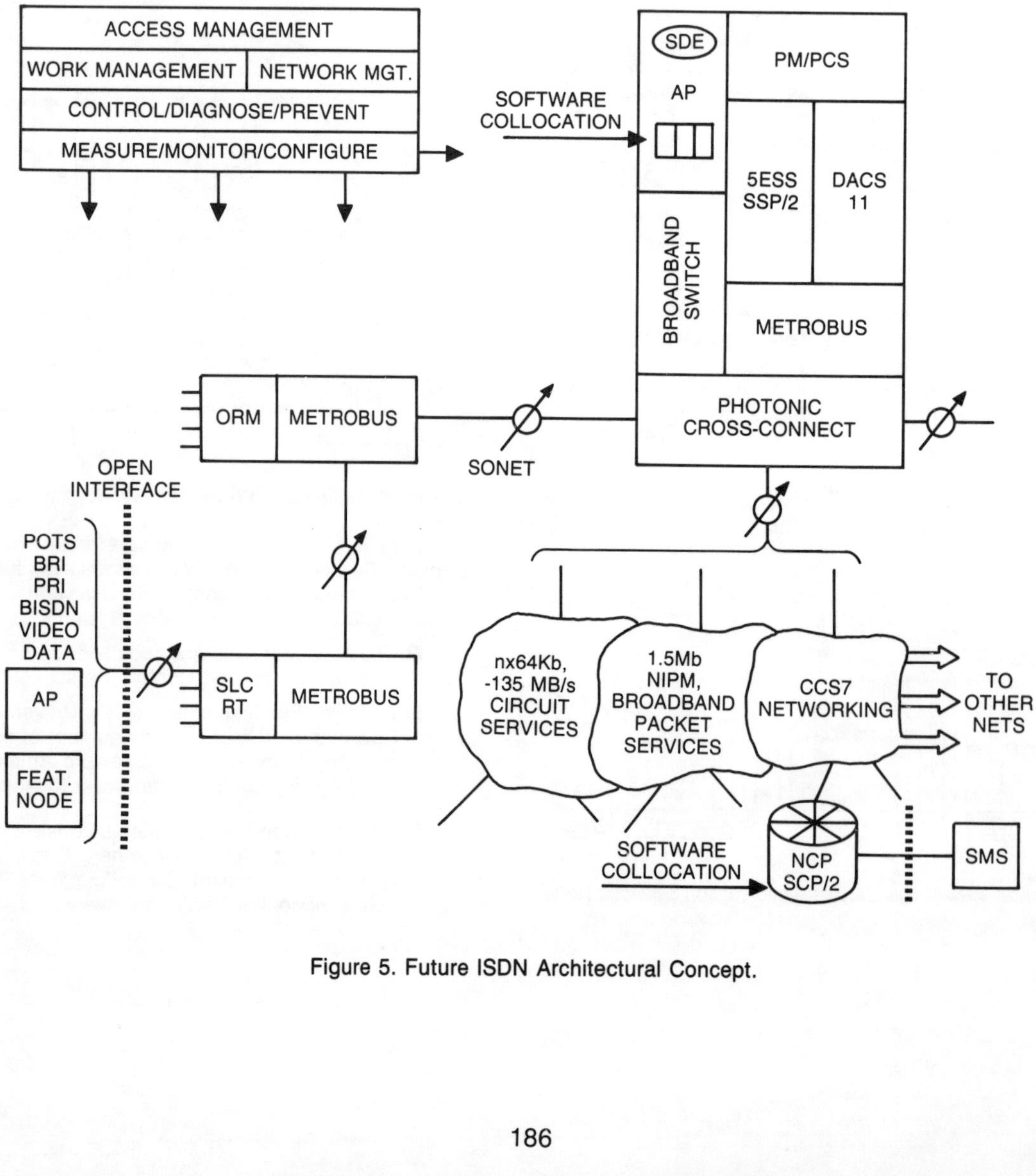

Figure 5. Future ISDN Architectural Concept.

carriers will continue to be the primary provider of information management and movement services to all network users.

Although some bypass may continue, the increased use of virtual private networks and their inherently more flexible, lower cost capabilities should replace and supplement private networks in many applications. Nonetheless, as we have shown, this target architecture will be able to access other networks as part of the Universal Information Services concept.

To reach that point in time when ISDN capabilities start providing that rich array of network-provided Universal Information Services, specific evolutionary steps must occur:

- Rapid Introduction of New Services must demonstrate the power of network intelligence to the subscriber base.
- Open Architectures Must be Developed, not only because they are required by the regulators, but because ONA will allow broader participation in enriched capabilities, stimulating further development of these capabilities by third party vendors, in turn leading to further network evolution.
- ISDN Must be Evolved to Non-channelized Packet Mode and Asynchronous Transport Mode to provide the flexible broadband transport needed for integrated voice and image as well as for information productivity capabilities.
- Virtual Private Network Features Must be Developed to offer the twin benefits of customer control and premium revenue potentials for network providers — benefits that can be implemented quickly at relatively little cost.
- Services and Facilities Provisioning Must be Automated, as a generalized aspect of virtual private networks and broadband capabilities. The more services and facilities are provisioned automatically, the more attractive the networks will become for end users, network providers, enriched services providers, and equipment manufacturers.

PETER K. BOHACEK is Universal Information Services Planning and Systems Engineering Vice President in AT&T Network Systems. He began his career with Bell Telephone Laboratories in 1964 with a variety of assignments on advanced radar systems. Between 1974 and 1983 he worked in switching systems engineering where his responsibilities included toll and local systems engineering and network services planning. In 1983 he moved to the marketing area of AT&T Network Systems. He assumed his present position in 1986 and is currently responsible for long range evolution of the network architecture, standards, systems engineering and end user market research.

Prior to his work at AT&T, he was a Ford Postdoctoral Fellow and Assistant Professor at the Massachusetts Institute of Technology. He attended Yale University and received a B.E., M.Eng., and Ph.D in electrical engineering.

ISDN IN 1995 – A NYNEX VIEW

Baily M. Geeslin
Casimir S. Skrzypczak
NYNEX Corporation

This paper provides a view of the Integrated Services Digital Network (ISDN) in the 1995 time frame. It takes a look at ISDN today as seen by users, network providers, and vendors. It then presents a NYNEX view of the future of the network architecture and services for the mid 1990's. It also examines market forces which will drive residence and business customer needs, as well as how the existing and emerging technologies of the time may evolve and coexist.

1. INTRODUCTION

In the year 1995, the Integrated Services Digital Network (ISDN), in the NYNEX metro LATAs, will provide the cornerstone for an architecture that will provide quality, innovative, and cost efficient services for the 21st century. In this paper we will discuss a view of that architecture and describe a set of elements that will either be implemented or in the final planning stages by that time. It will also provide a view of the marketing forces which will be influencing the residence and business customer.

2. ISDN TODAY

Before projecting ourselves into the year 1995, it may be worthwhile to take a short look at where this new architecture is today and review some of the significant issues confronting it.

2.1 The Market Push

ISDN has been recognized by technologists as a major step in the evolution of the public network architecture for many years. Recently, it has been embraced by the marketing segments of both equipment vendors and network providers and enthusiastically presented to the telecommunications user as a means to a variety of ends. In recent years, there has been a plethora of information and speculation regarding the benefits and capabilities of ISDN. This has generated a high level of curiosity among end-users as to what ISDN may be able to do for their telecommunications/data needs. Much of the available information has come from potential service or equipment providers who often project their own business/product applications into their descriptions of ISDN, leaving the potential end-user wondering, "what can ISDN do for me?"

2.2 The User Confusion

With the business telecommunications/data user community exposed to a heavy dose of ISDN hype along with much good factual information, the potential ISDN users are confused as to the benefits, scheduled availability, and price/performance relationships of this new architecture. With telecommunications playing an ever increasing role in their business strategies, users are concerned about being "left behind" their competitors technologically, yet they are not convinced ISDN will provide any significant additional service capabilities for the next several years. Many are therefore unwilling to include ISDN as part of their plans, and are taking a "wait and see" approach.

2.3 The Next Five Years

The next five years will prove crucial to the network evolution to ISDN. After the initial applications of ISDN by leading edge business customers, additional customers will need to be convinced of the visible benefits to their business operations. ISDN's success or lack thereof during this time, will largely depend on its ability to demonstrate to customers the operational and economic advantages it offers or the improved method of access to new and existing network transport capabilities. This improved access can be demonstrated in several ways, depending on the specific customer application. However, there can be little doubt that substantial savings in cost and/or productivity will be required by the majority of business customers to convince them to migrate to ISDN. The specific services themselves will be discussed later in this paper. It is mentioned here only to link their importance to the early success of ISDN. Initially, there will be a few truly new services offered on ISDN. What NYNEX will attempt to do is clearly demonstrate to potential ISDN customers, both the near and long term benefits that the architecture can bring to their specific business applications.

2.3.1 The Network Provider Challenges

The above scenario leaves the network providers with a considerable challenge. To continue to meet an increasing competitive challenge, they must explore new and advanced technologies for their potential benefits. Initially, network providers will deploy an access technology that will allow customers flexible transport to a host of services and data bases. NYNEX views access technology as a major challenge: to demonstrate to customers the abilities of the public network to access private or public information, and do this in a user friendly, ergonomic fashion. However, the full benefits of ISDN will only be realized when they are ubiquitously available, much like telephone service today. With many potential users playing a cautious waiting game, network providers will need to demonstrate convincingly, the future benefits of the architecture. They will need to enlist aggressive, leading-edge customers in exploring this new frontier. Other challenges the local exchange providers will be faced with are; regulatory issues, both local and federal, both national and international standards issues, as well as the need to establish successful relationships with embedded and new switching system vendors.

2.4 The Trials

Today, one only needs to scan the telecommunications trade publications to become aware of the numerous ISDN trials being conducted and planned throughout the country. NYNEX has also undertaken a number of trials with several vendors, both embedded and new. The trials provide valuable opportunities for the network provider to gain experience with the new technology and to better understand the operations and support issues which will impact the cost of ISDN supported services. The trials also provide the opportunity for business customers to see first hand, the types of service applications that can be supported with ISDN, both today and in the future. The network provider will need to nurture these trial relationships, for they are vital to demonstrating to the "wait and see" players, that a real and long term commitment is being made to ISDN. That it will not change communications overnight, but provide a robust architecture with far greater service capabilities to meet customers' changing needs in the future.

2.5 Uncertainty of Services

The new services ISDN will provide in the future are still unclear today. But it is certain that for ISDN to succeed it will have to provide subscribers with an added ingredient; a benefit or enhancement which will be viewed as a necessity. The new services will be used as a tool to give the end user a competitive edge. Businesses will be the first customers targeted. This new tool will affect the way companies do business. ISDN may not change the company's end product, but it will affect the internal and external daily workings of the company. A tire company would still manufacture tires, but communications between the factory and the distributors may change significantly. Information which today is sent over a private network or through the mail or given verbally over the phone, would with ISDN, be transmitted electronically over the public network. Distributorships around the country would communicate with the factory on a real time basis. Daily sales, shipping status, warehouse stock availability, and current pricing would be at distributorships' and factory personnel's fingertips. Information in a uniform format available to all personnel via the public switched network, can improve the daily operation of the business and change the way a company manages its business.

2.6 The Promise

Perhaps the true state of ISDN today is a promise of integrated solutions to customers continually changing telecommunications needs. Just as many years ago when the first copying machine was seen by several marketing surveys as "not needed, since everyone used carbon paper", so the true benefits of ISDN may be unclear to many potential users today.

3. ISDN IN 1995

We will look at two basic elements that make up the network in the 1995 time frame; the network architecture and the services.

3.1 The Network Architecture

The network architecture of 1995 will, as its foundation, provide ISDN access from local digital central office switches. The embedded base of analog stored program control (SPC) switches will either be capped with digital switches or employ digital adjunct devices to provide ISDN access. ISDN access, both basic and primary, will provide end users with a wide range of voice, data, and image services via a limited set of standard interfaces. Local loop plants will continuously be upgraded to meet evolving customer requirements.

Common channel signaling (CCS) based on the international standard CCITT-CCS7 will be an essential element in the 1995 network structure. One or more pairs of Signal Transfer Points (STP) will be located in each metropolitan LATA. The CCS network nodes, Switch Control Point (SCP), will be connected to the STPs via high capacity optical fiber or conventional T-Carrier links, see Figure 1.

ISDN/INTELLIGENT NETWORK

Network of the Future

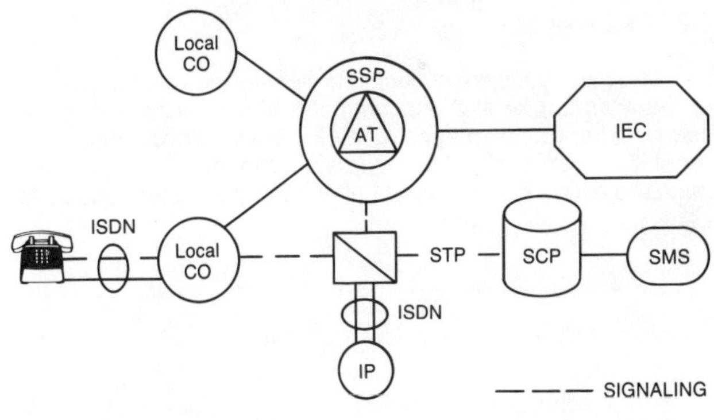

CO - Central Office
AT - Access Tandem
IEC - Interexchange Carrier
SSP - Services Switching Point
SCP - Switching Control Point
SMS - Service Management System
STP - Signal Transfer Point
SS7 - Signaling System #7
IP - Intelligent Peripheral

Figure 1

This robust out-of-band signalling capability will form the backbone for a new network architecture which is referred to as the Intelligent Network (IN). Initially deployed to support such data base driven services as 800 and Credit Card validation, this architecture will evolve, opening many opportunities for new services and user capabilities that will be data base driven. This network transport capability will provide for the proliferation of these services. Although these services may be accessed via traditional means, the availability of ISDN access will improve their flexibility and utility. Compatibility will exist between the out-of-band signalling provided by the ISDN "D" channel and the CCS-7 network out-of-band signalling. Together they will form the access linkage to the new Intelligent Network services that are being planned.

This network intelligence will allow the establishment of centralized ISDN customer "Profiles" to assist users in gaining access to network services quickly and easily. Network transparent services are what the intelligent network will bring to the user. Customers accessing a service via the intelligent network will interact with the service as if it resided within the user's own P.C.

The intelligent network will set up and transport information quickly through the network and the ISDN profile will provide the network service interface with information to facilitate effective communications between itself and the end user. This linkage of ISDN to the Intelligent Network via the ISDN "D" channel and Signaling System #7 (SS#7) will be a keystone to the acceptance of ISDN. ISDN will provide the most efficient means to access the vast potential of IN services due to the powerful set of "User-to-Network" signaling capabilities. ISDN customers wil be able to manipulate IN services more conveniently with far greater results than they could without the benefits of ISDN access. A valuable synergy will develop between these two separate, yet interworking structures, that will make ISDN the premier access architecture offered by exchange service providers.

In addition, wideband services will be emerging in the network. Wideband, as separate switching entities, or as features added to existing digital switches, will show its presence in the public switched network and add a new dimension to the architecture.

3.2 The Services

The mid-1990s are projected to witness the widespread use of many computer and telecommunications services that are just in experimental stages today. However, developments in regulation, competition, standards, and technology will determine the extent of deployment of new residence and business services.

3.2.1 Business

In the business market, ISDN will be well established, following significant deployment of ISDN/Centrex and Primary rate access to ISDN-capable PBXs. This success can only be realized if ISDN delivers benefits to the business customer in three key areas.

1.) Cost
2.) Simplified Access
3.) New Services

There will need to be significant improvements in the cost and simplicity of access to entice a large customer to consolidate several separate networks into an ISDN. The ISDN customer will realize economic benefit from the versatility and power of two 64Kbps channels and one 16Kbps packet switched channel delivered to the desk top. With user control over the bearer services, a variety of voice, circuit and packet switched data, and image services can be invoked by the user. Thereby offering the ISDN user the necessary capabilities, alone or in combination, when they are needed and where they are needed. A significant challenge to the ISDN provider will be to overcome the "S" Curve referred to by Foster [2] in which all new technologies, products, and services have a similar relationship between effort and performance as shown in Figure 2. Initially a large amount of effort is needed to realize little results. That is the present day situation with ISDN. By the mid-1990s there should be significant results from the effort and expense invested which would correspond to the middle of the S-curve. Later in the cycle, the upper limits of performance are reached at which time we again experience limited results from larger amounts of effort to improve performance. The position of ISDN on its S-curve in 1995 will be a key to its success in the marketplace in dealing with the customer's concerns relating to cost. The S-curve can also illustrate a challenge to realize benefits and user cost advantages into the mid-1990s as ISDN supported services displace other embedded analog services. Shown in Figure 3 with a pair of S-curves, Centrex service is used as an example of a service "Discontinuity". Rarely does a single technology meet all customer needs, which often requires two technologies to coexist in parallel. Effective management of their S-curves is often a key to economic success. Analog Centrex service has provided the end user with a rich and evolving set of service capabilities for over 25 years. However, in the 1980s it has reached a point in its technology life cycle where adding new capabilities requires increasing

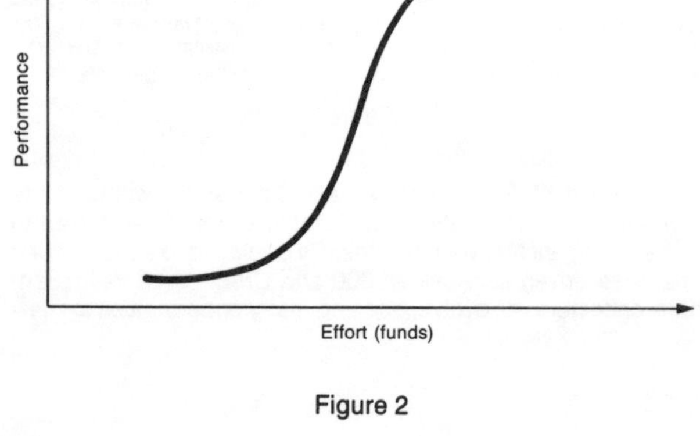

Figure 2

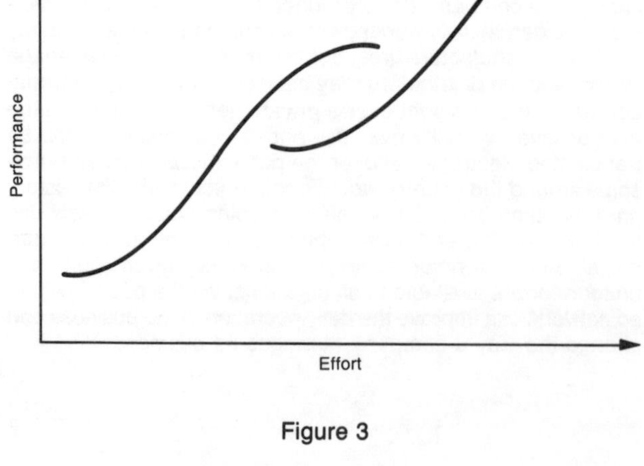

Figure 3

amounts of effort (funds). For example, much effort was required to provide analog Centrex switches with limited data capabilities (i.e., analog data above voice up to 19.2 kbps) via several non-standard interfaces. For this reason, NYNEX has begun to offer INTELLIPATH℠, and Digital Centrex services as a natural pre-curser and basis for evolving to ISDN. At the same time ISDN is also requiring relatively large amounts of effort for limited performance due to the large start up costs associated with a new and unique technology. However, the potential of this evolving technology appears superior to that it is displacing. A key to success for the ISDN provider will be in successfully matching these two curves to derive maximum performance from both while minimizing wasted efforts at the beginning of the ISDN curve and at the end of the Analog curve.

In the area of new services, simplified access to a variety of service capabilities will provide the ISDN customer with the needed flexibility to deliver virtually any service to any user, via a limited set of standard interfaces. ISDN will need to provide more than digital versions of today's favorite analog services. For example, ISDN Centrex services will need to provide ISDN specific elements which are not available through other services. The following are some application scenarios of ISDN business features which illustrate this flexibility.

Integrated Call Forwarding Terminating (ICFT):

This feature would combine several of today's features such as; Call Forwarding Variable, Call Forwarding Don't Answer, and Call Forwarding Busy Line [3]. It adds to these the ability to inform the calling party that their call has been forwarded and return the "forwarded to" directory number to the calling user. It can also inform the "forwarded to" party that the call they are receiving has been forwarded to them and include the directory number that forwarded it. ISDN, together with an Intelligent Network, will allow a served user to set up a customer profile with the local telephone company to "personalize" the feature. The following scenario may be one way customers "personalize" this service:

A medical supply company receives orders from three sectors of the medical community: hospitals, drug store chains, and small pharmacy stores. The company has divided its agents into groups corresponding to the three medical communities.

The medical supply company will give out one number to all customers for ordering purposes. They will utilize ICFT to enhance their order processing: a network profile will be set up which contains the directory numbers of various customers along with specific instructions to route the incoming calls to specific agents. Incoming calls which are not defined will be forwarded to an agent pool for appropriate handling.

A large hospital may have several employees purchasing medical supplies, each hospital employee's directory number or a block of directory numbers would be identified in the medical supply company's network profile. A hospital employee purchasing supplies dials the supply company's ordering number while ICFT is activated. The calling party's directory number is screened and if the directory number is identified in the profile, the call will be forwarded to an agent responsible for the hospital's account.

If the agent's line was busy the call would be forwarded to a group secretary. The calling party's directory number and the directory number that forwarded the call would be displayed on the secretary's phone.

If the agent does not answer the phone, the call would be directed to another agent. Again the calling directory number and the directory number that forwarded the call would be displayed on the agent's phone.

In the case where the calling party's directory number is not in the network profile the call would be forwarded to an agent pool. New accounts would have their directory numbers added to the supply company's network profile and an agent's directory number would be associated with these new numbers. Regular customers calling from unidentified directory numbers would also be forwarded to the agent pool where the new directory number would be associated with the agent who usually handles that account. The subscribers to ICFT would have control over the network profile to update and change routing as required.

Hospitals with computer systems may employ an electronic ordering system. They would, on their computer system, fill out an electronic form to order supplies from the medical company. The completed form would be sent to the supply company via ISDN. The call set up message, containing the calling party's directory number, would identify this call as a data call. ICFT would identify the appropriate agent, but since this is a data call, the "forwarded to" directory number would be the terminal on the agent's desk. If the terminal were busy or turned off, forwarding scenarios similar to the voice all would be invoked.

Facsimile service:

This service, available today, will be greatly enhanced with ISDN. High speed facsimile is available in limited areas. Text, graphic, and visual pages are received in black and white within a few seconds. Today's high speed facsimiles, 9.6 kbps to 19.2 kbps, must be connected to data circuits in the network in order for them to communicate with other facsimile machines. Computers in 1995, will be able to send text, graphics, and visuals to distant facsimile machines over ISDN. Bandwidth increases and flexible assignment of bandwidth will allow network providers and equipment manufacturers higher quality services. Facsimile machines operating at 384 kbps or 1.536 Mbps may be able to print out high quality color pages and may compete with color graphic plotters in the future. Optical readers in the future will provide another type of facsimile service; paper text, graphics or visuals may be scanned optically, and the information that resides on the paper would be processed, formatted and filed into a computer or data base. Once the paper information is entered into the data base, users will have full access to the file. The following is a possible application of facsimile features in a future business environment:

Law firm A, located in New York, subscribes to ISDN services. Each lawyer's desk is outfitted with a workstation and multi-button programmable phone. The law firm has several optical readers which enables them to enter paper documents into their data base. They also have a facsimile machine.

Law firm B, located in Boston, is a small firm and does not have ISDN capability. Their office has a high speed facsimile machine, some PCs, a printer and a LEXIS® machine.

Law firm A has received a request from a client to prepare a contract. The contract is technical, and graphics will be included. Law firm A prepares a contract on its workstation and utilizes the optical reader to enter the graphics into the contract document. Law firm A contacts law firm B by phone and wants to transfer the contract to their location. Law firm A utilizes ISDN service to transmit the contract document from its workstation to law firm B. Law firm B's high speed facsimile machine receives the contract while the firms are conversing.

Law firm B reviews the contract and contacts law firm A. Law firm B has amended several clauses in the contract and wants to send and review the amendments with law firm A. Law firm B calls law firm A's ISDN facsimile machine and sends the amendments. The amendments, received while they are talking, are minor and both firms agree on the changes. Law firm A, the originator, wants to expedite the process and asks law firm B to hold on while his secretary updates the contract. Law firm A accesses the contract file, updates the file, stores it and now sends the final contract to law firm B's facsimile machine. They review the changes again and law firm A states that a signed facsimile will be accepted by his clients to get things started.

In addition, a range of higher bandwidth services will find acceptance and usage, such as the following:

1. Conference services - Various types of teleconference service will be possible, at different bit rates in this time frame. At the lower bit rates, audio can be combined with high resolution graphics over switched channels to add overhead transparency capability to long distance conferences. At bit rates up to 45 Mb/s, switched video transmission may be appropriate for interconnection of business offices for informal communications, augmenting normal telephone service with dynamic allocation of bandwidth.

2. High resolution engineering graphics - This service would support transmission between centrally located image processors or supercomputers and personal workstations for applications of computer aided design.

3.2.2 Residence:

In the residence market, home computers will have access to 250,000 to 400,000 software packages and more than 100,000 data bases in this time frame. About 70% of the population will have access to a PC in the office or at home [1]. Many more people will work, shop, and bank at home. Improved and expanded Centrex capabilities, increased transmission capacity for home-to-office computer links, and enhanced videotex, and wideband services will support these capabilities. Several economic factors will play a major role in the residence marketplace for ISDN. Customers' needs for second access lines at home, to support PCs, both for personal and work at home use, must be met more economically by ISDN access for it to be greeted with acceptance. This will include the added value provided to the ISDN customer of integrated access to a variety of data base and enhanced service vendors. It will be the need for information and convenience services which will help drive ISDN into the home. In this time frame, we can expect to see customers looking for efficient and easy access to the above mentioned data bases, as well as the time saving advantages of banking and shopping at home. Home security, energy management, and high quality, color videotex services will also fuel this market need.

3.2.3 Ergonomics:

In order for these and all other ISDN services to be widely accepted they must address the human factors of the end user. The ergonomics of a service may play a larger role in gaining market acceptance than the service itself. The service must be packaged and addressed by the ISDN subscriber in a way that facilitates user confidence without tremendous user education. The ISDN subscriber must view the new service as a tool that can be called upon without reservation. For tools of this nature to develop much effort must be devoted to standards. Network providers and customer premise equipment manufacturers are involved with standard bodies today and must continue to support and interact with one another. Hopefully this will insure that the user will be able to access the ISDN and intelligent network from a variety of equipment with the same confidence that the phone is used today.

4. SUMMARY

This paper has presented a brief overview of where ISDN is today, and provided a glimpse at the possible future of the architecture in 1995 along with some of the factors which NYNEX feels will shape that future. Much of the future of ISDN will be in the hands of non-technical players in the telecommunications arena, such as the federal and state regulators, vendors business motivations, and the potential ISDN customer's perception of this new architecture. Major changes in the projected behavior of these key players could cause a significant deviation from our projected scenarios. The one element of our business which remains constant in an industry driven by change, is the challenge to manage change and yet achieve our goals.

Casimir S. Skrzypczak (SCRIP-ZAK) was appointed Vice President of NYNEX's Science and Technology department when it was formed on September 1, 1985. He is responsible for the formulation of NYNEX's strategic technology plans and for the direction of NYNEX's research and development programs, including the operation of the NYNEX Corporate Research and Development Center. Prior to joining NYNEX, Mr. Skrzypczak was Vice President of Network Planning at Bell Communications Research Inc.

Baily M. Geeslin was appointed Vice President - Planning - NYNEX Service Company in 1986. Mr. Geeslin holds an MBA from Northeastern University and a BS degree from the U.S. Coast Guard Academy. He is responsible for developing regulatory strategy and insuring regulatory strategy integration with business and market strategies. Prior to joining NYNEX, Mr. Geeslin was an Assistant Vice President at New England Telephone Company.

REFERENCES

[1] Bell Communication Research, Inc. . . Special Report SR-BEL-000390, Network Services Architecture Issue 2, October, 1985.

[2] Foster, Richard, Innovation: The Attacker's Advantage Summit Books, New York, 1986.

[3] Bell Communications Research, Inc. . . TM-NPL-004075, Integrated Call Features, November 11, 1986.

MIGRATING THE PRIVATE CORPORATE NETWORK TOWARD AN ISDN FUTURE

Donald K. Melvin

Intel Corporation
Santa Clara, California

1.0 MIGRATION TOWARD ISDN OF THE '90's

The long-awaited benefits of an integrated service digital network (ISDN) which allows voice, data, graphics and video to be transmitted over the same digital lines, can be achieved today. Not all the hardware or transmission facilities needed for a full, universal offering of ISDN services in the United States is here today. However, enough technology is available today to implement a private ISDN-migratable network to serve present corporate needs for a high-performance, cost-effective network, and permit basic ISDN-like services to be achieved, and their benefits realized now, while migrating toward ISDN of the '90's.

Intel Corporation, headquartered in Santa Clara, California, is participating in the ISDN frontier of technology from a number of aspects, thus proving that ISDN products and services can be a reality today. Intel's ISDN involvements include:

- Production of ISDN chips for digital subscriber-line terminals and switch interface circuits. The ISDN four-wire line "S" and "T" interfaces, which comply with CCITT Recommendation I.430 are supported by these available chips. (References 1, 2, 3, 4, and Figure-1)
- Installation of a private intersite digital corporate communications network which is migrating toward an ISDN future. (Reference 5)
- Participation in an ISDN trial in Chandler, Arizona, jointly with AT&T and Moutain Bell. (Reference 6)
- Publishing and presenting papers at conferences to convey the message that ISDN-compatible products can be developed, manufactured, sold, and applied in digital telecommunications networks today. (References 5, 7 and the following sections.)

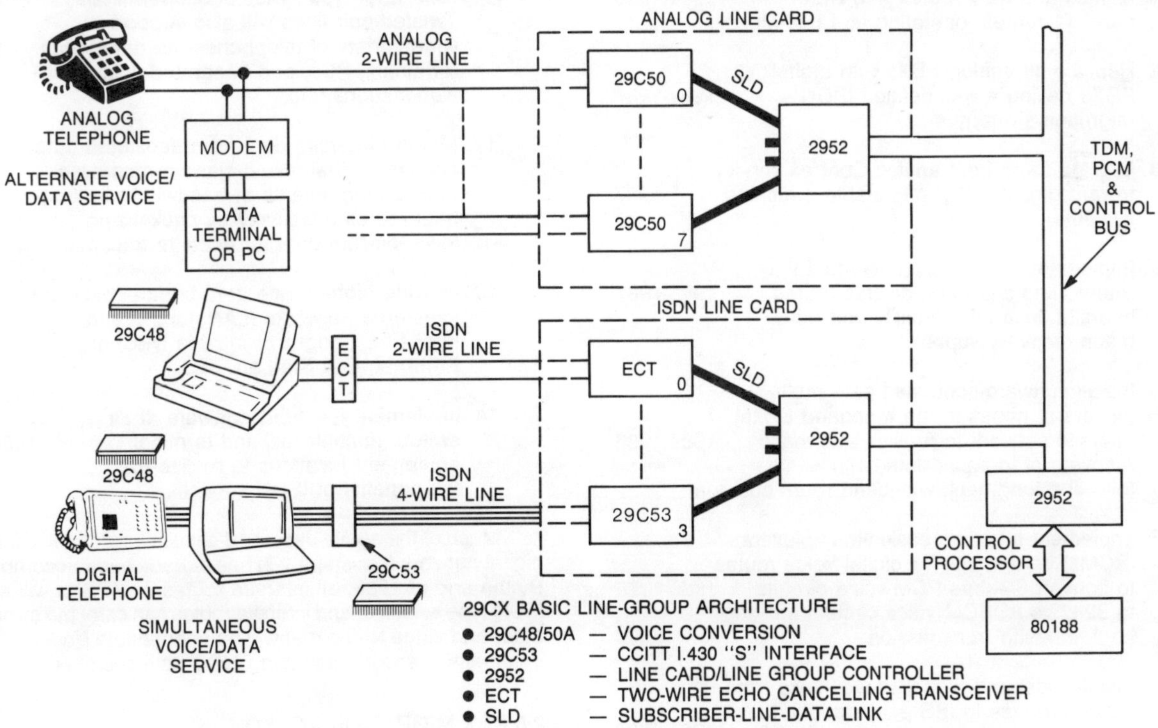

Figure 1. Example of Digital Switching System Line Group for Servicing Conventional Analog and ISDN Digital Subscriber Lines, Based on Intel's iATC Architecture and ISDN Chip Set

2.0 PLANNED STEPS IN MIGRATING A PRIVATE NETWORK TOWARD ISDN REALITY

As an example of how a private corporate network can be smoothly migrated from an analog technology to full ISDN compliance over a five-year period, the Intel corporate network can be described as a model. This paper covers the time span of ISDN conversion from 1983 to 1988, and will extrapolate the future ISDN vision for the period of 1989 to 1995.

Strategies outlined in this paper remain flexible for the period of 1988 to 1995, and the ISDN migration strategy proposed herein is not fully "bought into" by Intel, and is subject to future modification as new hardware and requirements become known. This paper does, however, indicate the general direction of Intel's ISDN migration strategy, and represents a goal toward which today's implementation designs are directed. Much of this strategy can be applied to both domestic and international corporation sites, since ISDN service is forseen as a worldwide offering within the next decade.

Intel's existing domestic digital-backbone network is composed of voice, data, and planned video routes between sites in California, Arizona and Oregon, with a network control center (NCC) in Folsom, California (near Sacramento). Major campuses are in the vicinities of Santa Clara, CA.; Phoenix, AZ.; and Portland, OR. An international digital network is presently planned, or being implemented to Europe, Pacific Basin, and the Far East. A digital satellite route from California to Puerto Rico has also been in service for several years.

A migration process was developed to convert a primarily analog network into a fully-digital ISDN network, and includes the following activities:

1. Replace all analog intersite voice tielines and data routes with wideband T1 routes, operating at 1.544-Mbps. — 1984-1986

2. Replace all analog PBXs with digital PBXs having a well-defined ISDN migration strategy. — 1984-1987

3. Migrate away from analog Centrex service toward digital, ISDN-migratable switching systems. — 1987

4. Implement closed-loop intersite T1 routes, and alternate, diverse-routed intersite links for reliability and automatic recovery support. — 1985-1987

5. Install software-controlled data multiplexers at nodes in the wideband digital intersite network to provide automatic recovery of intersite data circuits following long-haul, wideband route outages. — 1984-1986

6. Implement bit-compression-multiplexers (BCMs) on intersite T1 digital voice routes to convert 64-kbps PCM voice circuits to 32-kbps ADPCM voice circuits for long-haul, intersite transmission. — 1985-1986

7. Install wideband digital access links from network nodes to IBS earthstations to support international extension of ISDN services via 64-kbps clear-channel routes, or multiples of this bandwidth. International fiber routes are to be used when available and cost effective. — 1986-1988

8. Implement on-premise, IPX fast-packet, digital-crossconnect switches to reconfigure and reroute intersite traffic on wideband digital routes between private switching nodes, and on public network access routes for transmission common carrier diversity, and to provide 2:1 and/or 4:1 digital voice bandwidth compression, on the private network, as necessary. — 1987-1988

9. Append site ISN fast-packet data switches to major PBXs to support high-volume data switching needs, and provide switched gateways to the clustered protocol converters for giving terminals access to diverse data networks, and for LAN bridging and intersite data compression. — 1987-1988

10. Modify building wiring to support ISDN services using four twisted-pair wiring and modular jacks. Implement fiber riser and intercloset and interbuilding routes for data concentrator access to centralized packet switches, and for other high-speed data and compressed-digital video services. Local PBX nodes will also be fiber-interconnected to support high-speed information flow. — 1987-1988

11. Migrate coaxial-cable connect data terminals to twisted-pair access lines to reduce wiring complexity and cost. This includes some IBM 327X-type synchronous terminals. Twisted-pair lines will also support a wide variety of telephones, asynchronous terminals, PCs, and integrated voice/data workstations, etc. — 1987-1988

12. Migrate intersite data from dedicated data circuits to shared-wideband routes. Use 64-kbps increments of bandwidth on long-haul routes effectively by packetizing lower-bit-rate data for intersite transmission. — 1988

13. Provide high-bit-rate data bridges between local-area networks (LANs) intra- and inter-site. Bridges to include DECnet, Ethernet, and StarLan, etc. — 1988-1989

14. Implement full-ISDN software at all switch, multiplexer, and terminal equipment locations to realize the full benefits of ISDN services. — 1988-1989

Much of the above-described conversion of Intel's corporate digital network (i.e. steps 1-7) has already been accomplished. By the end of 1988, all intersite routes and PBXs will support ISDN-type services, and intersite routes can carry mixes of voice, data, and video traffic that will be changeable from day to day by on-site, network-managing NCC staff members.

2.0 PBX MIGRATION TO ISDN

In 1983, essentially all of Intel's PBXs were analog switches of the Dimension D400, and D2000 types, except for a few

Rolm CBXs in Oregon, and a #1ESS Centrex in Santa Clara, California, serving corporate headquarters and the Santa Clara campus. All switches were linked together by intersite analog voice tie lines. By the end of 1987, all domestic Intel sites in California, Arizona and New Mexico will be served by AT&T System-75 or System-85 digital PBXs, and a Rolm CBX-9000 will be the major Oregon site switching system, with the original smaller CBX's remaining in the smaller Oregon sites.

The PBXs selected for major sites were selected for their ability to switch 64-kbps increments of digital bandwidth between user lines and intersite and public-network access lines. Intel will proclaim its first ISDN-type service when a 64-kbps clear-channel circuit can be switched between two on-site and/or two different site locations over digital subscriber lines which support at least two 64-kbps B-channels, and via digital switches and internodal routes which support end-to-end bit integrity for clear-channel 64-kbps circuits. Figure-2 shows a proposed migration of Intel's Santa Clara switching systems to full ISDN functionality over 25 years from 1970 through 1995.

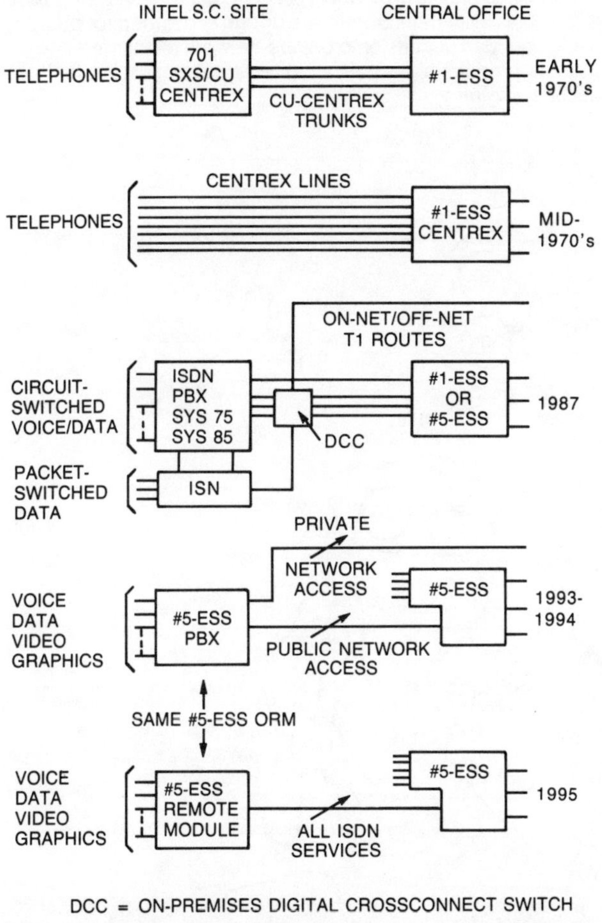

Figure 2. A Proposed Migration of Intel's Santa Clara, CA Site-Communications Network to Full-ISDN Service

3.0 T1 DIGITAL ROUTES USED BETWEEN MAJOR SITES

Independently of the switching technology used at network nodes, wideband digital T1 lines were installed between sites in the 1984 to 1986 timeframe, to set the stage for later ISDN-migration steps. Because of the high volume of intersite voice traffic, a separate T1 voice network was installed using M44 multiplexing with a combination of AT&T D4 and BCM32000 multiplexers used for point-to-point digital voice routes. Each voice circuit is carried intersite on a 32-kbps ADPCM channel.

Timplex Link-1, software-controlled multiplexers were used at nodes of the T1 data network. Only data traffic is carried on the data network, because some of the data-route endpoints are at locations not convenient for voice-grade circuit termination.

3.1 Intersite T1 Voice Routes

By using AT&T's BCM32000 bit-compression multiplexers on DS1 links between PBXs which switch voice circuits on a 64-kbps PCM basis, and the T1 lines which can carry digital voice in private networks at a 32-kbps rate by compressing and sub-multiplexing the T1-line bitrate, Intel was able to provide intersite digital trunking at 44 voice circuits per T1 route. This M44 multiplexing almost doubles the number of circuits at much less cost than two of the usual D4-multiplexed-only circuits that are used in public networks today.

Intersite voice tie-line requirements were sufficient to fill entire M44-multiplexed T1 routes with voice traffic, thus it was not initially necessary to integrate voice and data traffic on the same routes in Intel's private-digital network. Also, the termination points of the voice and data networks were at different locations, and thus it made sense to provide separate T1-based intersite voice and data networks.

The initial T1 voice network was a point-to-point network between sites, with no automatic alternate routing feature to allow the voice network to recover from momentary T1 route outages. At sites which utilize analog PBXs, industry-standard D4 voice multiplexers were used to convert from analog trunk interfaces on the PBX (4W/E&M) to 24 x 64-kbps voice circuits on a T1 line. These lines would then connect to the above-mentioned bit-compression-multiplexers along with another similar DS1 line from another local D4 multiplexer to provide the 44 x 32-kbps T1 voice route between sites.

3.2 Intersite T1 Data Routes

Even though Intel's initial ISDN-migration strategy called for the integration of intersite T1 voice and data routes in 1983, it was not practical at that time because of the early stage of development of compressed voice technology, and the volume of intersite voice traffic, which could support a cost-effective, separate digital intersite voice network.

An automatic intersite circuit recovery capability and an intersite route diversity requirement were more important factors in the data network design than in the voice network design, since voice calls could still be placed by DDD service if a private T1 route failed. This would not be so easily accomplished for a point-to-point data network.

4.0 INTEGRATING VOICE AND DATA ROUTES WITH IPX FAST-PACKET SWITCHES

Recently available fast-packet technology applied as an on-premise, privately-owned digital-crossconnect switch can provide a previously missing element needed for the graceful migration from Intel's present separate T1 voice and data networks.

The Integrated Packet Exchange (IPX), produced by Stratacom Corporation, of Campbell, California, has been used in feasibility studies and cost analyses of future integrated voice, data, and compressed-digital video services, utilizing fast-packet transmission for these services.

Up to 4:1 compression of bandwidth required to carry intersite digital voice traffic can be cost-effectively achieved by use of the IPX in this application. This IPX serves as a digital crossconnect for T1 (DS1) lines which pass through it to on-premise switching systems, or to telco and common-carrier access lines. The IPX also provides a form of alternate routing to automatically recover from T1 line outages, and also provides for centralized network-management functions of the integrated voice/data network.

Functioning as a combined digital crossconnect switch and a voice compression and decompression facility, up to a 4:1 compression of intersite bandwidth can be achieved for voice traffic, by sequentially applying PCM, ADPCM, digital speech interpolation (DSI), and packet assembly and disassembly processes on voice circuits carried on the fast-packet T1 lines.

This type of fast-packet transmission of intersite digital traffic can reduce transmission costs, provide automatic route recovery from momentary or extended-period T1 line outages, and provide automatic reconfiguration and sophisticated network analysis service which produces histograms of past network performance over periods of hours, days, or weeks.

A variety of fast-packet products are being introduced onto the telecommunications market in this time frame, and the use of compression ratios up to about 10:1 are already in development or in the field-trial stages of deployment. Examples are the products designed to carry several compressed voice circuits on a single 56-kbps or 64-kbps circuit, such as the Republic RLX-8 product, and others that can provide up to 120 or more voice circuits over a single T1 line.

Vocoder products which can compress voice to 9.6-, 4.8-, or even 2.4-kbps for transmission over conventional data circuits, are also available for use on costly international routes. Products marketed by TSP, Centigram, Entropic and others fall in this category, and provide adequate-quality voice transmission over low-bit-rate data circuits. Intel uses 4.8-kbps voice circuits to its Far East and Pacific Basin sites, and uses the TSP product. This bit rate was selected after a trade-off analysis of voice-quality versus bandwidth cost on international circuits.

International low-bit-rate-voice circuits hub on Intel's NCC at which a digital teleconference bridge is planned for bridging these decoded circuits onto private-network teleconference circuits, to produce high-quality digital teleconferencing for international digital voice calls.

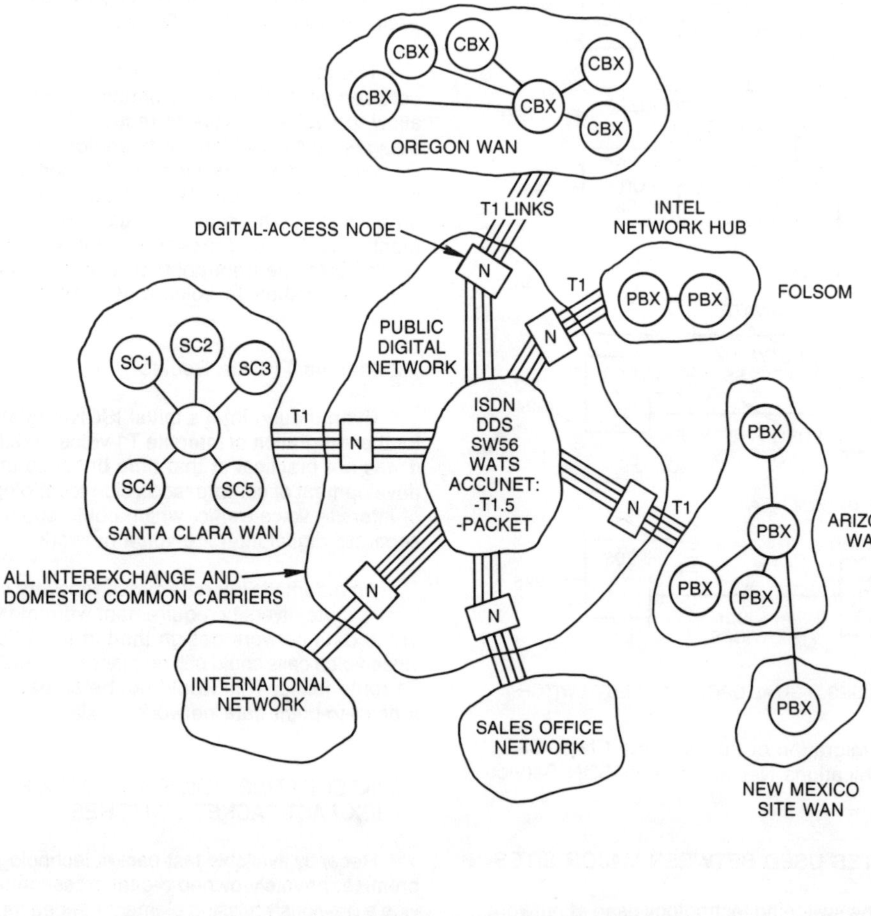

Figure 3. Plan for T1 (DS1, PRI) Access to Public Digital Network

5.0 ISN FAST-PACKET SWITCHING OF ON-SITE AND INTERSITE DATA TRAFFIC

For high-speed, high throughput, on-site switching of data within and between buildings on an industrial campus, or intersite via T1 lines, other fast-packet products are evolving. These can serve as bridges between local-area networks, and support fiber building distribution for a wide variety of asynchronous and synchronous terminals, and provide statistical-multiplexing capabilities for traffic applied to intersite digital links in increments of 64-kbps.

The AT&T Information System's "Information System Network" (ISN) product family, for instance, provides 8.6-Mbps fiber links between ISN concentrators in wiring closets, and ISN centralized packet-controller nodes which serve as the hub of a site packet network. The fiber-connected concentrators provide asynchronous data ports at rates up to 19.2-kbps, and synchronous ports for transmission rates up to 2.3-Mbps for terminals such as the IBM 3278, etc. Even 10-Mbps Ethernet and DECnet ports are available for switched bridging service between like LAN segments similarly connected at the same or a distant site.

Intel is currently evaluating the AT&T ISN fast-packet technology for potential use in its corporate network which is to be administered from a single network-control center, or a backup NCC. An ISN site packet network is planned for deployment at Intel's corporate headquarters site in Santa Clara in 1987 and 1988. Figure-4 shows the ISN functions.

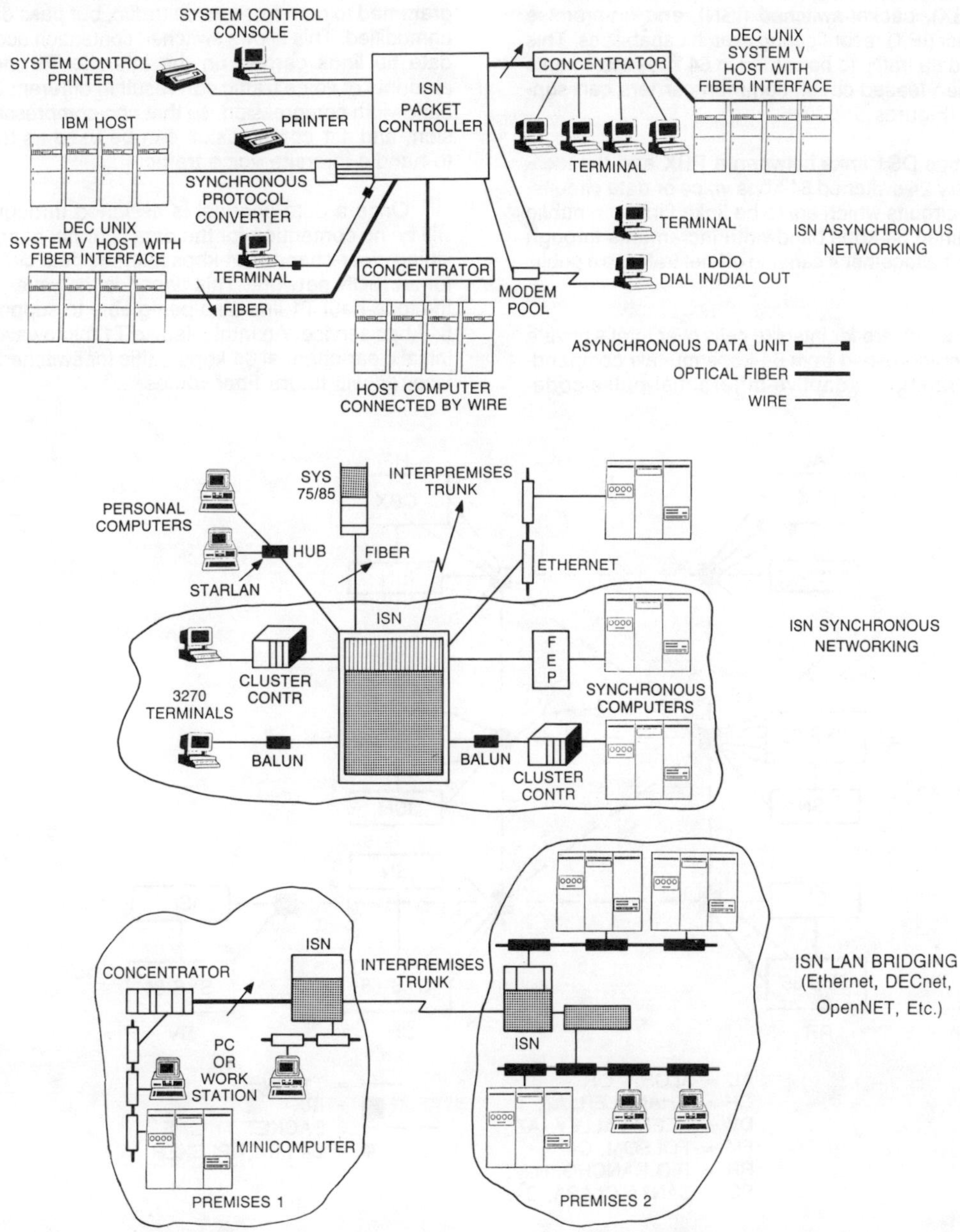

Figure 4. AT&T's "Information System Network" (ISN) Networking Functions

All intersite private-network traffic leaving an Intel site that is configured using the previously-described circuit- and packet-switching technologies, would be T1 (DS1) connected to the IPX digital crossconnect switch, so that voice and data traffic can be interspersed and assembled into large 192-bit packets, on any of the intersite T1 routes terminating in the IPX, to accommodate Intel's changing requirements. This will also greatly improve intersite network usage efficiency by mixing and "repackaging" T1-route traffic before applying it to intersite routes.

6.0 DYNAMIC MIXING OF VOICE, DATA, AND VIDEO TRAFFIC ON T1 ROUTES

Initial private digital network implementation is planned to interconnect sites which are equipped with a combination of circuit switched (PBX), packet-switched (ISN), and on-premise digital crossconnect (IPX) reconfiguring switch capabilities. This will allow intersite data traffic to be carried in 64-kbps increments over T1 lines, when leased-circuit common carriers can support this service. (Figures 5, 6, 7)

The 1.544-Mbps DS1 links between a PBX and the local IPX switch will carry 24 switched 64-kbps voice or data circuits, in any mix. Voice circuits which are to be linked into the public network will remain in 64-kbps bandwidth increments through the IPX, or bypass it altogether if carrying off-net traffic to a public central office.

Voice circuits which are for intersite calls over Intel's private network must be compressed from 64-kbps mu-law companded PCM into 32-kbps adaptive-differential-pulse-code-modulation (ADPCM) format, then by use of DSI, when necessary, into a packetized voice format for intersite transmission. Even though the voice channels are encoded into ANSI (T1Y1) 32-kbps format, the net result of all voice-circuit processing is that voice circuits occupy the equivalent of 16-kbps of bandwidth (average), thus allowing up to 96 voice circuits to be carried on a single intersite T1 line.

Other voice-compression algorithms have been developed to compress over 100 voice circuits, and perhaps over twice this number, over a single T1 line, however products based on these algorithms have not yet been validated or perfected for domestic toll-quality circuits for general application.

If 64-kbps time slots on a DS1 line which connects between a digital PBX and a DCC are dedicated to either voice or data traffic, then the above-mentioned Stratacom IPX can be programmed to compress voice traffic, but pass digital-data traffic unmodified. This allows switched contention access to voice and data tie lines carried on the intersite T1 lines, and varying amounts of voice traffic can result in different degrees of voice bandwidth compression, so that non-compressed, 2:1 compression, and 4:1 compression can be used as traffic loads vary, to handle intersite voice traffic.

Once a data channel is assigned through the IPX, there will be no contention for the data channel bandwidth, thus providing clear-channel 64-kbps intersite data circuits, as required for an ISDN network. This depends, however, on the ability of the long-haul T1 line also being able to support clear-channel 64-kbps service. An Intel intersite T1 microwave route can carry initial clear-channel 64-kbps traffic for switched data circuit service, as will future fiber routes.

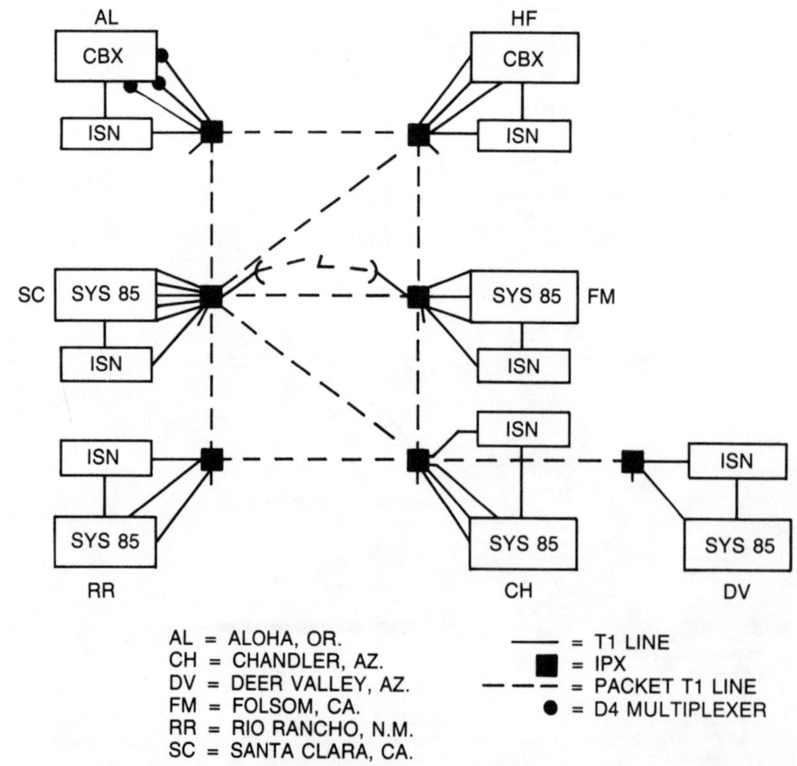

Figure 5. Integration of Voice and Data Networks into A Unified Fast-Packet Network

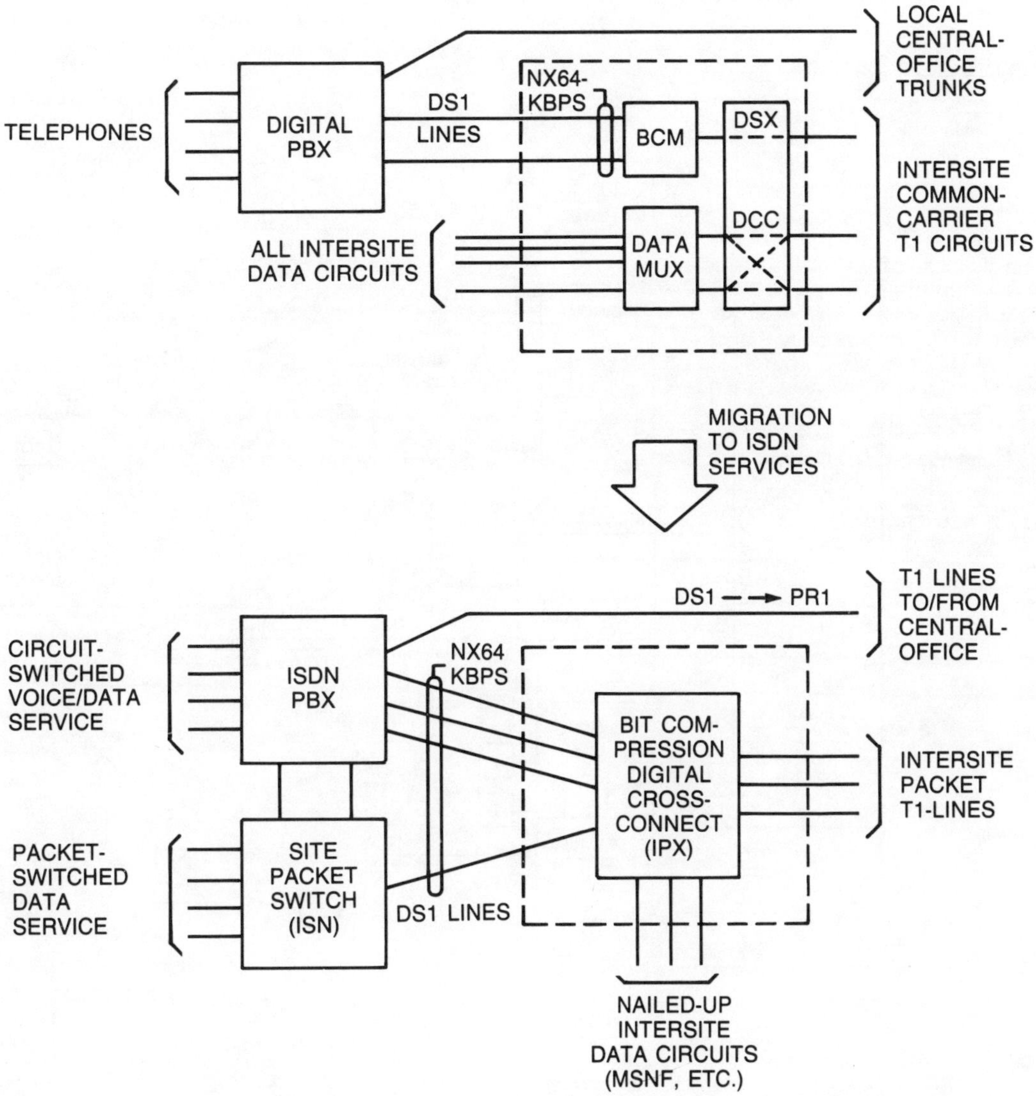

BCM = BIT COMPRESSION MULTIPLEXER FOR 32/64-KBPS CONVERSIONS
DCC = DIGITAL CROSSCONNECT SWITCH PART OF EXISTING DATA MUX
DS1 = DIGITAL SIGNAL LEVEL 1 24 x 64-KBPS, T1 LINE
DSX = DIGITAL CROSSCONNECT FRAME MANUAL PATCH
IPX = INTEGRATED PACKET EXCHANGE, COMBINES BCM, MUX, DSX, DCC

ISDN = INTEGRATED SERVICES DIGITAL NETWORK
KBPS = KILO-BITS PER SECOND
MSNF = IBM MULTI-SYSTEM NETWORK FUNCTION
MUX = MULTIPLEXER
PBX = PRIVATE BRANCH EXCHANGE
PRI = ISDN PRIMARY-RATE INTERFACE, 1.544-MBPS (23B + D)
T1 = DIGITAL TRANSMISSION AT 1.544-MBPS

Figure 6. Migration to Fast-Packet Digital Crossconnect and T1-Line Management Technology (IPX)

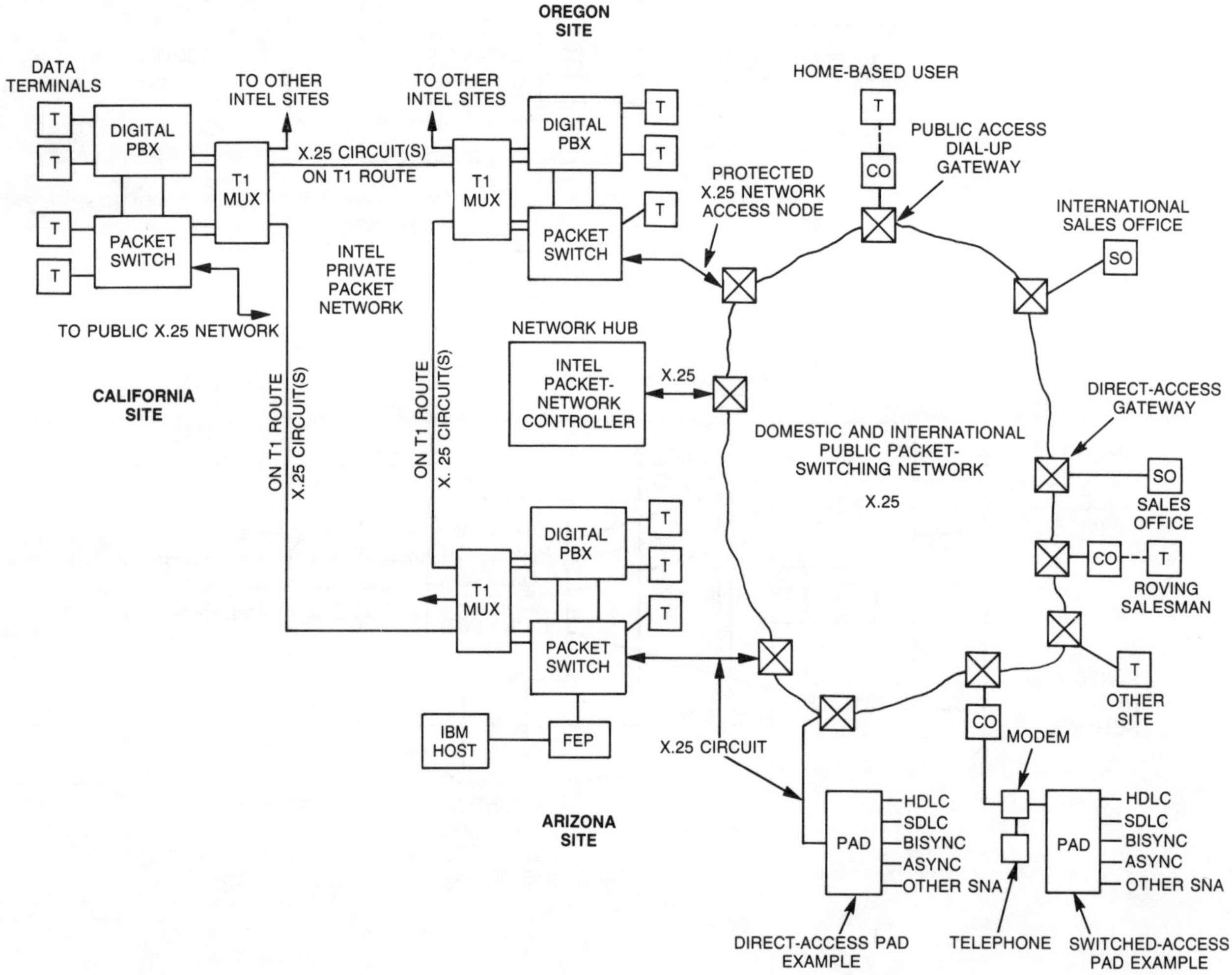

Figure 7. A Planned Domestic and International X.25 Packet Network For Migration to X.25 on Future ISDN Network

7.0 ISDN PRIMARY RATE INTERFACES (PRI) IMPROVE T1-LINE EFFICIENCY

As ISDN Primary-Rate Interfaces (PRI) for PBXs and digital central-office switches become available in the 1988 time frame, they can be used first on primary-rate links between PBXs and ISDN-equipped central offices. No immediate change would be required in the previously-mentioned private ISDN-like network that uses DS1 ports to link a PBX and an on-premise IPX switch.

A future evolution of on-premise digital crossconnect switches should support Primary-Rate Interfaces for routes to central-office switches and PBX switches. When this occurs, the dynamic assignment of voice and data traffic to any portion of the primary-rate line bandwidth will permit even more efficient use of the intersite T1 lines, since a PBX can signal the IPX via its ISDN D-channel signaling facility to identify just which portion of the T1 line bandwidth carries voice traffic that can be compressed, and which portion carries data traffic that must not be compressed or be in a contention mode. The 64-kbps data channels, for example, must be passed through the network with full end-to-end bit integrity, as required for ISDN B-channels.

It is important to recognize that ISDN services based on intersite switched 64-kbps clear-channel circuits for private network applications, can be achieved today, even before primary-rate interfaces and services are available from some common carriers. Intel will initially be able to support intersite clear-channel 64-kbps circuits over new private and public microwave and optical-fiber intersite routes, installed with this requirement in mind.

8.0 ARIZONA ISDN TRIALS DEMONSTRATE IMPROVED NETWORK PERFORMANCE

As a joint trial with AT&T and Mountain States Telephone and Telegraph Company (Mt. Bell), Intel has been utilizing AT&T's ISDN services since early 1987. A #5ESS switching system has been installed in the Chandler-West central office building to serve Intel's Chandler and Williams Field buildings, a few miles from the ISDN-equipped central office.

Following extensive reviews of Intel communications requirements, by holding meetings with existing-network users and designers, and by diagraming potential applications of ISDN services that would match Intel's specific needs, and improve existing services, a number of ISDN applications were planned and implemented. Using the 50 ISDN basic-rate interface (BRI) lines between the Chandler #5ESS digital-switching system and the Intel buildings, the following voice and data service applications were included in the Arizona trials:

1. ISDN PC Networking.
2. ISDN Asynchronous Data Networking.
3. Compressed, digital, color, two-way, personal video teleconferencing (not part of the ISDN trial, but served by ISDN BRI lines for customer-premise equipment).
4. Modem pooling between ISDN-digital and conventional analog lines, for extending data circuits into the public analog network.
5. ISDN coaxial-cable elimination for data terminal lines.

Each of the above-listed trials was conducted in conjunction with BRI line connections to the #5ESS digital central-office switch. Each of the trials provided experience in the effectiveness and application appropriateness to the Intel network. Since Intel had already started a migration plan toward private-network ISDN service prior to the Arizona ISDN trials, the PBX approach to a corporate ISDN network had already been brought into and proven to be the fastest way to implement a unified, cost-effective private ISDN network for Intel.

Intel's initial network will be ISDN-like, and when basic-rate and primary-rate interfaces become available in 1988, the network can be made ISDN compliant where appropriate. Most of the ISDN service features that are useful to Intel will be available with the initial ISDN-like network. Digital subscriber lines, interswitch trunks, and public-access lines may all support simultaneous digital voice and data services, and T1 route bandwidths can carry varying proportions of voice and data traffic.

The ISDN trial experience gained by Intel is first being applied to a corporate PBX-switched network. Essentially all the applications trialed on the #5ESS in Chandler can be directly applied to Intel's PBX network in the near future. A later evolution of Intel's network may likely utilize central-office ISDN services in various forms, starting in the 1988 and later timeframes, when primary-rate interfaces become available. Several #5ESS central-office switches are already in place near some Intel sites, but without the ISDN Centrex software.

8.1 ISDN-PC Networking

A digital-subscriber line interface on a printed-circuit board which is inserted into a PC expansion slot, gives the PC access to a 64-kbps B-channel that can be PBX switched under D-channel control to another similarly equipped PC, either locally, or at another site having clear-channel 64-kbps connectivity to the intersite corporate digital network. Switched 64-kbps access to public ISDN network data channels, and to public ISDN network data channels, and to on-site, and on-net computer networks, is also possible. Here, switched concentration can reduce computer ports to only enough to handle peak traffic loading, instead of one computer port (cluster controller port) per terminal.

Switched PC networking will be provided for 64-kbps file transfers via the digital PBX. Lower bit-rate networking will be provided by the site fast-packet switch, and higher bit-rate networking via StarLan (1-Mbps LAN), or Intel's OpenNET (10-Mbps LAN), or DECnet (10-Mbps) for high-speed VAX networking. Switched intersite access to terminal servers on LANS, and faster LAN bridges can be provided, as shown in Figure-8. The ISN site packet switch can provide same-type inter-LAN bridging.

8.2 ISDN Asynchronous Networking

Asynchronous terminal networking can be supported by either dedicated lines between terminals and the computer network access point, or by a switched digital connection to the access point (data switch, data concentrator, or cluster controller, etc.), switched access also provides switched concentration of the asynchronous data traffic, as well as permitting terminals to be located at great distances from computer network access points, without degradation of response time.

Integrated voice/data subscriber (IVDS) terminals and data module access units (DM in Figure 8), which terminate a digital subscriber line, and into which a terminal can be connected with a digital modemless RS232C connection, for PBX or central office switching over a B-channel, can support asynchronous rates up to 19.2-kbps. While a 64-kbps B-channel is used to transport

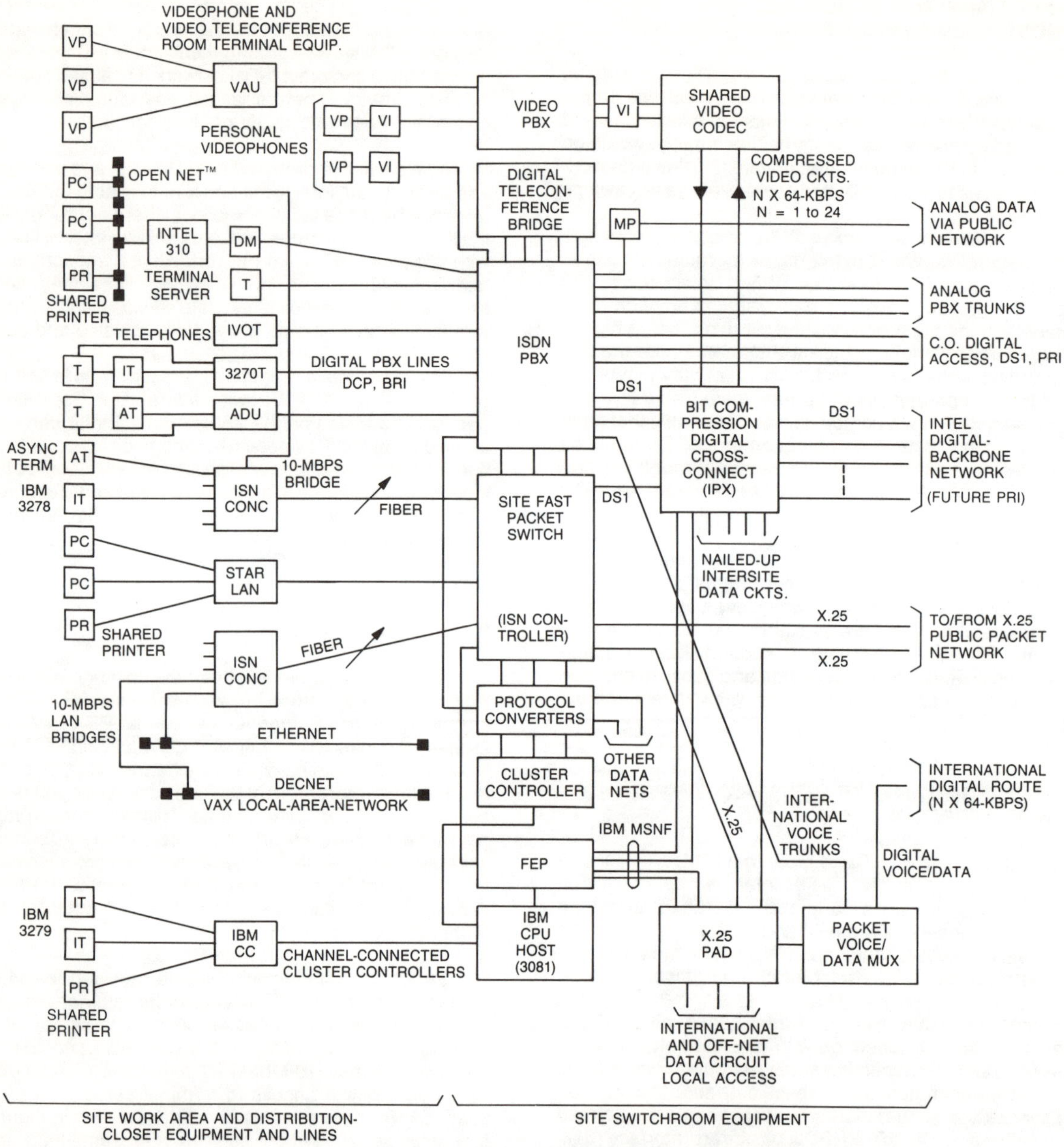

ADU = ANALOG DATA UNIT
AT = ASYNCHRONOUS TERMINAL, OR PC IN ASYNC. TERMINAL MODE
BCM = BIT-COMPRESSION MULTIPLEXER
BRI = ISDN BASIC-RATE INTERFACE (2B-D LINE INTERFACE)
CC = IBM CLUSTER CONTROLLER
CPU = CENTRAL PROCESSING UNIT (COMPUTER HOST)
DCC = DIGITAL CROSSCONNECT FACILITY (SWITCHES DSO INCREMENTS OF DS1 LINES VIA TIME-SLOT INTERCHANGE)
DCP = AT&T's DIGITAL COMMUNICATIONS PROTOCOL. AT&T's 2B+D DIGITAL-LINE PROPRIETARY PROTOCOL
DM = DATA MODULE, INTERFACE WITH DIGITAL SUBSCRIBER LINE
DS1 = DIGITAL SPEECH INTERPOLATION USED FOR 2:1 BANDWIDTH COMPRESSION
FEP = FRONT-END PROCESSOR
IVDT = INTEGRATED VOICE/DATA TERMINAL
IPX = INTEGRATED PACKET EXCHANGE (COMBINES BCM, DCC, AND DATA MUX)
IT = IBM DATA TERMINAL SUCH AS IBM-3278
LAN = LOCAL-AREA NETWORK
MP = MODEM POOL
MUX = MULTIPLEXER
PRI = ISDN PRIMARY RATE INTERFACE (1.544-MBPS)
T = SINGLE-LINE TELEPHONE (ANALOG OR DIGITAL)
VAU = VIDEO ACCESS UNIT (VIDEO TELECONFERENCE NETWORK ACCESS UNIT)
VI = VIDEO INTERFACE, VIDEO CODEC, FIBER INTERFACE, ETC.
VP = VIDEOPHONE, PERSONAL VIDEO TERMINAL, ETC.
X.25 = A LAYERED DATA COMMUNICATIONS PROTOCOL, COMPATIBLE WITH CCITT RECOMMENDATION X.25

Figure 8. Composite Diagram of ISDN-Supportive Switches, Terminals, and Wiring for Proposed Intel Site-Architecture Migration to ISDN (1988-1989)

the lower-rate (lower than 64-kbps) data traffic, it is desirable to packetize intersite asynchronous digital data to obtain bandwidth usage efficiency on intersite T1 lines.

This is done by switching the asynchronous-data-carrying B-channel into a site packet switch which is colocated with a PBX or an ISDN central-office switch before applying the asynchronous traffic to the intersite digital route. High-usage asynchronous terminals may be more appropriately served by lines connecting directly into computer access nodes, or into a fast-packet switch or its remote concentrator, such as an ISN concentrator, as also shown in Figure 8.

8.3 Modem, Protocol Converter and Modular Data Module Pooling

With modemless digital data access available at most user locations when an ISDN environment exists at a corporate site, low-cost asynchronous terminals can be switched by central office switch, PBX, or packet switch to shared protocol converters for extended access into different computer networks which support other protocols, such as bisync, SNA, etc., and also access Local Area Networks (LANs) such as Ethernet, DECnet, StarLan, etc., via RS323 ports on terminal servers attached to these LANs.

Pooled modems permit switched asynchronous data access into analog data networks and into the public switched telephone network for dial-up data links to value-added networks (VANs) or to other modem-equipped terminals and hosts. Pooled data modules, or modular data modules (MDMs), serve as an interface between a digital (2B + D) subscriber line (DCP or BRI) and an RS232C link to another-network access point, such as a terminal server which provides access to an Ethernet, DECnet, or other LAN network.

8.4 Compressed Digital Video Teleconferencing Over ISDN Lines

Intel has conducted intersite video teleconferencing trials, both over the Arizona ISDN-Trial network (between Phoenix and Chandler, Arizona), and over Intel's dedicated intersite T1 data network, between Santa Clara and Folsom, California. The Arizona trial used ISDN-switched B-channels for intercity transmission, and the California trials used vendor-supplied video-switching nodes for dial-up personal video service over Intel's private intersite T1 data network, at rates up to 224-kbps.

Recent improvements in video-compression algorithms have stimulated Intel's interest in trialing switched, personal, 2-way, color, video teleconferencing with integrated sound channels. About a year ago, video that was compressed to a 768-kbps transmission rate appeared to be the most cost effective "acceptable" rate for producing a full-motion video image, with good quality motion-handling capabilities, for Intel. More recently, however, rates as low as 128-kbps have proven adequate for minimal-motion video picture transmission, and rates between 192-kbps and 224-kbps support acceptable small conference-room video teleconferencing picture quality, if rapid movement of the subject is minimal.

From the experience gained from early dial-up, compressed, color video trials at Intel, the following bit rates have been determined to be useful for on-net, terrestrial route, video transmission in Intel's private, domestic digital network.

Video Technology	Low-Motion Security Video Applications	Moderate-Motion Personal Video (Videophones)	Small Conference Room Video	High-Quality Corporate Video Apps.
1987 Technology	56-kbps to 64-kbps	192-kbps to 224-kbps	224-kbps to 768 kbps	768-kbps to 1.544-Mpbs
Post 1988 Technology	56-kbps to 64-kbps	56-kbps to 128-kbps	128-kbps to 384-kbps	384-kbps to 1.544-Mbps

Table-1: Video Teleconferencing Bit-Rate Matrix by Application

Several video codecs have been identified, and the general picture quality, apparent resolution, and motion-handling capability at various bit rates compared by Intel. Codecs produced by Compression Laboratories, Inc. (CLI), PictureTel Corporation, GEC, Ltd., and Mitsubishi have been compared by judging picture quality at various vendor demonstrations, and video-tape comparisons of video-codec performance. Intel's video trials used the PictureTel C-2000 Video Codec with its MCT-200 software. This codec uses the Motion Compensated Transform (MCT) image compression technique, and was operated at a transmission line bit rate of 128-kbps for two-way color, real-time, video-teleconferencing demonstrations at the International Switching Symposium (ISS- '87) in Phoenix, Arizona in March, 1987, and at bit rates up to 224-kbps in an intersite demonstration between Intel's Santa Clara and Folsom, California sites.

Full deployment of switched, compressed, digital video will not take place until interbuilding fiber links are installed, and a full-duplex, video-synchronized audio (delayed for accurate voice/picture lip synchronization of talkers) capability is trialed and determined to be of suitable quality for Intel's terrestrial route video-teleconferencing needs.

Videophone terminal equipment produced by VideoTelecom, Inc., and Datapoint Corp. have been considered for use in Intel's network, however the video trials to date have been based on Video Telecom "Vision Plus" videophones, and PictureTel video codecs. Transmission rates between 56-kbps and 224-kbps have been demonstrated at the ISS-'87 and/or on Intel's private-digital network. Intel's ISS-'87 video demonstration operated over switched ISDN basic rate interface (BRI) lines, and utilized two BRI 64-kbps B-channels that were switched via an Optical Remote Module (ORM) and its host #5ESS ISDN-configured switching system was located in the Chandler-West exchange building.

In Intel's California video trials, inter-videophone switching was accomplished at the video level, utilizing Video Telecom's video-switching nodes which function much like a "video PBX". Two switching nodes were linked together by dedicated Santa Clara-Folsom intersite digital circuits carried on an Intel private T1 data route. Several offices and small conference rooms were linked by the switched intersite video teleconferencing trial network in California.

It is expected that in the near future, as more of the video compression task can be handled under software control, and less by physical circuitry, the costs and space factors for the video codecs will decrease, and picture resolution and motion-handling capabilities will improve. In fact, one video-codec producer believes there will be a 70% improvement in picture quality within a year, for any given transmission line bit rate within the range of their product.

8.5 Coaxial-Cable Elimination

Perhaps one of the most readily apparent cost-saving features of Intel's ISDN trials was the coaxial-cable elimination trial. These trials were used on lines which supported IBM 327X types of terminals, and other IBM devices which usually connect to cluster controllers by cables carrying 2.3-Mbps bursts of data. Data transmitted from a computer host to a cluster controller frequently utilizes 9.6-kbps or 56-kbps links between a front-end processor (FEP) and the cluster controller, if the latter is not located close enough to the CPU to be "channel connected" for very high rate transmission of 1.5-megabytes per second, for a parallel "Bus and Tag" connection. When a cluster controller which supports up to 32 terminals is connected to a FEP via a 56-kbps digital link today, such as a DDS circuit, this bandwidth is shared by contention by all 32 terminals it serves. (Figure-9)

The ISDN methods for connecting the above-mentioned IBM terminals utilizes twisted-pair, unshielded, telephone wire for the link between the terminal and the cluster controller. Terminals then may be direct connected via a dedicated twisted-pair line, or be switch connected via a PBX, packet switch, or packet-data concentrator. Switched access permits switching concentration and permits faster terminal response times.

Once an IBM terminal is switched into a cluster controller, there is no contention on that access line, and the cluster controller can then be located near the CPU and be "channel connected". This results in each such terminal being connected by a line having more bandwidth (one 64-kbps B-channel) than the entire cluster controller line which linked it with its FEP under the older method which required the coaxial cable. While the bit rate on the twisted-pair line serving a bisync terminal may be lower than with the coaxial-cable method, the display response time can be faster because there is no contention on the twisted-pair line, and the cluster controller has much more throughput when channel connected to a CPU than when connected by a much lower bit-rate data line to a FEP.

Intel users have experienced greatly increased response times on 3278 and other IBM terminals, and there is no longer a limitation on the distance a terminal may be located relative to its serving channel-connected cluster controller. In fact, in Intel's Arizona ISDN trials, an IBM 327X terminal was located in the residence of an Intel employee several miles from an Intel building or the #5ESS host switch. This same technology is applicable to PBX-based private ISDN networks, such as Intel is implementing, to reduce cluster-controller port counts by perhaps 40%.

Depending on the distance between the 327X terminal and the ISDN network access point, several optional methods of line connection can be used. Table-2, below shows these methods.

Method	Distance Between IBM-327X-Type Terminal & Channel-Connected Cluster Controller	Type of Access Node Equipment Used	Comments
1.	5000-feet, maximum	coax and IBM cluster controller	Usual connection method
2.	10,000-feet, maximum	coax and IBM 3299 concentrator, and cluster controller	Max. distance doubled for dedicated-ckt. IBM terminals
3.	Thousands of miles. Virtually unlimited.	twisted pair and ISN controller or ISN concentrator	Max. distance improved by ISN coverage and switched access
4.	Thousands of miles. Virtually unlimited.	AT&T 3270T & 3270C interface to DCP or BRI digital line	Max. distance virtually unlimited

Table-2: IBM 327X Terminal Distances from Serving Channel-Connected Cluster Controllers

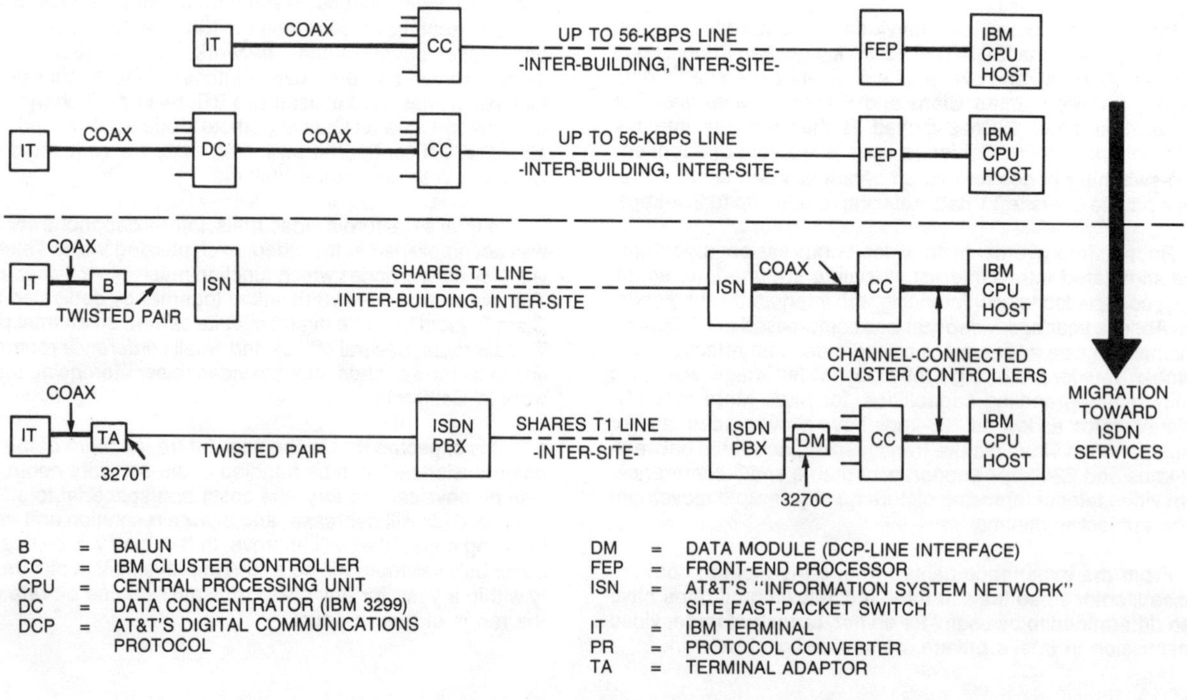

Figure 9. Alternate Methods of Connecting IBM 327x-Type Terminals

The 327X-terminal wiring methods 3 & 4, shown in Table-2, utilize twisted-pair telephone wire instead of the usual coaxial cable run between the terminal and its serving channel-connected cluster controller. This coaxial-cable elimination technology is applicable to both PBX and central-office ISDN network implementation methods. (Figure-9)

The AT&T 3270T throttle-down interface device, used between a short coaxial-cable from an IBM terminal and an ISDN BRI or DCP B-channel, includes sufficient buffer storage to store incoming data momentarily, then transmit it in 2.3-Mbps bursts to produce the rapid screen refresh that users expect.

9.0 A VISION OF THE CORPORATE PRIVATE ISDN NETWORK IN 1995

Depending on the pricing of central-office ISDN services, the private ISDN network, based on leased, dedicated intersite lines, may be partially or totally replaced by the public ISDN offering. Starting in the 1988 time frame, however, a private ISDN network can be realized to improve network performance, transmission efficiencies, computer terminal response times, and cost effectiveness by utilizing a migration strategy similar to that described in this paper. Over the period from 1988 to 1995, the following evolutionary changes in ISDN offerings are foreseen by the author, and are listed by category of technology below.

9.1 PBX Versus Central-Office Remote Switching Modules

With time, it is anticipated that both the ISDN PBX and the Central-Office ISDN switches will evolve to a similar distributed architecture, so that remote-switching modules of the central-office ISDN switch will take on more and more of the same features of the PBX, approaching each other in architectural design, functionality, interface specifications, and application. Both will support copper line and fiber links with the central-office host switch, and both will also provide basic-rate and primary-rate interfaces to support a wide variety of customer-premise equipment. (Figure-2)

A merging of both circuit-switching and packet-switching capabilities into the same switching unit appears to be feasible, with fast-packet technology being part of the PBX and central office switch capabilities. Looking farther out in time, it is even likely that all voice, data, graphics, and video traffic will be switched with fast-packet switches, and that switches will become more modular and less costly as the methods for switching various types of traffic become more alike, and switching units for most types of traffic can be located in distribution closets instead of centralized in a building switch room.

9.2 Building Wiring Migrates to Heavier Use of Fiber-Optic Lines

In addition to the optical-fiber links between a central office and the corporate customer premises, and the inter- and intra-building routes as well as inter-wiring-closet and riser runs, which are being implemented with fiber today, the extension of fiber lines into user work areas and offices is also foreseen. Lower fiber costs, fiber node units which can convert a star fiber building network into a logical bus, ring, or star architecture, may drive the office wiring plan to support a mix of twisted-pair wire and fiber outlets by 1995.

Fiber-optic office outlets with dual-fiber plug-in sockets, and flexible dual-fiber line cords, which plug into both the wall outlets and the terminals, can extend extremely high-bandwidth capabilities to a wide variety of terminal devices. Fiber patch panels, for use in wire- and fiber-distribution closets, extend a similar kind of distribution flexibility to fiber that we have today with twisted-pair building wiring. Early hardware devices are available for the fiber building distribution systems today. Fiber is expected to become more-and-more practical and cost effective with time, and may be in much more general use by 1995 within buildings, and for wideband service drops to work areas.

9.3 On-Premise, Fast-Packet Switches May Combine Data Switching and Digital Crossconnect Functions with Voice Switch Functions

Fast-packet switching systems which may have backplane transmission rates, and fiber concentrator link rates up to perhaps a ten megabit rate range today, may evolve over the next decade to operate at rates well in excess of 100 Mbps. The volatility of requirements, terminal interface and transmission-line interface specifications today make separate packet and circuit switches, and T1 network multiplexers and digital-crossconnect switches appropriate today. However, as the functions and interface specifications become more stable, an integrated circuit-switch and fast-packet switch, which combines digital crossconnect and T1 transmission line multiplexer functions, seems inevitable and perhaps may be available by 1995, as part of a very distributed switching system for customer-premise applications.

Extension of T3 (DS-3) transmission line bit rates of 44.736-Mbps (nominally 45-Mbps) to major corporate-user network nodes will bring the low-cost transmission capability needed to support new high-bandwidth terminal applications, particularly video technology. By 1995 however, photonic switching, by an add-on module to a #5ESS, may allow switched video or computer-to-computer fiber links to be established under BRI D-channel dial-up control, for rapid reconfiguration of a corporate fiber network at a moment's notice by the corporate user.

A digital-access-crossconnect-system (DACS) function, located both in central office and customer premises installations can terminate other 45-Mbps fiber routes, and permit wideband digital bandwidth to be subdivided, reconfigured, and redistributed to site switching, transmission, video, and computer nodes, also under corporate-user control.

9.4 Digital Video Teleconferencing

With the present rapid improvements in digital video compression algorithms, it appears certain that low-bit-rate color video for personal videophones will be well developed and in extensive deployment in the 1995 time frame. Although it is not here yet, full-motion color video at 64-kbps should be available for now emerging dial-up personal video applications. With present-day video codec prices in the $70,000 price range, most near-term video conferencing will be based on shared video codecs which can be accessed by on-premises video switching systems. As the video codec technology improves, and prices drop, each video terminal will be able to have a self-contained video codec, resulting in dial-up 64-kbps or 128-kbps compressed digital video service over a basic-rate access line to the ISDN switched digital network.

Standards are already being formulated to make the 384-kbps ISDN HO-channel rate an international standard for good-quality compressed digital video. A standardized 384-kbps video-compression algorithm seems practical, since this rate can be carried and partitioned in an ISDN network with appropriate switch and digital-crossconnect capabilities. This rate is seen as an internationally-evolving video rate around which standards will evolve, and international connectivity of different-vendor-produced video codecs may be possible in the 1995 time frame, with picture quality expected to be comparable to that obtained with 768-kbps compressed video today.

Small videophones that are either stand-alone, or adjuncts to personal computers are already available. This technology will quickly evolve into a major market once the video codec quality and price improvements, that are already seen on the horizon, actually materialize, perhaps around the 1990 timeframe.

The success of video teleconferencing will be heavily impacted by the ability to transmit high-resolution video images, even if by freeze-frame, in addition to real-time video images. Both computer-generated images, and camera viewed copy must be able to be transmitted over a video-teleconferencing network so that users can refer to a diagram or text presentation. By 1995, cost-effective, comfortable-to-use, video teleconferencing in a variety of forms should be available for deployment.

9.5 Interswitch Signaling to Migrate toward CCITT Q.931 Protocol

PBXs produced by various vendors today do not all support the same interswitch signaling protocols. This situation must be resolved near term for large corporate networks based on PBXs produced by more than one vendor to be able to function together with maximum feature transparency between switches, particularly in the ability to switch intersite data at rates up to 64-kbps.

PBX suppliers that retain only older or proprietary interswitch protocols, or that don't provide interfaces with other major-supplier switches for full-feature control functions, lead corporate customers toward single-vendor switch selection choices, during early ISDN-migration days. Otherwise, an early ISDN corporate network may have different feature functionality on different sites, particularly in the area of switched-data services.

Fortunately, the stand-alone, fast-packet switch that can be added to a digital PBX at each corporate site, allows many of the same switched-data services to be provided with different-supplier PBXs. This separate packet-switch concept "saves the day" during early ISDN migratation in a mixed PBX-suppler network, such as Intel's. The ISN packet switch described in Section-5 can serve as the different LAN and PBX "integrator" in Intel's early private ISDN network.

Three major interswitch signaling systems are evolving along with ISDN migration. These have already been implemented in varying degrees. These include AT&T's DMI, the CCITT's LAP D and signaling system number 7. These ISDN standards rely on the basic concepts of the open systems interconnection (OSI) model for communications protocols.

The DMI (Digital Multiplexed Interface) protocol is a technical specification that is expected to evolve in line with the evolution of the CCITT's Q.931 ISDN recommendations for communicatiaons between a PBX and another PBX, a private or public network, or a host computer. DMI also supports both local and remote host-to-host and PBX-to-PBX communications. More than 100 companies have obtained DMI licenses from AT&T for DMI connections, including Intel, ITT Corp., Advanced Micro Devices Inc., Amdahl, Control Data Corporation, NCR, Rockwell International Corp., and many others (Figure-10)

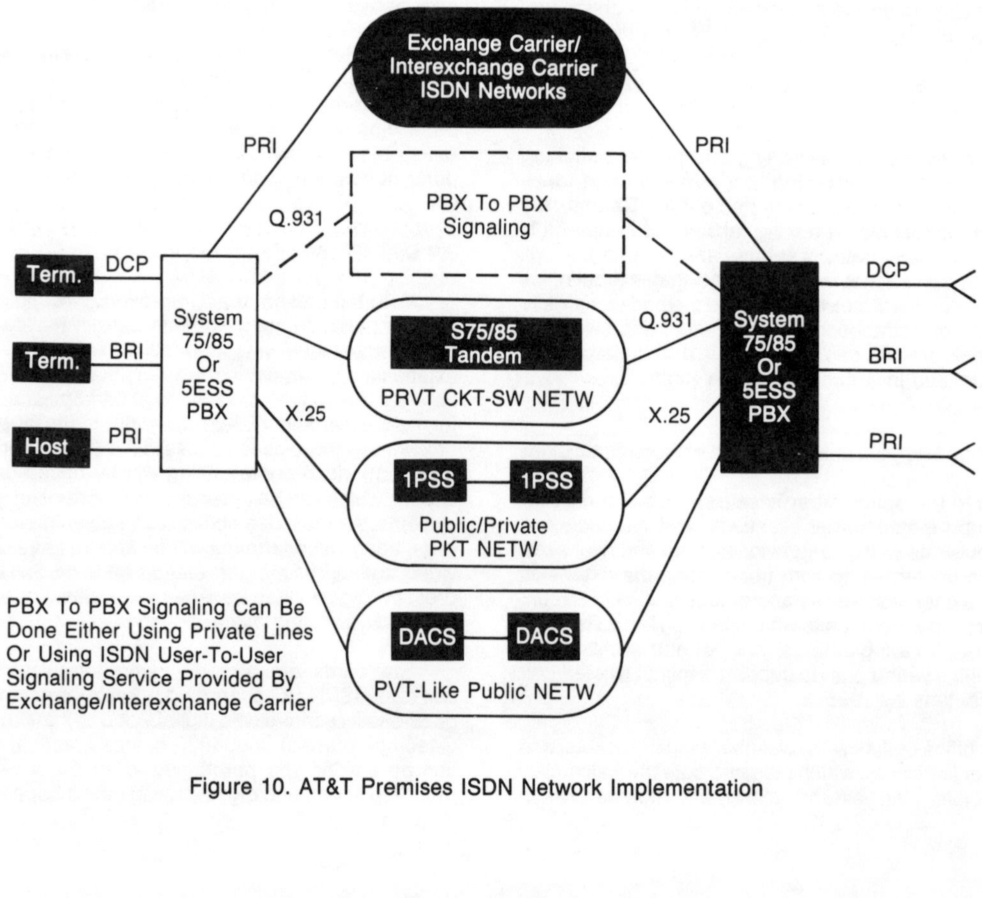

Figure 10. AT&T Premises ISDN Network Implementation

Intel's evolving network is planned to utilize DMI protocols for interswitch signaling in California and Arizona initially. As the ISDN standards evolve, the network signaling system will migrate toward the new standards. In the meantime, Intel will be enjoying many ISDN benefits before the standards are totally finalized.

The status of basic-rate and primary-rate protocols may be summarized as follows:

- AT&T's DCP subscriber-line protocol will evolve into the future ISDN basic-rate interface protocol BRI. CCITT Recommendation I.430 specifies this protocol, and Intel's chip set for ISDN "S" and "T" interfaces support the I.430 protocol.

- AT&T's Digital Multiplexed Interface (DMI) is a precursor to the ISDN primary-rate protocol. When the CCITT Recommendation Q.931 is ratified in 1988, the primary-rate interface (PRI) specification can be finalized. (DMI + Q.931 = PRI)

The benefits Intel expects to achieve by utilizing DMI interswitch signaling near term, are that DMI will allow a digital PBX to route incoming calls directly to a host computer. In this configuration, the PBX acts as if it were a digital-data switch. Thus, data may be transmitted between remotely located hosts, and through one or more digital PBXs without modems. In effect, DMI creates a private data link that bypasses the local telephone operating company. This technology dramatically will decrease the costs and error rates of data transmission. Also, several hosts connected to a single PBX may be accessed individually, at different times through a PBX. Older telephone signaling systems do not support this level of digital-data switching and routing sophistication between different-supplier PBXs.

10. ELECTRONIC-TANDEM NETWORKS UTILIZE NEW PACKET ROUTING OVERLAY

A side benefit of the fast-packet digital crossconnect technology described in this paper is that they can also provide an alternate-routing function that is independent of PBX routing functions. The proposed IPX packet switch can readdress packets of voice, data, and video following failures of previous packets to reach the originally-addressed destination. If this technology is also used as the sole method of alternate routing to recover from intersite line failures, then a large number of additional ports may be required on the served PBXs to group circuits into destination-defined trunk groups, thus increasing the number and cost of PBX trunk ports.

The planned future deployment of the packet-switched DCC calls for the establishement of two types of PBX trunk groups, primary routes, and secondary routes. Primary routes will be direct routes from one site to another that is either nearby, or has heavy traffic requirements relative to the local PBX. The secondary routes will carry tandem traffic to a centralized electronic-tandem PBX for switching traffic on to low-traffic requirement sites relative to the local PBX, or to very distant locations having a low community of interest with the local PBX users. This combination of primary and secondary trunk groups matches the original corporate network strategy, and provides a cost-effective mix of tandem-switching capabilities whereby the IPX and PBX back-up each other in alternate-routing capabilities to speed recovery from T1 route overloads or outages.

11. OPTICAL SWITCHING MAY BRING SWITCHED FIBER ACCESS BY 1995

Photonic switching, the switching of very high bit rate signals which traverse optical fibers, is now heavily in the trade press. In the corporate site network, photonic switching may open the way to switch a campus fiber network from one form of communication to another under electronic control.

Perhaps video teleconferencing can utilize a fiber at one time, then very high bit-rate computer linkages can be established at a later time, then internodal routes can be rearranged in a distributed PBX network to support switch growth, etc. Optical switching systems can be categorized in three groups—space division, time division, and wavelength division. As each of these switching technologies are developed, a variety of new optical product offerings will become available to change forever the way communications networks can be designed.

As optical-switching technology matures, and fiber routes are implemented between cities and into buildings, the cost of bandwidth should dramatically drop. When this occurs, less effort will be devoted to high-compression technology, which is needed today to make ISDN migration a feasible and cost effective undertaking, and more effort will be directed toward the design of lines, switches and terminals that can utilize high bandwidths for useful services.

The new optical-transmission services can support dramatically improved, high-resolution video services and quality videophone services that will benefit business and the home user of the evolving new telecommunications networks. At the heart of much of the controlling electronics for the universal-communications network of the future will be the new semiconductor devices that will perform even more rapidly, and with even higher levels of integration than those of today, to make the future very bright for the telecommunications network implementers and users of the '90's.

12. CONCLUSIONS

While real ISDN services, based on CCITT recommendations, and agreed upon by national and local telephone operating companies in the U.S.A., are not yet fully available, enough of the ISDN infrastructure is available today to make it possible to implement a private ISDN-like network with many of the important ISDN features. Features such as switched 64-kbps data circuits, with end-to-end bit integrity. Also, packetized voice and data circuits can be routed by fast-packet technology for efficient intersite switched communications links. Fast-packet digital crossconnect switches allow an ISDN deployment strategy to evolve today, and to remain valid and be upgraded as further CCITT Recommendations and ISDN standards become available to guide additional product and service offerings.

Fortunately, the present lack of a finalized CCITT primary-rate recommendation does not hinder the implementation of a private ISDN-like network today. The voice and data compression hardware needed to make a private ISDN network technically feasible and cost effective today is not dependent on the CCITT recommendations for primary-rate lines, planned for resolution in 1988. Present DS1 network interfaces to intersite T1 routes are sufficient, and appropriate for early ISDN migration plans, since the added primary-rate functionality is not a hinge factor in providing intersite, private-network, ISDN-like services based on highly-compressed voice and data technology.

Intel's private domestic network will use 64-kbps, 32-kbps, and equivalent 16-kbps voice circuits in 1988, and 8-kbps and 4.8-kbps digital voice on international routes. The primary-rate line protocols are not optimized for handling the various forms of higher-compressed digital voice circuits, particularly if they are carried intersite in packet-transmission form to achieve the desired cost advantage needed to make early ISDN-like services implementable in the 1988-1989 timeframe. As long as the private switch and private network approach serves the corporate need, and is cost effective, it will be used. When public ISDN offerings become more cost effective, the private ISDN-like network may vanish, but this seems unlikely over the depreciation period for the equipment and services being installed in this time frame.

Some of the benefits Intel expects to obtain from its early private ISDN network include:

- Avoidance of coaxial-cable lines to terminals
- Faster terminal responses than at present
- Fewer cluster controllers (about 40% less expected)
- Inexpensive terminals can access a variety of diverse computer networks
- Digital quality voice circuits
- Lower cost intersite routes
- Higher-feature telephones
- Twisted-pair building wiring for voice and data
- Centralized maintenance and management of network
- Higher-features PBXs
- Lower switching costs
- Rapid intersite provisioning under Intel's control
- High-Quality digital voice teleconferencing
- Future dial-up, compressed, digital-video teleconferencing
- Equivalent voice and data services at most sites
- High speed file transfers between PCs at same and different sites
- System that is easily migratable as ISDN evolves
- Fewer intra- and inter-building wires and cables
- Integration of private LAN, WAN, and public ISDN network functions
- Fast network recovery from intersite route failures
- Getting in step with an ISDN future

The Arizona ISDN trials mentioned in this report were conducted in the February to August, 1987 time frame, and provided the basis for much of the ISDN Migration strategy described in this paper.

REFERENCES

1. Intel Corporation, "Telecommunications Handbook". Available from Intel Corporation, Santa Clara, California 95051

2. "Microcommunications Handbook---Components, Boards, Systems, Software for Microcomputer Communications". Intel Corporation, Santa Clara, California 95051

3. Stout, David F., and Melvin, Donald K., "Microprocessor Applications Handbook". McGraw Hill. Chapter 9: "Microprocessor Applications in Telephony", 1982.

4. Bergstroem, Curt--Editor, Intel marketing brochure: "ISDN--Making it Happen". Published by Intel in Europe, 1987.

5. Melvin, Donald K., "Development of ISDN Products and Services as a User and Manufacturer." Wescon/86, Anaheim, California, November 1986, Session 10, Conference Paper 4.

6. Feldman, Robert; "New Demos Link ISDN Sites Over Long-Haul Lines", Management Information Systems Week, Monday, March 16, 1987, Page 27.

7. Melvin, Donald K., "Fast-Packet Transmission Trialed in Intel's Digital Network", Data Communications (McGraw Hill), Late 1987. (Issue date not known at publication date of this paper.)

8. AT&T Network Systems, "International Switching Symposium--- ISS-'87-Phoenix, Arizona---AT&T Technical Papers", March 15, 1987.

9. M. Iwama, J. W. Timko, J. F. Day, "AT&T ISDN Architecture", published as pages 18 through 36 in reference 8, above.

10. Personick, Stewart D., "Photonic Switching: Technology and Applications". IEEE Communications Magazine, Photonic Switching issue, May 1987, Vol. 25, No. 5, Pages 5 to 8.

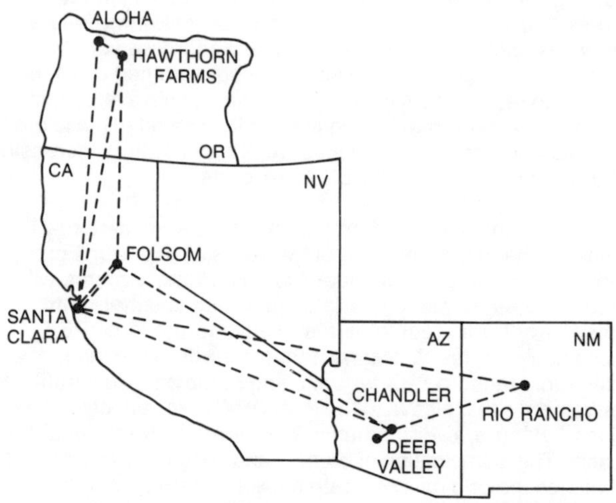

Figure 11. Proposed Final Intel Fast-Packet Corporate Network for the Digital Transmission of Voice, Data, and Compressed Video

BIOGRAPHY OF DONALD K. MELVIN

Donald K. Melvin is Corporate Telecommunications Senior Strategist for Intel Corporation, Santa Clara, California. He has been with Intel since 1975, initially in a product planning capacity for integrated circuit chips for digital switching systems, including codecs, PCM filters, and ISDN subscriber line interface chips. Since 1982, he has been developing and pursuing the strategy for migration of Intel's worldwide communications network into a fully-digital private network, to heavily support ISDN services in the 1990 timeframe. Prior to working with Intel, Mr. Melvin was employed by GTE Laboratories Lenkurt Electric Company, Stanford Research Institute, and was President of Pacific Telecommunications Laboratories, Inc. He has 196 patent claims, and has published heavily in trade and professional journals, has authored and coauthored several books on digital telecommunications, and has frequently lectured at college, professional society and technical seminars. He is also a senior member of the IEEE, and has received a best technical paper award from the AIEE.

July 22, 1987

ISDN - The Competitive Advantage

J. F. Luby

Illinois Bell Telephone Company
Chicago, Illinois

This presentation will discuss the progress so far in turning up ISDN applications at the McDonald's Corporation ISDN trial in Oak Brook, Illinois. The discussion will deal with how this customer views the emergence of ISDN based services combining voice and data as an opportunity to improve the competitiveness of their business, from the ISDN based Centrex service at the home office in Oak Brook expanding to include the individual stores.

1. Introduction.

I think it's appropriate that the subject of this final seminar should focus on the new business opportunities ISDN makes available. Many of the discussions about ISDN to date have focused primarily on technical questions -- and certainly the technology of ISDN is one of extraordinary significance.

However, it's the application of ISDN technology that I want to discuss today. That's because ISDN is much more than just a means of upgrading a company's voice and data communications system or reducing its costs.

To illustrate this point, I want to cite Illinois Bell's experience in implementing ISDN-based services for the McDonald's Corporation. From the beginning, McDonald's has viewed ISDN as a means to increase the productivity -- and thus, the profitability -- of their operation. McDonald's sees effective communications as a strategic resource. Thus, the implementation of ISDN is part of the company's overall corporate strategy and a way to enhance their business opportunities.

2. Communications and Business Opportunities

Although a specific business opportunity may vary from company to company, and from industry to industry, it can be described simply. The most basic definition of a business opportunity is a way to make a profit, either by increasing revenues or decreasing costs.

McDonald's has stated many times that they expect ISDN to help it do those things and more. They see ISDN as a means to improving employees' productivity and creativity. They see it as a means to free up employees' time so they can concentrate on actually managing the business. They even see ISDN as a tool that, in the long run, will help them serve better hamburgers and develop new products.

Now, as far as any business opportunities go, the traditional way to look on a telecommunications system is in terms of how it can contribute to decreasing a company's fixed costs. Now, certainly, ISDN compares favorably in this area.

To take just one example: it's been estimated that the average direct cost of rewiring a coaxial cable system when a person moves is approximately $1000. That's just in direct costs. The hidden cost of having employees temporarily cut off from the company's information flow is much greater.

With ISDN, such a move can be made at substantially less cost -- our studies indicate less than $200. And since the time it takes to make the change is minimal, so is the time that employees are affected.

Take another example. The time when a system is down can be disastrous to a company. This is especially true in the area of data communications. As a result, it's imperative that customers have the control and support systems that allow them to identify a problem and take the necessary steps to correct it themselves.

As part of the McDonald's ISDN trial, we are testing a prototype system that can give the customer an unprecedented degree of control -- and thus provide an unprecedented way of controlling service and costs.

But the area where McDonald's believes that ISDN has its greatest potential is in the area of generating new revenues. To help you understand why, I want to give you a brief picture of McDonald's corporate structure, our experience to date with our trial, and then describe some of the future applications they see ISDN providing.

3. McDonald's at a glance

McDonald's U.S.A. has its home office in Oak Brook, just west of Chicago. The company has nearly 9,100 restaurant outlets. Its international division is rapidly expanding, with locations in 44 foreign countries. Everyday, McDonald's serves some 19 million customers.

The current ISDN trial is taking place in the Oak Brook complex, where more than 1600 employees work in five locations in an area covering approximately one square mile. The Oak Brook locations are served from a No.5 ESS switch at our central office in this area. McDonald's currently has more than 2,000 Centrex lines, and a number of data circuits.

Eventually, McDonald's plans to extend their ISDN network to their field offices and to the restaurant locations both here and overseas.

4. Trial Progress

How has the trial progressed since it began last December? To date, we have turned up well over 100 lines at the Oak Brook complex. One application includes IBM 3270 coax elimination using the B channel to communicate between the terminal and the cluster controller.

In other areas of data communications, we are using the packet transport capability of the ISDN line, particularly the "D" channel packet, to access four different host computers, accomplish PC to PC communications, and achieve local area network functionality via the ISDN network.

In addition to the data applications, compressed video, Group IV facsimile, and imaging (interactive graphics and text) applications have been successfully turned up and have now become commonplace capabilities within the trial.

It is very important to insure that the Oak Brook implementation is not an island unto itself. To this end, we have successfully interconnected the ISDN network to our Public Packet, our high speed digital transport and our voice switched networks. We even successfully tested these connections from the Chicago area to Phoenix, Arizona during the ISS 87 Symposium in March.

5.1 Oak Brook Applications

At Oak Brook, McDonald's is already seeing benefits from ISDN. The fact that they can provide uniform wiring, especially in a building now under construction, is making moves and changes easier and less expensive.

Thanks to the increased connectivity that ISDN provides, McDonald's already foresees the day when they can reduce the number of computer terminals, and gain access to several different computer systems from a single terminal, rather than today's two or three terminals.

5.2 Field Applications

McDonald's plans to first extend ISDN to its 68 field offices, and it sees considerable benefits that should result.

For example, in addition to being the largest fast-food company in the country, McDonald's is also the largest owner of retail land and they are also still growing. This growth requires significant coordination between the field and home office as blueprints are reviewed and schedules updated. McDonald's feels that ISDN's high bandwidth capability will improve the company's ability to manage this real estate process.

ISDN is also expected to improve the effectiveness in the introduction of new products and special seasonal promotions. In addition, ISDN is expected to play an important role in the company's marketing and advertising reviews and, in general, to improve overall administration.

5.3 Restaurant Applications

McDonald's restaurants are, of course, its most numerous and familiar locations. Extending ISDN to the restaurants will take some time, but its eventual importance to the performance of the entire operation cannot be minimized. In fact, McDonald's already foresees applications in at least four critical areas: training, market research, store promotions and data collection -- functions in which the restaurants play key roles.

Training of employees is of paramount importance to McDonald's. Today, much of the training is through videotaped instruction, with each store being equipped with several training tapes. Gauging the effectiveness of the training is somewhat limited because no interaction is possible with the tapes.

McDonald's thinks that the data transport capability of ISDN will make it possible to upgrade the training process so that employees can interact with a regional or possibly centralized computer. Under the system, employees would receive immediate feedback about their progress. Just as important, the training department would be able to quickly identify any training materials that needed to be updated, improved or localized.

Market research plays a vital role in McDonald's success.

One demonstration of how ISDN can aid market research took place in Phoenix last March. In this demo, customers were able to ask a computer about the nutritional aspects of different McDonald's products. This provided health-conscious customers information that could aid them in making their selections and gave McDonald's valuable data based on the selections the customers actually made. The company sees similar applications of ISDN as a way to gauge the probable success of potential new products.

Special store promotions help attract customers, and McDonald's sees ISDN as improving the effectiveness of such promotions in the years ahead. With an ISDN-based network, each store could have access to a centralized computer system with updated information during the promotion. At the same time the home office could keep track of inventories of special items and keep abreast of how the campaign is being received by the public.

ISDN also offers significant business opportunities in the area of data collection. Each store must submit sales and inventory information -- a critical function, but one that, ideally, should not impede a manager's ability to deal with customers and manage the operation of the store. ISDN offers McDonald's a user-friendly network that can help the field people spend most of their time with their customers.

Finally, with a mature American market, McDonald's sees the greatest opportunities for growth in its overseas operations. The fact that ISDN has the promise of international compatibility and standards made it the system of choice for McDonald's.

5. Conclusion

Let me leave you with a few additional thoughts.

First, ISDN customers can realize a significant benefit because of the flexibility of ISDN-based networks. Business needs for information are dynamic. An office area that is an administration center one day might be needed as a computer center the next.

Second, ISDN-based networks allow such changes to happen quickly and with significantly lower costs than are possible with today's dedicated networks. Such flexibility is being furthered by the variety of terminals and telephones that are being developed by many suppliers.

In general, only software changes are necessary for a location's capability to be altered from an attendant console, to a multiline key system, to a high speed facsimile, to a data center, to a message desk application, or to an executive office. This flexibility and variety offers to the end user a complete new set of options in managing a business.

Finally, the ability of an ISDN-based network to provide a combination of communication services simultaneously and efficiently provides numerous business opportunities to users who see communications as a resource, and not merely as an expense. ISDN provides a competitive advantage to those firms that find communications an essential component of their operation.

ISDN: A SOURCE OF NEW BUSINESS OPPORTUNITIES

S. I. McLeod

BANK OF MONTREAL
TORONTO

Within the context of today's banking environment, this presentation assesses the potential of ISDN for addressing current business issues and new market opportunities. Critical success factors from the business user's perspective are identified.

Introduction

- The Bank of Montreal is a subtantial international Bank
  - extensive branch network throughout Canada
  - units in all major commercial centres around the world
  - services cover wide spectrum of banking activities
- Communications play a key role in delivering services
  - large central computingfacilities (IBM based)
  - T1 digital backbone network
  - extensive private voice network
  - transatlantic satellite system
  - public packet-switched network used
  - interconnection with other financial private networks
  - PSTN
  - telex and other messaging services

Business Concerns

- Banking is now intensely competitive at the national and global levels
- Traditional service offerings no longer adequate to increase or even retain market share.
- Customers at all levels demand more personal service and better responsiveness to their needs.
- Deregulation means competition from non-Bank's, and opportunities for new business
- Apart from a thrust to increase revenues, cost cvontrol remains a high priority.
- Overall result is a demand from the Bank's marketeers for the application of technology to enhance the competitiveness of our products by:
  - differentiating products/services from those of competitors
  - improving customer service
  - reducing production costs
  - enhancing management decision-making

ISDN and Business

- ISDN is relevant to the Bank's business only if it addresses the above concerns, principally:
  - cost containment
  - increased revenues through
    - new products/services
    - international market expansion
    - improved quality of service
- Focus of this presentation is to examine the potential of ISDN to enahnce the Bank's product offerings and the level of customer service.

. Although shown in a banking context, most of the considerations apply to other industries, particularly those in the service sector.

Potential Opportunities

1. At the Branch
   . Increased data bandwidth will allow improved information exchange between the branches and central locations to enhance customer service across the counter.
     - Reduced lead time for service provision (single network)
     - Reduced response times for full-screen online applications
     - electronic report distribution
     - high speed facsimile
     - image processing (cheque truncation)

   The result will be more up-to-date and complete customer, account, and product information at the branch, in keeping with our belief that quality of information will be a major market factor.
   . Improved Telephone Operations
     - PC interface for director, call logging, statistics gathering for operational and marketing support.
     - Enhanced call handling based on automatic incoming call identification, eg. prioritized queuing, selective routing, message taking.

2. International Financial Dealing
   . Trading is done primarily by voice, with settlement laTER VIA SWIFT, Telex, and other messaging networks.
     - opportunity to unify the trading and settlement networks through ISDN voice/data integration
   . More generally, inter-Bank combined voice and data communications through a single international standardized communications utility will greatly aid the Bank's ability to service the growing number of customers with international requirements.

3. At the Customer's Business Location
   . Cost of current offerings limits market acceptance
   . Security of the Bank's database and customer privacy are major considerations.
   . Many clients have international requirements, but diverse national comms services make international expansion difficult. (only partly addressed by X.25)
   . Market expansion depends on cost reductions and international applicability of product designs
   . ISDN is expected to provide
     - cost-effective dial-up security through use of the D-channel for user signalling
     - improved response times and high-speed batch transfer of data (high bandwidth)
     - concurrent access to different services, eg. Bank and Service Bureau via B or D channel packet-switching, or voice enquiry while online.
     - standardized communications interface to permit global application of a single design
     - LAN interface for multi-user business support

   Applications include:
     - online access to Bank accounts and related cashmanagement services
     - securities traing
     - import/export financial services
     - financial planning support
     - consumer loan application processing eg. car loan at auto dealer
     - payroll

4. EFT POS
   . Full service to the merchant demands the capture of all transactions, irrespective of the means of payment (cash, credit, debit, cheque or coupon)
     - including complete sales data
     - without imposing a service delay
     - at a reasonable cost
   . Service opportunities include
     - credit authorization
     - debit card processing
     - cheque verification, and even cheque processing (i.e. cancelled cheque is returned to customer at the time of sale
     - cash drawer balancing in the back office
     - inventory control, order placement
     - sales clerk performance and payroll
     - statistical data on sales, etc. for marketing support
     - combination of other banking services with EFTPOS delivery
   . ISDN can support the high volume high speed data transfer requirements of the above applications
     - sub-second connect time
     - alternate routing
     - automatice switching for voice authorization of dubious or high-value transactions

5. In the Home
   . Minimal automated home to business communication today
     - typically by combined touch tone and voice response
     - has limited capability
   . Major expansion under current technology depends on widespread deployment of terminals and modems, with the provision of adequate security mechanisms
   - Cost is a prohibiting factor

- ISDN may overcome the cost barrier
  - modemless environment
  - smart phone/terminal
  - data connections possible without interfering with normal voice usage or requiring an extra line
  - use of the D channel for cost-effective security and low-volume data
- Home banking can be seen as a market extension of commercial online banking
  - ease of use of the terminal device/PC is critical, but so is security
  - opportunity for the Bank to provide extensive financial services to the individual in his/her home, eg.
    - income tax preparation support
    - expert system for budgeting and investment planning
    - stock trading
    - electronic mail for queries
    - ABM in the home - smart card technology will permit the implementation of electronic "cash" withdrawals

Critical Success Factors

- Standards
  - ease of implementation and support is crucial to the success of products and services
  - international agreement on comprehensive ISDN standards within the OSI framework is vital
  - as is adherence to these standards in a uniform manner by PTT's, Telco's, and manufacturers
  - addressing is a particular concern: singular addressing within the whole network is required, not just the ISDN segment
- Cost
  - 2 times the business rate is expected; any increase should be justified by an increase in functionality
  - Usage charges of public ISDN facilities will bear directly on the feasibility of using the technology for many of the applications. Today, long distance telephone charges price PSTN out of the market for many products (B and D channel packet-switching may resolve this problem)
  - Customer premises equipment has to be priced competitively with alternative technologies
- Equipment
  - The availability of functional, cost-effective equipment at both ends of the network is essential
  - a new approach for connecting the network to the central processing systems appears inevitable; current FEP's will not adequately address ISDN bandwidth requirements
  - adapters are needed for non-ISDN devices to allow an orderly migration from the pre-ISDN environment
- Flexibility
  - given the inevitable lengthy period between the introduction of ISDN and its global availability, flexible interworking between ISDN and existing services is critical
  - Users must be able to match bandwidth usage to bandwidth requirements; B channel packet-switching and access to the D channel address this requirement
  - the availability of virtual private ISDN facilities and the public networks is another requirement, ie. without the need for multi-step addressing
- Network Management
  - Users of private and public ISDN services must be given the ability to manage this use within the context of their total network management function, i.e. integration of ISDN and non-ISDN management at the central site
  - the above implies also direct customer involvement in the management of certain aspects of the public network
  - the D channell is the apprpriate bearer of user network management traffic
- Security
  - the cost of providing adequate security on today's PSTN is prohibitive for many potential applications; security maintenance is also a problem
  - ISDN needs to facilitate the provision and management of global access control based on a central security database
- Availability
  - ISDN should provide a high level of availability, with automatic recovery of interrupted connections without jeopardizing security

Conclusions

- Unless there is wholehearted adoption of a sufficiently comprehensive set of standards ISDN will present a new set of problems without solving those we face today
- The comms industry must pay careful attention to the requirements of end-users if ISDN is to be embraced by the business community. The appropriate allocation of bandwidth is a concern (eg. D channel for security and network management)
- The cost of ISDN is critical to its success
- Computer vendors must work closely with the Comms industry to promote true end-to-end connectivity and network management within an open architecture. Vendors that fail to do so will ultimately limit their own market opportunities
- Members of the user community must actively provide input to the industry to promote the development of ISDN that meets user needs as well as those of Telco's.
- Many of the opportunities identified above can be addressed by current technology.

Cost and management are the constraining factors

- To some extent this may be a reflection of the radical thinking that is required to invent unique ISDN applications. This challenge is open to all participants in the ISDN arena.
- Only time will reveal the full impact of ISDN, but it is already apparent that it has the potential to change significantly the way business is conducted

ISDN - The Business Network Key

James G. Schram

Ottawa
Bell Canada

This presentation provides Bell Canada's assessment of the Business Opportunities afforded by ISDN from both a customer and Telco perspective. Successful implementation will be dependent on developing new applications and satisfying existing applications in a cost effective and efficient manner.

1. Introduction

ISDN as a concept is no longer new and has recently been the subject of many synposiums and seminars including this important IEEE conference here in Dallas.

Until recently ISDN has been a question of technology and standards. These issues have mostly been resolved or are well on their way to reasonable conclusions. The challenge facing us now is a marketing challenge - the business opportunities for ISDN.

I intend to look at these opportunities from both the telephone company and the customer perspective for if there is not a business opportunity for both with ISDN then probably an opportunity does not exist for either.

I entitled my talk ISDN - The Business Network Key - The unlocking of powerful network capabilities for business customers. Let me review with you Bell Canada's views on these potentials.

To begin, let me tell you a little bit about Bell Canada. We are the largest telephone company in Canada, with 50,000 employees, we serve most of the provinces of Ontario and Quebec, plus a large portion of the Northwest Territories, a total of more than a million network access lines. Our total operating revenues in 1986 were $6.3 billion (Canadian). We are also a member of Telecom Canada, an organization of the nine major telephone companies, and Telesat, Canada's domestic satellite carrier, which together provide telecommunications services on a nationwide basis.

ISDN is now a technical reality. What now needs to be accomplished is for the telephone companies in conjunction with customers to unlock the business opportunities.

2. The Service Concept Evolution

To look at ISDN from a marketing perspective let's view the services the marketing people see emerging in Bell Canada.

Bell Canada's strategic service planning process envisions that ISDN is not one single technology nor is it one service but an evolution.

It started with digital transmission facilities - both interoffice and intercity facilities: These were employed by Bell Canada in the 1970's to introduce Dataroute a fully digital private line data network - offering digital capabilities from 2.4 to 56 Kbits from customer premises to customer premises. This was followed by Datapac - a fully digital packet switched network, also offered from customer premises to customer premises.

Next our engineers embarked upon a modernization project to upgrade central offices with digital switches. This allowed us to offer Centrex III - a full featured central office located customer switch switching both voice and data based upon intelligent capabilities within the switch. This was also combined with propriety telephone sets that provided intelligent capabilities in the terminal on the customer premises.

Next we employed data bases and centralized processing in the network to offer services such as Envoy - a text messaging store and forward service available network wide - inet a standardized method to access data bases both from dumb and intelligent customer terminals. This has been enhanced now with Exten a voice store and forward messaging service.

We also capitalized on the capabilities of digital cross connects to offer our customers our Megaplan series of services - starting with Megaroute - a customer premises to customer premises 1.5 mb service. It proved very popular and is now being offered in 1.5 mb increments up to 44 mb capacity. This is both a local and intercity service and increasingly uses fibre optics as the transmission medium. Megastream is a 64 Kb service that uses DCC's and both network and customer located intelligence to offer full featured dedicated integrated service capabilities.

We have recently used central office based data base capability to offer call management services to our residential customers. These are a few examples of the Marketing applications of technology.

As can be seen from above, we have progressed from voice services to digital data services to digital integrated voice data and image services. However, each of these services was based on a discrete access technology and each service required our customers to purchase a unique terminal and subscribe to a network service with its own loop access.

The transport and signalling capabilities which ISDN brings to the local loop will modify this "end-to-end" service architecture considerably. Multi-functional terminals will access a variety of network services through switching centre gateways using a common access arrangement. The same facility will be shared to provide connection to dedicated,circuit switched and packet switched services.

3. Initial ISDN Services in Bell Canada

We intend to offer 2B+D and 23B+D to our customers as digitally integrated link services that connect to other digital centrex switches.

It makes marketing sense to offer our newest access arrangement as an integral part of our most intelligent switch.

Thus we now have an integrated access service to a network link, that can be used as a node to all of our network services.

The evolution of our services, we believe will be to a lower number and variety of interoffice and intercity network transport services. Differentiation of these services will be less dependent on performance and voice - data considerations in the future. We foresee there being fewer transport options - in fact we are evolving to:

- a digital integrated dedicated service
- a digital integrated circuit switched service
- a digital packet switched service

all being accessed by one ISDN basic rate and primary rate access.

The signalling capability of the D channel extended from the customer premises to network nodes and data bases has exciting potential and we are working to provide these as feature options to our customers. Thus we can see our network services being redefined - fewer transport options but more functionality on these services.

Now let me dispell a myth - the question of which is best for large business customers ISDN or private networks. Our ISDN based access services will have link functions at the central offices to both public and private networks - one time a call over the ISDN link service may be a public network voice or data call and the next call may be a connection to a customers intercity private 1.5 mb network. Both can be handled by ISDN. So as today we can expect customers to have a mix of services to meet their individual needs. ISDN will not change this, it may however change the cross over point where the public and private networks offer the optimum solution for each customer's needs.

Now we have our first business opportunity - to bring the integrated voice, data and image capacity to the desktop. This is our first thrust. To date we have universal data services but in Canada as in the U.S. their deployment has been dedicated to specific applications that concentrate on data processing.

The widespread availability of our ISDN link services that offer both efficiency and speed of provisioning to us and to our customers will we believe make higher speed data access a desktop service. This is the first business opportunity for all of us. We expect the provisioning and ongoing maintenance and administration to enable the cost to both customer and ourselves to be less than that of the equivalent services. This should lead to more uses of these services and we should both benefit. Do the increased number of integrated connections to our network mean greatly increased number of calls? For voice probably not, we don't think the voice access costs have inhibited voice users, however for data there may well be a different answer.

Thus the business opportunity is to enable the customer to be more efficient and effective with increasing data uses and to increase our network utilization. Here ISDN access is the key.

The next business opportunity is a major step for us. To date, aside from the specialized packet services, most of our services have been application independent. That is, pure telecommunications services that were not primarily designed to specifically improve our customers business problems but rather to permit them to communicate more quickly, more accurately and with a lower unit cost, of course this will still be an objective of our evolving services.

Today customers specific total solutions to their business needs have not been met by the telephone companies. We have long met his complete needs for voice services, and the data processing industry have been dominant in meeting his data needs, using our services as a transport system solution. In the integrated communications we plan to work with our customers to offer a solution to their business needs. Many times this will be based upon generic solutions but tailored to that specific customer or it may be a solution unique to that specific customer. We believe many of our solutions in its future will be based upon the integrated ISDN access as the key to our network capabilities.

The success of ISDN has been said by many to rest on successful applications and this is our thrust. We intend to offer our customers improved efficiency and cost effectiveness but as a major business opportunity to offer total solutions to our customers based upon ISDN. One aspect of these solutions will be to provide customers with competitive advantages in their market place so communications leads to increased revenues. At this point communications solutions become a strategic asset to our customers. This is both the promise and the challenge of ISDN.

Bell Canada has started our "in house" ISDN Trial on the principle we should use it first before we pass it to our customers. Our customer trial begins in November of this year.

Another feature of our ISDN program is a capability demonstration facility that will be opened very shortly.

In closing let me introduce another business opportunity, this time not involving its customer directly but between Bell Canada and terminal and data processing manufacturers. The combining of terminal intelligence with increasing network intelligence and data bases over a separate signal channel undoubtly is a source of good potential. By close association between ourselves and software and hardware vendors, we believe we can enhance both offerings that will use ISDN.

Authors of ICCC - ISDN '87

Executive Committee of the International Council for Computer Communication

President

Dr. Philip H. Enslow, Jr.,
Georgia Institute of Technology,
United States of America

Executive Vice President

Dr. M. Clayton Andrews,
IBM Corporation,
United States of America

Secretary General

Dr. Ronald P. Uhlig,
Northern Telecom, Inc.,
United States of America

Executive Secretary

Dr. Pramode K. Verma,
AT&T Information Systems,
United States of America

Treasurer

John D. McKendree,
Defense Communications Agency,
United States of America

Deputy Secretary General

Professor John M. Bennett,
Basser Dept. of Computer Science,
Australia

Vice Presidents

Professor Alexander S. Douglas,
London School of Economics,
United Kingdom

Mr. Yasuo Makino,
The Nippon Telecom Consulting Co. Ltd.,
Japan

Dr. Louis Pouzin,
CNET,
France

Dr. Martin E. Silverstein,
Georgetown Center for
Strategic and International Studies,
United States of America

Dr. Prem P. Gupta,
CMC Limited,
India

Past Presidents

Dr. Stanley Winkler,
IBM,
United States of America

Dr. Peter E. Jackson,
Northern Telecom Canada Ltd.,
Canada

Mr. Douglas F. Parkhill,
Department of Communications,
Canada

ICCC Governors

Mr. de Aguiar
Mr. J. Alvey
Dr. M. Andrews
Mr. S. Barbera
Prof. S. Beer
Dr. M. Benedetti
Prof. J. Bennett
Mr. H. Bown
Mr. E. Boyar
Dr. K. Chon
Dr. P. Claringbold
Mr. J. Cook
Dr. I. Cotton
Mr. A. Coulter
Mr. J. Curtis
Mr. C. Dalfen, Esq.
Prof. A. Danthine
Dr. D. Davies
Prof. A. Douglas
Prof. P. Enslow, Jr.
Dr. E. Fredriksson
Mr. R. Fukutomi
Mr. H. Gassmann
Mr. E. Goldstein
Mr. A. Goldsworthy
Mr. J. Granado
Dr. P. Green, Jr.
Dr. P. Gupta
Dr. C. Hammer
Dr. G. Handler
Mr. K. Harris
Mr. A. Hartung
Mr. K. Hirota
Mr. D. Horton
Mr. T. Housley
Mr. P. Hughes
Mr. F. Israel
Dr. P. Jackson
Dr. R. Kaenel
Dr. F. Kamoun
Dr. M. Karnaugh
Dr. M. Kato
Dr. J. Kella
Mr. P. Kelly
Hon. Justice M. Kirby
Prof. L. Kleinrock
Dr. P. Kuehn
Mr. H. Kunze
Dr. S. Kyong
Mr. T. Larsson
Mr. H. Leroux
Mr. G. Lindberg
Dr. R. Lyons
Dr. N. Macon
Mr. Y. Makino
Dr. Y. Masuda
Mr. O. McAleer
Mr. J. McKendree
Mr. T. Meyers
Dr. M. Miller
Dr. Y. Nakagome
Mr. W. Newport
Mr. D. Parkhill
Dr. R. Parodi
Mr. J. Pernin
Mr. R. Pertuiset
Dr. D. Phillips
Mr. P. Picard
Mr. L. Pouzin
Dr. S. Ramani
Dr. J. Raviv
Dr. A. Reid
Dr. F. Reza
Dr. J. Salz
Dr. S. Samoylenko
Dr. E. Silverstein
Dr. T. Szentivanyi
Mr. G. Thompson
Dr. R. Uhlig
Dr. P. Verma
Mr. R. Wagner
Mr. S. Wernikoff
Mr. R. Weingarten
Mr. J. Whyte
Dr. S. Winkler
Mr. J. Yglesias
Mrs. R. Winkler